The Logica Yearbook
2014

The Logica Yearbook 2014

Edited by

Pavel Arazim
and
Michal Dančák

© Individual authors and College Publications 2015
All rights reserved.

ISBN 978-1-84890-177-3

College Publications
Scientific Director: Dov Gabbay
Managing Director: Jane Spurr

www.collegepublications.co.uk

Original cover design by Laraine Welch
Printed by Lightning Source, Milton Keynes, UK

All rights reserved. No part of this publication may be reproduced, stored in a retrieval system or transmitted in any form, or by any means, electronic, mechanical, photocopying, recording or otherwise without prior permission, in writing, from the publisher.

Preface

The book that you are holding in your hands is a further entry in a series of volumes that aspires to make some of the ideas presented at the annual international symposium Logica permanently accessible to both the conference participants and the wider public. The symposium, which took place at Hejnice Monastery in the Czech Republic from June 16 to June 20, 2014, brought together logicians from many different countries. This volume contains a representative sample of the contributions made at the conference.

The Logica symposium is an event with a tradition that began in the late nineteen eighties. Since that time, it has evolved into a respected conference which possesses a firm place in the annual schedule of the international community of logicians. Though the symposium is open to researchers of both a mathematical and a philosophical bent, its audience traditionally consists mostly of logicians with philosophical interests. The informal atmosphere provides a space for a stimulating exchange of ideas among logicians of all generations, including students. As the editors of this volume we are proud that we can contribute to the successful completion of the annual symposium cycle by presenting this collection to you.

Last year's Logica was – as were all previous Logica symposia – organized by the Department of Logic of the Institute of Philosophy of the Czech Academy of Sciences. More than thirty lectures were presented during the conference, including those given by a distinguished list of invited speakers: Dorothy Edgington, Dag Prawitz, Graham Priest, and Göran Sundholm. As happens every year, the conference was enriched by a social programme that provided room for friendly debates concerning professional topics as well as for starting and developing personal friendships. The proceedings, which are traditionally published within one year of the conference, unfortunately offer only a very limited record of the topics discussed and cannot hope to even partially convey its atmosphere. In spite of that, we hope that you will find this book worthy of your attention.

Both the Logica symposium and The Logica Yearbook are the result of a joint effort by many people to whom we would like to express our gratitude. We are, of course, very grateful to the Institute of Philosophy for all the important support that made the event possible. We express our thanks to the staff of Hejnice Monastery for their hospitality and friendly assistance. Special thanks from the organizers and also, we believe, from the guests go to the Bernard Family Brewery of Humpolec, which has traditionally sponsored the social programme of the symposium by providing three barrels of its excellent beer. We owe thanks to the Czech Science Foundation, which provided significant support for the meeting and for the publishing of this volume with the funding of the grant project no. 13-21076S. We would like to express our gratitude to Petra Ivaničová, who is a key member of the organizing crew. We would also like to thank College Publications and its managing director, Jane Spurr, for their very pleasant cooperation during the preparation of this book. Last, but not least, we would like to thank all of the authors for their exemplary collaboration during the editorial process.

Prague, May 2015

Pavel Arazim and Michal Dančák

Contents

Jesse Alama:
  Without E   1

Thomas N.P.A. Brouwer:
  Prospects for a Cognitive Norm Account of Logical Consequence   13

Massimiliano Carrara:
  Preliminaries to a Logic of Malfunction   33

Filippo Casati and Naoya Fujikawa:
  The Totality and Its Complement   49

Roberto Ciuni:
  Conjunction in Paraconsistent Weak Kleene Logic   61

Michal Dančák:
  The Intuitionistic Robinson Arithmetic(s)   77

Marie Duží:
  Quantifying into Hyperintensional Attitudes   91

Noah Friedman-Biglin:
  Carnap's Tolerance and Friedman's Revenge   109

Luis Estrada-González:
Fifty (More or Less) Shades of Logical Consequence 127

Yacin Hamami:
Poincaré and Prawitz on Mathematical Induction 149

Geoffrey Hellman and Stewart Shapiro:
Points, Gunk, Boundaries, and Contact 165

Hidenori Kurokawa:
The Principle of Reflection via Nested Sequents 185

Pavel Materna:
Two Approaches to Philosophically Analysing Language 203

Peter Milne:
Structures, Homomorphisms, and the Needs of Model Theory 215

Martin Pleitz:
Curry's Paradox and the Inclosure Schema 233

Emma Ruttkamp-Bloem, Giovanni Casini, and Thomas Meyer:
A Non-classical Logical Foundation for Naturalised Realism 249

Igor Sedlár:
Action Frames for Weak Relevant Logics 267

Vítězslav Švejdar:
On Strong Fragments of Peano Arithmetic 281

Jacek Wawer and Leszek Wroński:
Towards a New Theory of Historical Counterfactuals 293

Alexander Yates:
Clarifying Sense: Frege's Commentary on His Symbolism 311

Dmitry Zaitsev:
Logics of Generalized Classical Truth Values 331

Without E

JESSE ALAMA[1]

Abstract: Dialogue games are a two-player game-based semantics for a variety of logics. The Proponent (**P**) asserts an initial formula φ; the Opponent (**O**) disputes that φ is valid. The players take turns, attacking or defending logical formulas. There are a few ways in which dialogues can be defined. There seems to be no unique formalization, though one can clearly see family resemblances among them. One rule that comes up often is the so-called E rule, which says that **O** must respond to **P**'s immediately preceding move. The E rule is regarded as awkward and lacking a clear significance in any ordinary sense of "dialogue". Its sole value appears to be that it facilitates proofs of adequacy (soundness and completeness). While acknowledging its utility, one might naturally wish to eliminate the E rule, or at least replace it by a rule that has intuitively greater plausibility. In some contexts, it is known that E is in fact redundant (a set of rules with E has the same logical force as the ruleset without E), while in other contexts, E seems to be essential. Toward that end, this note investigates a dialogical characterization of classical logic in which the E rule is relaxed.

Keywords: dialogue games, classical logic, dialogical logic, dialogues

Dialogue games (or, for short, simply "dialogues") provide a two-player game-based semantics for various logics. A Proponent (**P**) asserts an initial formula, the validity of which is disputed by the Opponent (**O**). The players take turns, attacking or defending various claims as the game unfolds. Dialogues were originally bound up with intuitionistic logic (IL), but today dialogue-based semantics can be given for a range of logics. The logical significance of dialogues shows up at the level of winning strategies: a formula is valid if **P** has a winning strategy.

Although there is broad agreement about the underlying principles that govern how dialogue games ought to go, there is no unique formalization of the notion, that is, there are a number of ways in laying down precise rules that the players are to follow, and of specifying when one or the other player wins a dialogue. Are dialogues allowed to be infinite? Are the players

---

[1] Supported by FWF grant P25417-G15 (LOGFRADIG).

allowed to repeat themselves? What are the rules, precisely, that capture various logics in a dialogical way? One can find in the literature various approaches to these issues, even though, indeed, dialogues can be traced back to a single source (Lorenzen, 1958). (For a historical overview of dialogical logic, see (Lorenz, 2001) or (Keiff, 2009). Krabbe (1985) also treats a variety of options for formalizing dialogues for intuitionistic logic.)

Of interest here is one rule that is often found across various formalizations (e.g. Felscher, 1985; Fermüller, 2003): the so-called E rule, which requires that **O** must respond to **P**'s immediately prior move.

The E rule is rightly regarded as intuitively awkward and hard to justify. It lacks a clear justification in any ordinary sense of "dialogue". Why, after all, must **O** be so reactive? (A symmetric form of the the E rule can also be considered (Rahman, 1993), but we do not consider that rule here. The symmetric form is, in any case, only marginally less awkward than the asymmetric E rule that is in focus here.) Apparently, one postulates the E rule only for its instrumental value, namely, because its presence simplifies proofs of adequacy (soundness and completeness) of dialogues when showing that they capture various logics. Felscher (1985), for instance, formalizes dialogue games for intuitionistic logic (IL) both with and without the E rule and shows that the set of dialogically valid formulas (formulas for which **P** has a winning strategy) is the same in both cases. In other words, one can capture IL dialogically without the E rule. Yet when one follows Felscher's formalization of dialogues with the aim of dialogically characterizing classical logic (CL), one finds, perhaps surprisingly, that the E rule cannot be dropped (Alama, Knoks, & Uckelman, 2014). The result is that an unintuitive, hard-to-justify restriction on **O** suffices yet seems to be unavoidable. In (Fermüller, 2003), the set of rules of IL is streamlined, yet nonetheless the E rule is among them; removing E leads to logical chaos (IL is no longer dialogically characterized). Naturally, one wishes to do away with E while still acknowledging that some of its "effects" must be retained (at least in settings such as Felscher's or Fermüller's). Our problem is: can one give a more intuitively justified substitute for E with which one can capture classical logic?

The problem has, in another setting, already been solved. Clerbout (2014) gives a soundness and completeness proof for a dialogical characterization of first-order classical logic. In Clerbout's formulation, the E rule is not present at all. This does not contradict the previously mentioned result of (Alama et al., 2014) because the formalization of dialogues in the two is slightly different. The main difference is that Clerbout adheres to the tradition of dialogues where all possible developments of all dialogue games (that is, any particular play of the game, starting from any initial formula) are finite. The finiteness is assured with the help of so-called repetition ranks, which bound the number of times either player can repeat previous moves. By contrast, in the Felscher-style formalization of dialogue games, dialogues are permitted to be infinite even though winning strategies for **P**, whose existence provides the notion of validity in dialogical logic, are always finite.

By viewing Clerbout's result on classical dialogues from Felscher's perspective, one gets the idea that the E rule can be replaced by a rule that has the effect of somehow limiting potential infinity in dialogue games. In this note we give another dialogical characterization of classical logic where the E rule is relaxed. We are then able to give an account, in another dialogical setting, of Clerbout's important Proposition 2 which expresses that it is enough to restrict the number of times that **O** may repeat himself to 1. We work in a reformulation of Felscher-style dialogues due to Fermüller (2003), which, for the sake of completeness, will be repeated below.

The main insight is to replace the E rule with a new rule, R, which says that **O** is not allowed to repeat himself. By removing E we give **O** to option to delay responding to **P**'s moves. Increasing the number of possible moves for **O** can potentially decrease the set of valid formulas, because **P** may not be able to always successfully respond to **O**'s new moves. The main result of this paper (Theorem 2 below) is that the relaxed "no repeat" rule maintains balance.

The basic setup of the dialogue games here is an adaptation of Fermüller-style dialogue games (Fermüller, 2003). Fermüller's games are themselves a streamlining of Felscher's dialogue games (Felscher, 1985). Fermüller's games were initially developed for intuitionistic logic as a stepping stone for a dialogical treatment of intermediate logics. Our work below assumes the results of Alama, Knoks, and Uckelman (2011), in which Fermüller's games were slightly modified for treating classical logic.

We now define dialogue games for classical logic. The main idea is that in dialogue games, the two players, **P** and **O**, attack formulas and defend against attacks. The ways in which attacks and defenses proceed are governed by the so-called particle rules, which are captured in Table 1 below.

The particle rules for conjunction $\wedge$ are to be understood as follows: if a player X has asserted $\varphi \wedge \psi$, the other player Y has two possible attacks, $\wedge_l$ and $\wedge_r$. The understanding is that X must then defend either φ (if Y's attack is $\wedge_l$) or ψ (if Y's attack is $\wedge_r$). For disjunction, there is a unique possible

Assertion	Attack	Response
$\varphi \supset \psi$	φ	ψ
$\varphi \wedge \psi$	$\wedge_l$ or $\wedge_r$	φ or ψ as appropriate
$\varphi \vee \psi$	?	φ or ψ
p (atomic)	?	—

Table 1: Particle rules for dialogue games

attack, which can intuitively be read as a question, or demand: "which disjunct do you mean?". There are, naturally enough, two possible defenses: one can assert either the left of the right disjunct. Atoms cannot be defended, though they can be attacked (the "?" attack on an atom p can intuitively be understood as: "I dispute p"). The symbols "?", "$\wedge_l$", and "$\wedge_r$" are called symbolic attacks.

The particle rules on their own have hardly any logical significance. In addition to the particle rules, one has so-called structural rules (not to be confused with "structural rule" in the setting of sequent calculus).

Definition 1 (Ruleset $\mathcal{E}$) *The following rules comprise ruleset $\mathcal{E}$:*

Start *The first move of the dialogue is carried out by* **O** *and consists in an attack on the initial formula φ.*

Alternation *Moves strictly alternate between players* **O** *and* **P**.

Atom *Atomic formulas, including $\bot$, may be stated by both players, but only* **O** *can attack them.*

E **O** *must immediately respond to* **P**'s *previous move.*

An $\mathcal{E}$-dialogue is a sequence of moves that adheres to this ruleset.

Finally, one needs to say when the game is over. We are interested only in what counts as a win for the Proponent; it does not interest us here what it means for the Opponent to win.

Definition 2 (Winning conditions (for **P**)) *There are two ways that* **P** *can win a dialogue:*

W^{CL} *The game ends (with* **P** *winning) if* **O** *has attacked a formula that he earlier asserted (or initially granted).*

Without E

$W_\perp$ *The game ends (with **P** winning) if **O** asserts $\perp$.*

W^{CL} refers to initially granted formulas. These are formulas that are considered as asserted by **O** before the game begins. One can think of them as assumptions. (The CL in name "W^{CL}" has to do with the relevance of this winning condition for classical logic. Fermüller also has two winning conditions, one of which is $W_\perp$, but the other is slightly different. This is no surprise because Fermüller is interested in IL rather than CL.)

With these ingredients in place, one can appeal to the notion of winning strategy coming from game theory. (Again, since we are interested only in winning strategies for **P**, by "winning strategy" we always mean "winning strategy for **P**".) One says that a formula φ is $\mathcal{E}$-valid if there exists a winning strategy for φ: a way for **P** to play in such a way that, no matter what **O** does at any stage, **P** can ensure that the game ends with a win for **P**.

We thus have the set of $\mathcal{E}$-valid formulas. It is known that $\mathcal{E}$-valid formulas are precisely the (classical) tautologies:

Theorem 1 φ *is a (classical) tautology iff there exists an $\mathcal{E}$-strategy for φ.*

(This is the main result of (Alama et al., 2011).) The ruleset $\mathcal{E}$ contains, as its name suggests, the notorious rule E. Since we cannot simply eliminate it, the best we can do is relax it.

Definition 3 (Repetition) *A move m' in a dialogue is a repetition of an earlier move m just in case m and m' have the same stance (that is, both are attacks or both are defenses), make the same statement (occurrence of a subformula of the initial formula, or identical symbolic attack), and refer to the same earlier move.*

Some kind of bookkeeping technique would be needed to keep track of whether a certain move in a dialogue is a repetition. We do not care to spell out such details; it suffices for us to work with occurrences of formulas, with the understanding that we can tell whether two formula occurrences are identical. This suffices for our purposes here because, by inspecting the particle rules, one sees that any formula that gets asserted in the course of the game is either a subformula of either the initial formula φ asserted by **P** or else a subformula of a member of the set Π of formulas initially granted by **O**. We may also view the symbolic attack "?" as a kind of formula. φ and Π thus give us a forest of formula occurrences; together with the symbolic attack "?", we have a finite set of all possible formula occurrences in any possible dialogue.

Definition 4 (Ruleset $\mathcal{R}$) *The following rules comprise ruleset $\mathcal{R}$: **Start**, **Alternation**, **Atom** from the ruleset $\mathcal{E}$, plus:*

R **O** *may not repeat himself.*

An $\mathcal{R}$-dialogue is a sequence of moves that adheres to this ruleset.

The winning conditions for $\mathcal{R}$-dialogues are the same as for $\mathcal{E}$-dialogues. The particle rules in Table 1 are understood to be present both for $\mathcal{E}$ and $\mathcal{R}$.

Every $\mathcal{E}$-dialogue is an $\mathcal{R}$-dialogue: if **O** must immediately respond to what **P** just said, there can be no issue of **O** repeating himself. Conversely, it should be intuitively clear that there are $\mathcal{R}$-dialogues that are not $\mathcal{E}$-dialogues: rule R is a genuine weakening of rule E. Thus, in general, $\mathcal{E}$- and $\mathcal{R}$-strategies are different. Nonetheless, we have:

Theorem 2 φ *is a tautology iff there exists an $\mathcal{R}$-strategy for φ.*

This is the main result of the current paper. The theorem follows from Proposition 1 and 2 below, which separately prove the left-to-right and right-to-left directions of the equivalence.

In the following, we shall restrict ourselves to formulas built from atoms, implication and $\bot$; extending the proofs to treat $\wedge$ and $\vee$ as well is a straightforward exercise. (Besides, $\supset$ and $\bot$ suffice to define all connectives in classical propositional logic, so we are on safe ground from a semantic point of view.)

To prepare for the proof of Proposition 1 below, we need the notion of a dialogue sequent. Following Fermüller, we may think of a dialogue as a state transition system, in which edges are moves that follow the dialogue rules. For us, the notion of state in the game will be a sequent $\Gamma \vdash \Delta$, in which Γ and Δ are sets of formulas. The game begins with a set of initially granted formulas by **O** (which we can regard as assumptions), and the initially disputed formula. For us, the state changes only when **O** makes a move. The left hand side Γ of a dialogue sequent $\Gamma \vdash \Delta$ is simply a recording of what **O** has said so far. Δ records what **O** has attacked.

Considering dialogues as state transition systems, we say that a node is a **P**-node if it is **P**'s turn (that is, the outgoing edges from this node represent moves by **P**). Similarly, an **O**-node is a node where it is **O**'s turn to move.

In Fermüller's treatment of intuitionistic logic, the notion of state in a game was slightly different. In his case, the right-hand sides Δ were always singletons, so the right-hand side could not be a recording of all the formulas that **O** has ever attacked. Our modification of Fermüller's idea parallels the way that classical sequent calculus differs from its intuitionistic counterpart. One sees that the notion of state in a dialogue game is rather simpler in classical logic than in intuitionistic logic.

The proof of Proposition 1 refers to the notion of a dialogical sequent $\Gamma \vdash \Delta$ being classically valid. By this we mean, as usual, that every Boolean truth assignment that makes every member of Γ true makes at least one member of Δ true.

Proposition 1 *If φ is $\mathcal{R}$-valid, then φ is a (classical) tautology.*

Proof. Let τ be an $\mathcal{R}$-strategy for φ. We show that classical validity is preserved when moving up from leaves to the root.

Every leaf λ of τ can be understood as an $\mathcal{R}$-dialogue in which **P** has won. Thus, at λ one of the two winning conditions W^{CL} or $W_\perp$ is satisfied. If **O** has attacked a formula that he either initially granted or asserted in the course of the dialogue, then the dialogue sequent at λ looks like $A, \Gamma \vdash \Delta, A$, which is classically valid. If we got to λ by **O** just asserting $\perp$, then the dialogue sequent at λ looks like $\perp, \Gamma \vdash \Delta$, which is also classically valid.

Let us now consider internal (non-leaf) nodes of τ. Let ν be an internal node. There are two possibilities:

ν is a P-node If ν is a **P**-node, then ν has exactly one child ν' (since τ is a winning strategy). Since the dialogue sequents are merely a bookkeeping device that records what **O** has said and what **O** has attacked, the dialogue sequent at ν is identical to that of ν'. Thus classical validity trivially transfers from ν' to ν.

ν is an O-node Let $\nu_1, \ldots, \nu_m$ be the children of ν. **O** might be responding to **P**'s immediately prior move (or attacking the initial formula of the dialogue), but the R rule allows the possibility that **O** might be responding to earlier moves by **P**. Interestingly, the increased ability of **O** does not matter. In other words, for the purposes of the theorem under discussion, it is enough to show that classical validity of the dialogue sequents corresponding to ν_k that do adhere to the E rule implies the classical validity of the dialogue sequent corresponding to ν. Consider what move **P** made to get to ν:

1. **P** defends $A \supset B$ and **O** immediately counterattacks. This means that **O** asserted A earlier and **P** has just asserted B, so the dialogue

sequent at ν has the form $A, \Gamma \vdash \Delta, A \supset B$. **O** now attacks B. There are two cases:

- If B is an atom, then the dialogue sequent the ν_k corresponding to this move is $A, \Gamma \vdash \Delta, A \supset B, B$. But

$$A, \Gamma \vdash \Delta, A \supset B, B$$

classically entails

$$A, \Gamma \vdash \Delta, A \supset B.$$

- B is an implication $B_a \supset B_c$. Then the dialogue sequent at the ν_k corresponding to this move is $A, B_a, \Gamma \vdash \Delta, A \supset B_a \supset B_c$. But

$$A, B_a, \Gamma \vdash \Delta, A \supset B_a \supset B_c$$

classically entails

$$A, \Gamma \vdash \Delta, A \supset B_a \supset B_c.$$

2. **P** attacks $A \supset B$ and **O** responds immediately against this attack. There are two ways for **O** to do so: by attacking **P**'s assertion of A, or by defending B. Let $A \supset B, \Gamma \vdash \Delta$ be the dialogue sequent at ν. Both moves are possible for **O**; we need to show how we can use the classical validity of the dialogue sequents for these two successors of ν to guarantee the classical validity of the dialogue sequent for ν. As in the previous case, we need to distinguish between the case where A is an atom from the case where A is complex.

- If A is an atom, then the dialogue sequents of the two relevant children of ν are

$$A \supset B, \Gamma \vdash \Delta, A$$

and

$$A \supset B, B, \Gamma \vdash \Delta.$$

But these classically entail

$$A \supset B, \Gamma \vdash \Delta$$

as desired.

Without E

- If A is complex, say, $A = A_a \supset A_c$, then the dialogue sequents of the two relevant children of ν are

$$(A_a \supset A_c) \supset B, A_a, \Gamma \vdash \Delta, A_a \supset A_c$$

and

$$(A_a \supset A_c) \supset B, B, \Gamma \vdash \Delta.$$

There may be other children of ν (which would be violations of the E rule), but we do not need to account for them; what matters is that there are two children of ν with the displayed dialogue sequents. Since these two classically entail

$$(A_a \supset A_c) \supset B, \Gamma \vdash \Delta,$$

the proof is complete.

□

The following definition anticipates the proof of Lemma 1 below.

Definition 5 *An* **O**-*node* ν *in a strategy* τ *is said to be* **bad** *if* ν *lacks a child node corresponding to a possible* **O**-*move violating the E rule yet still adhering to the R rule.*

An $\mathcal{E}$-strategy may have bad nodes. A strategy that has no bad nodes is an $\mathcal{R}$-strategy. The root node of an $\mathcal{E}$-strategy cannot be bad (because **O**'s initial move is, by definition, an attack on the initial formula).

Lemma 1 *Every $\mathcal{E}$-strategy for φ can be converted into an $\mathcal{R}$-strategy for φ.*

Proof. Let τ be an $\mathcal{E}$-strategy, and let n be the number of bad nodes in τ. If $n = 0$, then τ is already an $\mathcal{R}$-strategy. Supposing that $n > 0$, we shall show how to produce a sequence $\tau_1, \ldots, \tau_n$ of strategies, starting with τ, with the property that τ_k has $n - k$ bad nodes. In other words, we can repair all the bad nodes. We start with an $\mathcal{E}$-strategy and end with an $\mathcal{R}$-strategy. Along the way, the in-progress strategies we consider are neither $\mathcal{E}$- nor $\mathcal{R}$-strategies.

Let ν be a bad node in τ_k ($1 \leq k < n$). Without loss of generality, we may assume that ν is maximally shallow (that is, above ν there are no bad nodes). (If there are multiple maximally shallow bad nodes, the choice of which one to choose is arbitrary.) Let $\nu_1, \ldots, \nu_p$ be the children of ν. To say that ν is bad means that some of the $\mathcal{R}$-possibilities of **O** are being neglected at ν, or, in other words, that $\nu_1, \ldots, \nu_p$ is not a full listing of **O**'s possibilities under rule R. Simply extend ν with each of these missing possibilities; call the new nodes $\nu'_1, \ldots, \nu'_q$. Our task now is to show that **P** can continue playing in such a way as to force a win for himself. Some of the new nodes $\nu'_1, \ldots, \nu'_q$ might actually be terminal (that is, **O** might have just lost the game by making some of these moves). But for the sake of notational simplicity, let us assume that none of the new nodes are terminal. Each of the ν'_k, which are **P**-nodes, can be extended by simply having **P** repeat whatever move he made to get to ν. (There is such a move, since ν cannot be the root of τ.) Such a move by **P** is legal because rule R prevents only **O** from repeating himself, not **P**. Each of the ν'_ks has now been extended with a unique **P** move, and we return to the earlier situation. We now extend each of these extensions by all $\mathcal{R}$-possible moves of **O**. The crucial thing to notice is that each of these extensions has one fewer possibility, owing to rule R. We continue as follows: as long as there are bad nodes, have **P** simply repeat himself. Eventually, we force **O** into a situation where all his available moves are indeed responses to **P**'s immediately prior move. Thus, we can extend the tree, finally, by copying $\nu_1, \ldots, \nu_p$, as appropriate.

After all this work, we have produced a new strategy τ_{k+1} having precisely one fewer bad node. Repeating this process n times, we eventually reach a strategy τ_n having 0 bad nodes, which is an $\mathcal{R}$-strategy. □

Proposition 2 *If $\Pi \vdash \varphi$ is classically valid, then there is an $\mathcal{R}$-strategy for φ with initially granted formulas Π.*

Proof. By Theorem 2 of (Alama et al., 2011), if $\Pi \vdash \varphi$ is classically valid, then there is an $\mathcal{E}$-strategy τ for φ with initially granted formulas Π. By Lemma 1, τ can be extended into an $\mathcal{R}$-strategy τ' (for φ, with the same set Π of initially granted formulas). □

What has been gained by relaxing rule E to rule R? Is ruleset $\mathcal{R}$ more compelling, from a dialogical logic point of view, than ruleset $\mathcal{E}$, which is based on the inelegant E rule? Rule R is, in our view, more persuasive than rule E because it formally captures a kind of "focused conversation" principle. By allowing **O** to make any move that adheres to the particle rules subject only to the constraint that he not repeat himself, it would seem that we have eliminated at least some of the intuitively unwelcome aspects of rule E. **O** now has a wider range of possibilities, and is bound only by an intuitively acceptable constraint.

Nonetheless, rule R shares with rule E the curious property that only **O** is constrained. Looking at the proof of Lemma 1, we introduce potentially many repetitions by **P**. We have tried to shake off the instrumental attitude that justifies rule E, but we get our desired completeness proof at the cost of having **P** repeat himself without abandon. It would be worthwhile to investigate the extent to which one could restrict **P** at least somewhat more. But, as we often see in logic, there is a tradeoff that one must be aware of: if we constraint **P** too much (or if we give **O** more freedom), than we might fail to capture any logic at all.

References

Alama, J., Knoks, A., & Uckelman, S. L. (2011). Dialogue Games for Classical Logic. In M. Giese & R. Kuznets (Eds.), *TABLEAUX 2011 Workshops, Tutorials and Short Papers*.

Alama, J., Knoks, A., & Uckelman, S. L. (2014). A Curious Dialogical Logic and Its Composition Problem. *Journal of Philosophical Logic*, *43*(6), 1065–1100.

Clerbout, N. (2014). First-order Dialogical Games and Tableaux. *Journal of Philosophical Logic*, *43*(4), 785-801.

Felscher, W. (1985). Dialogues, Strategies, and Intuitionistic Provability. *Annals of Pure and Applied Logic*, *28*(3), 217–254.

Fermüller, C. G. (2003). Parallel Dialogue Games and Hypersequents for Intermediate Logic. In M. Cialdea Mayer & F. Pirri (Eds.), *Automated Reasoning with Analytic Tableaux and Related Methods (TABLEAUX 2003)* (Vol. 2796, pp. 48–64). Berlin: Springer.

Keiff, L. (2009). Dialogical Logic. In E. N. Zalta (Ed.), *The Stanford Encyclopedia of Philosophy* (Summer 2009 ed.). http://plato.stanford.edu/archives/sum2011/entries/logic-dialogical/.

Krabbe, E. C. W. (1985). Formal Systems of Dialogue Rules. *Synthese*, *63*, 295–328.

Lorenz, K. (2001). Basic Objectives of Dialogical Logic in Historical Perspective. *Synthese*, *127*, 255–263.

Lorenzen, P. (1958). Logik und Agon. In *Arti del XII Congresso Internationale de Filosofia* (pp. 187–194). Florence: Sansoni.

Jesse Alama

Rahman, S. (1993). *Über Dialoge, Protologische Kategorien und Andere Seltenheiten*. Pieterlen: Peter Lang.

Jesse Alama
Vienna University of Technology
Austria
E-mail: `alama@logic.at`

Prospects for a Cognitive Norm Account of Logical Consequence

THOMAS N.P.A. BROUWER[1]

Abstract: When some P implies some Q, this should have some impact on what attitudes we take to P and Q. In other words: logical consequence has a normative import. I use this idea, recently explored by a number of scholars, as a stepping stone to a bolder view: that relations of logical consequence can be identified with norms on our propositional attitudes, or at least that our talk of logical consequence can be explained in terms of such norms. I investigate the prospects of such a cognitive norm account of logical consequence. I go over the challenges involved in finding a plausible bridge principle connecting logical consequence to cognitive norms, in particular a biconditional principle that gives us not only necessary but sufficient conditions for logical consequence in terms of norms on propositional attitudes. Then, on the assumption that an adequate norm can be found, I consider what the philosophical merits of such a cognitive norm account would be, and what theoretical commitments it would generate.

Keywords: logical consequence, normativity, propositional attitudes, expressivism

1 Introduction

It is natural to think that when some P implies some Q, that fact should have some impact on our attitudes to P and Q. At a minimum: if I accept P but reject Q, my attitudes are not as they should be, in an objective sense of 'should'. This natural thought has motivated a small but significant literature in the philosophy of logic, exploring the idea that there is a characteristic sort of norm that relations of logical consequence give rise to. In this paper I want to build on that literature to explore the prospects of a bolder hypothesis: that relations of logical consequence not only give rise

---

[1] I would like to thank Ole Hjortland for first giving me the idea for this paper. This paper was prepared within the 2013–15 AHRC project The Metaphysical basis of Logic: the Law of Non-Contradiction as Basic Knowledge (grant ref. AH/K001698/1).

to cognitive norms, but that they are cognitive norms – or, at least, that our talk of logical consequence is to be explained in terms of such norms. That view has some challenges to overcome, I'll argue, but there is some mileage in it.

2 Looking for a bridge principle

The aforementioned literature on the normative import of logical consequence has focused on finding a suitable 'bridge principle', one which takes us from claims about logical relations between propositions to claims about how our propositional attitudes ought to be. Significant contributions to this research programme are (Field, 2009, Forthcoming; MacFarlane, Unpublished; Restall, 2005) and, from a more sceptical angle, (Harman, 1986). Here is a first stab at a bridge principle:

(1) $P_1, P_2, \ldots, P_n$ logically imply $Q \to$ if one accepts $P_1, P_2, \ldots, P_n$, one ought to accept Q.

Basically, the logical implication gives rise to an instruction to expand our acceptances in a certain way, given which things we've already accepted.[2] Before we discuss the merits of this principle, let me get a few clarifications out of the way.

First off, I use "P_1", "P_2", "P_n", and "Q" as names of propositions. The relation of logical consequence can also be understood as a relation between sentences, so it might be worth pointing out why I use the letters in this way. One obvious reason is that P_1, P_2, P_n, and Q are to be the objects of propositional attitudes and, as the term suggests, those are usually taken to be propositions. Another reason is that if logical entailments are ultimately cognitive norms, then presumably the reason to care about logical consequence is to find out what is required of us, cognitively. If so, the requirements must be pre-existent, not artefacts of our logical theorising. In other words, the relation of logical consequence we are looking at is not that of some particular formal system – it is rather the thing that such systems aim to capture. As such, it seems more appropriate to treat it as a relation between propositions, a relation which can then be modelled, in formal systems, by a relation between sentences.

---

[2]Throughout this paper I will use the phrase 'B is a consequence of A' interchangeably with 'A implies B'.

Prospects for a Cognitive Norm Account of Logical Consequence

Secondly, I am not going to be greatly concerned with the nature of the (bi)conditionals appearing in these bridge principles. I am going to suppose that they at least (i) detach and (ii) contrapose. Apart from that I take no view on the matter.

Now, what are the merits of principle (1)? It seems to get something right, but not all that it should. To begin with, it might be thought to be too strong. One reason to think so is that it imposes obligations to believe even when the implication relation between $P_1, P_2, \ldots, P_n$, and Q is very difficult to find out about. Presumably some things we already accept either imply Goldbach's Conjecture or imply its negation. Principle (1) tells us we are presently under an obligation to accept one or the other, though we can't tell which. That may appear overly strong.

It is not clear whether it is indeed too strong. One thing to observe is that the 'ought' in principle (1) is intended to be an objective one. Failure to comply with it may well be excusable and even to be expected in cases where our epistemic situation is less than ideal. Also, there is a certain amount of interpretation possible with regard to the force of the 'ought'. One might understand it in a very weak sense, as saying that we are 'committed' to accepting Q by what we already accept, where commitment could be such a deflated notion that we may well expect to sometimes end up committed to accepting things which we have no real business accepting. However, while it may be understood that way, it is implicit in the literature on the cognitive import of logical consequence that something more robust and genuinely demanding is intended. And that strikes me as right – commitment in this deflated sense doesn't seem all that worth caring about. But even on a robust reading of the 'ought', it is not clear to me that it is too strong, given that some failures to comply might be eminently excusable, such as in the aforementioned case of Goldbach's Conjecture.

Principle (1) has another clear failing, though. As Broome (1999) has pointed out, since logical consequence is very often thought to be reflexive, we get the result that I ought to accept anything I do in fact accept. And that seems inappropriate in cases where I have no business accepting something in the first place. This also shows us a way in which (1) is too limited: it only gives us instructions about how to expand our acceptances. And finding out that some P implies some Q can have other upshots too: if I happen to think that Q is a ridiculous thing to accept, the entailment shows me I shouldn't accept P either. Principle (1) does not capture that.

2.1 Improving the principle

One may fix this latter failing by giving the 'ought' wide scope:

(2) $P_1, P_2, \ldots, P_n$ logically imply $Q \rightarrow$ one ought to be such that if one accepts $P_1, P_2, \ldots, P_n$, one accepts Q.

This gives us the right sort of bi-directionality. If P_1 through P_n are good things to accept, then the right thing to do is to accept Q as well. But if Q is a bad thing to accept, not all of P_1 through P_n should be accepted. However, insofar as (1) was overly demanding, (2) is no better. MacFarlane considers restricting the antecedent to known entailment:

(3) One knows that $P_1, P_2, \ldots, P_n$ logically imply $Q \rightarrow$ one ought to be such that if one accepts $P_1, P_2, \ldots, P_n$, one accepts Q.

But he rejects this option, correctly in my opinion, because it muddles up the order of explanation between knowledge of entailments and the normative import of entailments. According to (3), logical entailments require things of me insofar as I know about them. But it seems, rather, that I have an interest in knowing about logical entailments precisely because they require things of me. In that case, these requirements had better be there already, before I find out about them. Furthermore, (3) might just be too weak altogether. In 'easy' cases of entailment, I do not get excused from complying with the norm simply by failing to spot the entailment. It's possible, though, to explain such cases by appealing to a further, separate obligation to know about certain entailments.

2.2 A proscriptive norm

Another different to deal with the alleged over-demandingness of (2) is to change the content of the obligations themselves:

(4) $P_1, P_2, \ldots, P_n$ logically imply $Q \rightarrow$ one ought not accept $P_1, P_2, \ldots, P_n$ whilst rejecting Q.

This principle, suggested by MacFarlane and defended by Restall (2005), changes the obligations from obligations to accept to obligations not to reject, and on this principle, logical entailments merely prohibit certain combinations of attitudes.[3] Whereas principles (1) – (3) give us prescriptive

---

[3] It should be noted that Restall defends a version of this principle which allows for multiple conclusions – i.e. some things entail some others when one ought not accept all of the former

norms, (4) gives us proscriptive norms. If one is worried about cases like Goldbach's Conjecture, these are now less worrisome: the fact that Goldbach's Conjecture is entailed by what I accept (let's say) does not require any action on my part. It merely requires that I do not reject Goldbach's Conjecture, which was already the sensible thing to do in my epistemic situation.

Principle (4) may strike some as overly weak, in that the requirements it gives rise to can be respected by simply suspending judgment on all matters. But I believe that the merely proscriptive nature of those norms is entirely appropriate to the role that logic plays in our epistemic lives. While we may well think that we are sometimes obliged to accept something, this can be understood as arising from factors extrinsic to logic. One might think, for instance, that there is a value to having true beliefs (and false disbeliefs) and so we have compelling reason to expand our beliefs and disbeliefs. Logic then merely gives us guidance on how to go about it. Such a division of labour gives us much more room to manoeuvre, too: for while one might believe that true belief is generally valuable, one might instead believe that its value depends on the subject matter (in certain matters it is not worth my while to expand my opinions) or on further extrinsic factors (e.g. true belief is only of full value if combined with justification, or somesuch). Principle (4) leaves such matters open, and that seems appropriate. It is the principle I favour, on balance.

This is not nearly the end of the matter. One might worry that even the relatively weak (4) still gives the wrong result with regard to cases like the Preface Paradox. In such cases it seems appropriate that I accept a bunch of propositions, yet reject their conjunction, even though the latter is entailed by the former. There are various ways to go; one might claim that the appropriate attitude to the conjunction is suspension of belief rather than rejection, which is compatible with (4). Or one might just reject adjunction. MacFarlane (Unpublished) favours diagnosing the paradox as a case where objective epistemic obligations come into conflict with subjective epistemic obligations; principle (4) gives us the objective one, the conjunction-rejecting attitude to the paradox is the one we are subjectively obliged to take.

But probably the most popular line on the preface paradox is to treat it in terms of partial belief: the appropriate stance is one in which one has a high but lower-than-one credence in the conjuncts, and a low but higher-

---

while rejecting all of the latter. I believe there is something to be said for a multiple-conclusion formulation, but in the present context it would be a distraction, so I restrict myself to single-conclusion principles.

than-zero credence in the conjunction. Given enough conjuncts, that seems to give the right result. Principle (4) is in principle open to being generalised to the case of partial belief, and (Field, 2009, Forthcoming) attempts to do this. I will not explore this option, however, as it would introduce many complications not germane to our topic.

It should also be noted that there are further options for bridge principles that might in principle be considered. One might choose not to cast the principles in terms of obligations at all; there are other normative notions around. One might cast them in terms of reasons to accept or not reject, an option MacFarlane considers, or in terms of the epistemic value of accepting or rejecting something, or in terms of what is correct to accept and reject. I think, all things considered, that casting them in terms of obligations is our best bet, but other options deserve examination. However, for the remainder of this paper, I will be building on principle (4).

3 Getting logical consequence from cognitive norms

Let us assume, for the sake of argument, that principle (4) gives us a plausible account of the normative import of logical consequence. There is a considerable step from there to the much bolder claim I announced above, that logical entailments are to be identified with norms on our attitudes. Can bridge principle (4), or any of the others for that matter, bear that weight? If it is to do so, it had better be the case not only that every case of logical entailment has the normative upshot (4) predicts, but also that there are no cases in which such norms arise without it being a case of logical entailment. In other words, principle (4) should plausibly give us not only a necessary condition for logical entailment but a sufficient condition as well. We ought to be able to replace the conditional in (4) with a biconditional:

(5) $P_1, P_2, \ldots, P_n$ logically imply $Q \leftrightarrow$ one ought not accept $P_1, P_2, \ldots, P_n$ whilst rejecting Q.

Is (5) plausible? Not as it stands; it gives us false positives. Plausibly, one ought not accept that the sky is green. Also plausibly, one ought not reject that snow is white. Combining the two, one ought not accept that the sky is green whilst rejecting that snow is white. But there is not the merest whiff of logical consequence to be detected here. We have, then, a false positive.

Prospects for a Cognitive Norm Account of Logical Consequence

3.1 Necessitation

A fix immediately suggests itself: we should necessitate the consequent:

(6) $P_1, P_2, \ldots, P_n$ logically imply $Q \leftrightarrow$ necessarily, one ought not accept $P_1, P_2, \ldots, P_n$ whilst rejecting Q.

Now the false positive is gone. But there are other ones much like it which meet this modalised requirement. Plausibly one ought not accept that the moon is red and green all over, and since it is necessarily not red and green all over, one necessarily ought not accept it – bearing in mind that we are speaking of objective epistemic obligations. It follows that necessarily, one ought not accept that the moon is red and green all over whilst rejecting that Q, for any Q. But surely this necessary falsehood does not logically entail everything. A further fix seems to be required.

3.2 Incoherence

Restall (2005) suggests that the cognitive requirements that logical entailments impose on us arise from the need to avoid a kind of incoherence in our mental states. Here's how one might build that in:

(7) $P_1, P_2, \ldots, P_n$ logically imply $Q \leftrightarrow$ one ought not accept $P_1, P_2, \ldots, P_n$ whilst rejecting Q, on pain of incoherence.

The thought is as follows: when some P entails some Q, it's because either accepting P would land one in a state of incoherence, or rejecting Q would do so, or the combination of the two would do so. If P were a contradiction, one could see how it would do so, and similarly if Q were a tautology. Hence contradictions imply everything and tautologies follow from anything. But the claim that the moon is red and green all over is not a contradiction, necessarily false though it might be, and so it is not incoherent to accept it, even though one ought not do so. Thus, to accept it whilst rejecting some other proposition is also not incoherent (unless the other proposition is a tautology).

Principle (7) is promising, but as it stands it is not quite satisfactory. It appeals to the notion of incoherence, and that in itself is not an unreasonable thing to do – we do have such a notion. But the problem is whether the notion of incoherence that is available to us pre-theoretically is circumscribed enough to get the job done here. And if it is not, can we specify the more precise notion of incoherence in a way that does not involve a prior account of what logically entails what?

First off, it is not clear to me that the claim that the moon is red and green all over is not incoherent, on a pre-theoretic notion of incoherence. Its truth is unimaginable, so the attempt to do so at the very least engenders some confusion in us. It is not clear to us what believing that the moon is red and green all over would amount to. So we cannot assume that incoherence, on a pre-theoretic understanding, is sufficient for being logically at fault. Secondly, it seems fair to me to describe a subject as incoherent whose beliefs are radically unconnected, i.e. seemingly chosen at random. But to believe entirely unconnected things is not sufficient for being logically at fault – as long as none of the things one believes are inconsistent. So we cannot assume that incoherence, on a pre-theoretic understanding, is sufficient for being logically at fault.

What we need, then, is a more precise and independently motivated notion of incoherence. Restall does sketch one. It boils down to the idea that incoherence is, in the first instance, a state where one accepts and rejects the same proposition – as state in which one, as it were, disagrees with oneself. Other states can then also be regarded as incoherent if they implicitly commit one to taking opposite attitudes to some particular proposition. For instance, the state of accepting A & B and rejecting A is incoherent because, in some sense, to accept A is to commit to accepting A, and hence to be committed to accepting something one rejects.

The state of accepting and rejecting the same proposition strikes me as a distinctive form of incoherence, and to that extent I think Restall's suggestion is attractive. What is not obvious is how the extended notion of 'implicit' incoherence is to be unpacked satisfactorily, i.e. without relying on the fact that A & B logically implies A. One way to do this, perhaps, is to claim that the very meaning of a logical connective like "&" is bound up with commitments or obligations to take (or not take) certain attitudes to its connectands, so that (for instance) by asserting a conjunction one expresses a felt obligation to not reject its conjuncts. I believe such a view has its attractions.[4]

3.3 Logical form

Another tactic for disposing of the false positive is to appeal explicitly to considerations of logical form. Here is how the principle might run:

---

[4] I have explored the merits of such a view in (Brouwer, 2013).

Prospects for a Cognitive Norm Account of Logical Consequence

(8) $P_1, P_2, \ldots, P_n$ logically imply $Q \leftrightarrow$ for any $\phi_1, \phi_2, \ldots, \phi_n$ and ψ of the same logical form as $P_1, P_2, \ldots, P_n$ and Q, one ought not accept $\phi_1, \phi_2, \ldots, \phi_n$ whilst rejecting ψ.

The thought would be as follows. Perhaps we necessarily ought not accept that the moon is red and green all over, but there is nothing about the logical form of that claim that makes it so. It is not the case, for instance, that we necessarily ought not accept that the moon is rocky and cold all over. This is different with contradictions: the very reason they are unacceptable (*pace* dialetheic views on contradictions) is that they have a particular logical form, and mutatis mutandis the same goes for logical tautologies.

So far I have been speaking solely about propositions, but the prototypical bearers of logical form are sentences. There is thus a theoretical choice at this point; either we understand principle (8) as somehow bringing sentences into the picture, or we adopt some account of propositions, one which they can be attributed to logical form. The latter is not at all unprecedented: views that treat propositions as structured entities can talk of the logical form of propositions, and such views can be independently motivated, for instance by the need to deal with the phenomenon of hyperintensionality (see for instance Soames, 1987).

If we do not want to treat propositions as bearers of logical form, we could understand principle (8) with reference to the logical form of sentences. Let us imagine a hypothetical language L, the consequence relation of which (a relation between sentences) perfectly models the 'true' consequence relation (a relation between propositions). When principle (8) speaks of propositions 'of the same logical form as' some proposition P, take these to be the propositions expressed by sentences of the same form as any sentence of L that expresses P. I say 'any', because if propositions are understood to lack logical form and hence as coarse-grained, there will be multiple sentences of L that express P.

But whichever way talk of logical form in principle (8) is understood, it had better be the case that the notion of logical form employed does not implicitly involve a prior conception of logical consequence. Otherwise, the account of logical consequence which we are exploring will end up being parasitic upon another. One option that fits with the spirit of the proposal we are examining is to understand the logical form of a sentence/proposition directly in terms of what the assertion/acceptance of it would require of us as regards to attitudes to other propositions (particularly the ones occurring in the sentence/proposition). In other words, something similar to the view floated above, in the context of Restall's notion of logical incoherence.

3.4 Generic consequence

An entirely different response to the false positives we've been discussing is to embrace them. Perhaps they are cases of consequence – just not cases of logical consequence. This would amount to shifting the goalposts: to claim that the more interesting notion or our account to capture is a more general notion of consequence, of which logical consequence is just one species.[5] This is not a crazy view. One might think that this more general kind of consequence is really of more basic interest to us than logical consequence: after all, what we primarily want is for our reasoning to take us from things worth believing to (only) other things worth believing. The logical validity of such reasoning, one might think, is only of interest to us secondarily: insofar as it confers particularly high grade of reliability on the reasoning.

There are a few ways to go. One might hold that the proper notion of generic consequence is captured by principle (6). That is: some propositions generically imply some other proposition when, of necessity, one ought not accept the former whilst rejecting the latter. However, once one has abandoned logicality as a target, it might be hard to motivate caring in particular about those inferences which are backed up by necessitated norms. One might as well embrace all of the purported false positives and go back to principle (5). After all, to reason from 'Fido is a dog' to 'Fido is a mammal' is a perfectly good thing to do, even though neither logical form nor any deep kind of necessity backs up the goodness of the inference.

3.5 Relevance

Nevertheless, even if one is willing to abandon logical consequence as one's particular target, one might be bothered by another aspect of the false positives discussed above: the fact that they constitute failures of relevance. To reason from the 'grass is red' to 'snow is white' is indeed bad – one might think – but not so much because the latter does not follow logically from the former, but rather because the latter has nothing to do, content-wise, with the former.

Relevance is usually discussed by logicians as a possible necessary condition on logical consequence. But there is no reason why one could not

---

[5]I would like to thank Göran Sundholm and Dorothy Edgington for drawing my attention to this option.

care about relevance in its own right, quite apart from whether it contributes to the *logicality* of a consequence relation. One might simply desire from one's inferences that they preserve truth or some other privileged status, and in addition that they be relevant. One might think that inferences meeting the standard of relevance are by that token more reliable, in a way different from but comparable with the way that logical validity makes inferences more reliable.

There are various ways one might build a condition of relevance into principle (5), and inspiration for these might be sourced from the various approaches to relevant logic that are already around in the literature. A popular view is that relevant consequences should at least exhibit *variable sharing*: the premises and the conclusion have some proper or improper part in common. This however, can only be built into the principle if one has a notion of logical form available, which one may not. A better option might be to go directly for a condition of a more semantic nature. At a first approximation, we might formulate it in terms of aboutness:

(9) $P_1, P_2, \ldots, P_n$ imply $Q \leftrightarrow$ one ought not accept $P_1, P_2, \ldots, P_n$ whilst rejecting Q and every proposition among $P_1, P_2, \ldots, P_n$ and Q is about the same thing or things as at least one other proposition among $P_1, P_2, \ldots, P_n$ and Q.

The notion of some proposition being about some thing or things would then of course need theoretical unpacking, but this is already an area of active research (see for instance Yablo, 2014). It should be noted too that if one is not attracted to the goal-post-shifting tactic, and one wants to pursue a notion of properly logical consequence, there is no reason why one could not use a relevance condition in addition to any of the other conditions above, albeit at the price of further complicating one's bridge principle.

4 Features of a cognitive norm account

The preceding section leaves a lot open, and is meant to. I have indicated a number of directions along which one might pursue a biconditional principle that captures necessary and sufficient conditions for logical or generic consequence. I will be happy if I have shown that there is at least some hope for finding such a principle. In the remainder of this article I will assume, for the sake of argument, that we have found ourselves a biconditional principle that does the work required. Let us see what a theory of logical consequence built on top of such a principle would be like.

4.1 Identity theory versus expressivism

The most straightforward version of a cognitive norm account of logical consequence will simply identify the relation of logical consequence between some P and Q with the norm that our preferred biconditional associates with it. Relations of logical consequence just *are* norms, on this version of the view. Call it the 'identity theory'. Is it an appealing view?

On the identity theory, to assert the existence of a relation of consequence is to assert the existence of a norm. There are advantages to that. Assuming that we were in any case committed to some norm arising from the consequence relation, we have now effected a theoretical reduction of two entities to one. And we have rid ourselves of the burden of explaining why relations of consequence have a bearing on our propositional attitudes – it is their very nature to have a bearing on propositional attitudes, given the sorts of things they are.

But one might also have metaphysical misgivings at this point. Normativity is and has long been a source of philosophical puzzlement. How can such things exist in a world that at bottom is supposed to consist of cold hard facts? How can an 'is' ever yield an 'ought'? If the identity theory identifies an entity which up until now was not a source of metaphysical puzzlement (though perhaps of other kinds of puzzlement) with an entity that is, then we might be worse off than before.

Norms are indeed ill-understood, but that is no reason for despair, for many important notions are. Meta-ethicists are on the case, and hopefully someday they will tell us how it is that norms fit in a world of cold hard facts. But, that being said, it might be that the task of explaining how cognitive norms exist can be avoided altogether. Hartry Field (2009, Forthcoming), who as far as I am aware is the only one presently defending a cognitive norm account of logical consequence, does not defend the identity theory. In fact he is a primitivist about logical consequence, and does not think it admits of any analysis. But he thinks our talk of logical consequence can be explained in an expressivist manner, in terms of cognitive norms.

An expressivist version of the cognitive norm account would say that, whenever we assert that some P implies some Q, we are expressing a felt obligation to have certain attitudes to P and Q. Perhaps, along the lines of principle (5), we feel that we ought not accept P whilst rejecting Q – and

Prospects for a Cognitive Norm Account of Logical Consequence

nor should anyone else. The necessary and sufficient conditions that principles (5) – (9) give us are, on this view, not conditions for the obtaining of a consequence relation, but conditions for the assertibility of a consequence claim. Talk of norms does not vanish on the expressivist version, but appears in a different, less metaphysically robust way. Statements of logical consequence do not state the existence of a norm, but express our commitment to a norm. A norm one so commits to need not have any mind-independent status – it can be a mere fiction, and nevertheless do all the work it needs to.

That, at least, is the hope. It is not altogether obvious that metaphysical misgivings about normativity will not reappear somewhere else in the picture. One might be tempted to say, for instance, that the cognitive norms that we are inclined to impose on ourselves are not entirely accidental. There are better and worse norms that one might impose on oneself, and that fact demands explanation of a metaphysical sort. And perhaps there our prior puzzlement will reassert itself. Perhaps – but it may be that a sufficiently serious expressivist will dismiss such further questions about better and worse norms as symptoms of an insufficiently exorcised realism. One can indeed – they would say – affirm that some norm is better than another; but to do so is nothing more than to commit to the former and not the latter.

The expressivist version of the account also helps with another potential worry about the identity theory. When we assert that some P implies some Q, we do not think that we are talking about obligations or attitudes. We think we are talking about whatever P and Q are about, and nothing else. Yet if the identity theory is correct, to talk about a relation of logical consequence is ipso facto to talk about obligations and attitudes. Isn't that the wrong result? To be honest, I think the force of this worry should not be overstated. It is perhaps a surprising result that talk of consequence is talk of cognitive norms, but the identity theory is an analysis of logical consequence, and analyses are allowed to sometimes be surprising. We did not think we were talking about obligations and attitudes, but perhaps it has turned out we were.

Nevertheless, if one does feel that the identity theory does too much violence to our intuitions of aboutness, the expressivist theory does better. On that view, talk of consequence isn't about anything more than the things that are stated to follow from each other. Something about obligations and attitudes is expressed, but not by means of a descriptive claim about obligations and attitudes. It is like saying 'hooray': doing so does not describe my elation – it expresses it.

Although it is not clear that the identity theory cannot overcome the above challenges on its own, I do think that the expressivist version of the cognitive norm account of consequence is preferable overall. It also strikes me – though I could not say precisely why – as more elegant.

4.2 Whither truth?

One striking feature of a cognitive norm account of (logical) consequence is that it does not appear to mention truth at all, despite the fact that this is a notion usually closely associated with consequence. Is this a feature or a bug? It might appear as a bug if one thinks that an adequate theory of consequence should have a close connection between truth and consequence drop out automatically. But it is not clear to me that the connection ought to be quite so automatic.

Let's first observe that we may still secure a general connection between truth and consequence by the expedient of equating the obligatoriness of accepting a proposition with its truth, and the obligatoriness of rejecting it with its falsity. True, we cannot allow truth and falsity to be the notions that explain why some things ought to be accepted and others ought to be rejected, for if we did that our cognitive norm account would effectively collapse into a truth-theoretic account of consequence – and such accounts are already widely available. So perhaps this will not satisfy the friend of truth, because they might demand an explanatory role for truth vis-à-vis consequence.

But perhaps they should not be so quick to do so. First off, many philosophers are attracted to deflationary accounts of truth, on which the entire content of the truth predicate is given by a principle of disquotation, and the notion of truth has no robust metaphysical nature of any sort. But if one is attracted to that sort of account, one should be hesitant about according truth any explanatory role vis-à-vis consequence or indeed anything else: can something with no nature of its own explain the nature of anything else? The cognitive norm account allows truth to drop out of the picture, as perhaps it should.

Secondly – and somewhat speculatively – there might be discourses on which consequence has a grip but truth does not. Suppose we think that moral discourse is not truth-apt. We might think that, and still feel that some moral claims should be accepted and others rejected, perhaps because they are appropriate in some non-alethic fashion. There is no reason to think that considerations such as consistency, coherence and relevance could not be just as applicable to a non-truth-apt discourse as to a truth-apt one, and if

Prospects for a Cognitive Norm Account of Logical Consequence

so, relations of consequence – conceived as cognitive norms – would obtain in these discourses. It would be good, then, if our account of consequence was not inextricably bound up with truth.

4.3 Non-classical logics and structural rules

Would a cognitive norm account of (logical) consequence be ecumenical with regard to the wide variety of formal systems of logic that are out there, or would it be compatible with only some? This depends on the conditions that our preferred biconditional principle imposes on the consequence relation. As it happens the principles considered above apply without any issues to a wide variety of logics, including classical, paracomplete and at least the more standard paraconsistent logics. This is not surprising: the main contributors to the literature on the normative import of logical consequence, on whose work we have built here, are keenly aware of non-classical logics, and have designed their principles appropriately.

The matter becomes trickier once we consider sub-structural logics, logics which omit one or more of the classic structural properties of the consequence relation: transitivity, reflexivity, monotonicity, permutability and contractibility. Restall (ibid.) has suggested that principle (4) does impose these properties on the consequence relation, but it seems that it does not really do so without the help of some further assumptions. Since we have mostly explored norms based on principle (4), it will be instructive to look at how much it tells us about the consequence relation. As a reminder, here is the principle:

(4) $P_1, P_2, \ldots, P_n$ logically imply $Q \to$ one ought not accept $P_1, P_2, \ldots, P_n$ whilst rejecting Q.

First we can observe that the principle does not immediately give us transitivity. Suppose that I ought not accept A whilst rejecting B, and that I ought not accept B whilst rejecting C. With regard to the first requirement, I can oblige by accepting A and suspending judgment on B. With regard to the second, I can oblige by suspending judgment on B and rejecting C. I thus may end up accepting A and rejecting C in full compliance with (4); thus it does not tell me that I ought not accept A whilst rejecting C. Nevertheless, transitivity might be secured by means of a further, quite plausible norm: that there are no propositions such that I ought not accept them and ought not reject them. This norm is to be firmly distinguished from the less plausible norm that there are no propositions such that I neither ought to accept them nor ought to reject them. The latter rules out, draconically, that there could be propositions concerning which I have no obligations; the former, more reasonably, rules out that there are propositions concerning which I am not allowed any attitudes. With this latter norm in place, the counterexample to transitivity above will be ruled out. A defender of non-transitive logic wanting to adopt the cognitive norm account therefore had better reject this additional norm, which they might motivate by claiming that paradoxical propositions, or whatever else they regard as proper loci for failures of transitivity, are indeed such that we are obliged not to have any attitudes towards them.

On the very plausible assumption that I ought not accept and reject P, for any P, principle (4) tells me that any P implies itself. In other words, reflexivity is fairly easily obtained, and quite hard to get rid of.

With regard to monotonicity (aka weakening), permutability (aka exchange) and contractibility (aka contraction) we might think that they are quite easily obtained. With regard to monotonicity: if it is the case that I ought not accept some P whilst rejecting some Q, then presumably I also ought not do that whilst doing any other particular thing, like accepting R or playing cricket. With regard to permutability: since there is no apparent sense in which my obligations with regard to some propositions in any way involve any ordering on them, there is also no sense in which a change in their order could affect my obligations. With regard to contractibility: there seems to be no difference between me merely accepting P and me accepting P and accepting P, and so it seems there could be no difference between the obligation to accept P and the obligation to do so twice.

We are assuming too much here, though. Implicitly, the reasoning I used to obtain monotonicity assumed that the obligations I have with regard to P and Q arise only from the nature of P and Q, i.e. that they are somehow internal to P and Q. If so, then it is indeed clear that my performing some further action cannot change that. But if my obligation to not accept P whilst rejecting Q arose specifically because of something else I was doing or (let's say rejecting R) then it is conceivable that my obligations with regard to P and Q might change if I stopped doing this further thing (rejecting R). So then it does not follow from my obligation to not accept P whilst rejecting Q that I have an obligation to not accept P and accept R whilst rejecting Q. Accepting R might be just the thing for getting me out of my obligation vis-à-vis P and Q. The plausibility of monotonicity thus depends on whether we think that our cognitive obligations concerning any given bunch of propositions do indeed depend only on the nature of those propositions. That claim strikes me as plausible, but it might be doubted, and it should be doubted by a proponent of non-monotonic logic.

Similarly, with regard to permutability, though we do not think that propositional attitudes, of themselves, have any ordering, one could imagine such orderings obtaining relative to some external factor (for example time). If my obligations involving some propositional attitudes depend in part on that external factor (i.e. the temporal order in which I adopted them) then a change in ordering could mean a change in obligations. So to uphold permutability is implicitly to deny the involvement of any such external factor. Similarly with contractibility: on the face of it there is no difference between me accepting P and doing so twice, but if there is some external factor relative to which my attitudes can be distinguished in a more fine-grained way, there may be such a difference after all. To uphold contractibility is thus to deny the relevance of any such external factor.

To sum up: it seems that all the structural rules can be called into question within the context of a cognitive norm account of (logical) consequence, though some more easily than others. A cognitive norm account need not restrict us to some particular subset of the available logics.[6]

4.4 Logical pluralism

At the beginning of this article I said that a cognitive norm account of (logical) consequence involves the thought that there is some particular relation of consequence that exists independently of our formal systems, a relation between propositions, which we try to model in those formal systems by means of a relation between sentences. One might assume this involves thinking that there is a single 'true' logic out there, i.e. logical monism. If so, then a cognitive norm account is of no use to those attracted to logical pluralism, the view that there is a plurality of logics, all of which are in some manner 'in good standing'. The assumption is natural, but I don't think a cognitive norm account is committed to monism in quite such a rigid fashion.

What a cognitive norm account is strictly committed to is the claim that there is one and only one consequence relation out there which corresponds to the norm which governs our propositional attitudes. It leaves open the possibility that there are other relations among propositions which, some-

---

[6]Which is not to say that no version of it will. It is entirely possible that some ways of embellishing principle (5) will indeed restrict us to a narrower set of logics. Imposing a relevance condition, for instance, will compromise monotonicity.

how or other, also deserve to be called consequence relations, and which are significant to us for some different reason. If we imagine for the moment that it is the classical consequence relation which captures our general cognitive obligations, it could for instance be that some other, stricter logic captures our obligations with regard to some particular intellectual pursuit or some particular subject matter, such as mathematics. A certain amount of pluralism is thus compatible with a cognitive norm account.

5 Conclusion

In this article I have sketched a possible view of logical consequence. It is a view, or family of views, on which consequence relations are to be identified with norms on our propositional attitudes, or, at least, our talk of consequence relations is to be explained in terms of such norms. I have explored some of the challenges involved in getting the basic machinery of such an account off the ground: to wit, a plausible biconditional bridge principle that connects consequence claims with normative claims. And I have explored, to some extent, what a cognitive norm account of (logical) consequence would be like and what further theoretical commitments it would generate.

I am by no means convinced that a cognitive norm account gets at the truth of things. No one should be; it is too early to tell. But I think that it may get something important right, and I think that there is enough mileage in the view to make it worth exploring in greater detail.

References

Broome, J. (1999). Normative Requirements. *Ratio, 12*, 398–419.
Brouwer, T. N. P. A. (2013). *The Metaphysical Commitments of Logic*. Unpublished doctoral dissertation, University of Leeds.
Field, H. (2009). What is the Normative Role of Logic. *Proceedings of the Aristotelian Society, 83*, 258–261.
Field, H. (Forthcoming). What is Logical Validity? In C. Caret & H. O. (Eds.), *Foundations of Logical Consequence*. Oxford: Oxford University Press.
Harman, G. (1986). *Change in View*. Cambridge, Massachusetts: M.I.T. Press.

Prospects for a Cognitive Norm Account of Logical Consequence

MacFarlane, J. (Unpublished). *In What Sense (if Any) Is Logic Normative for Thought?* Retrieved from http://johnmacfarlane.net

Restall, G. (2005). Multiple Conclusions. In P. Hájek, L. Valdes-Villanueva, & D. Westerståhl (Eds.), *Logic, Methodology and Philosophy of Science: Proceedings of the Twelfth International Congress* (pp. 189–205). London: Kings' College Publications.

Soames, S. (1987). Direct Reference, Propositional Attitudes, and Semantic Content. *Philosophical Topics, 15*, 47–87.

Yablo, S. (2014). *Aboutness.* Princeton: Princeton University Press.

Thomas N.P.A. Brouwer
Northern Institute of Philosophy, University of Aberdeen
Scotland
E-mail: tnpabrouwer@gmail.com

Preliminaries to a Logic of Malfunction

MASSIMILIANO CARRARA[1]

Abstract: Aim of the paper is to present some preliminaries to a new logic for reasoning about malfunction in (technological) artifacts. First, I show that 'malfunctioning' denotes a modifier rather than a property. Then I argue that – as modifier – there are two readings of it: *subsective* and *privative*. The two readings have a different logical behavior: If *subsective*, a malfunctioning F is an F; if *privative*, a malfunctioning F is *not an F*. The answer to the question: "Is a malfunctioning F, which fails to do what Fs do, nonetheless an F?" depends on whether the malfunctioning modifier is construed as being *subsective* or *privative*.

Keywords: logic of malfunction, property modification, subsection, privation

1 Introduction

Functions are used ubiquitously in some sciences as engineering, biology and social sciences. Let us concentrate on the engineering case: The term "function" is used everywhere in the engineering literature and it plays a central role in the description of a lot of processes, as for example, the process of designing. A general and crude characterization of designing in engineering is that it starts by a client expressing a specific goal, which is translated by engineers in a set of *functions* the artifact to be designed has to perform, setting the task for the engineers to then find a design solution to those *functions*. Given this role of function in designing, the term proliferates within engineering.

But, even if "function" is a key-term in engineering its formalization, semantics, and logic are just a recent challenge; this, even if formalizations of *function* in turn lead to the means necessary to build automated tools for engineering, such as tools for supporting reasoning in engineering designing

---

[1] Thanks to Bjørn Jespersen, Vittorio Morato, and Elena Pariotti for insightful objections and helpful suggestions to previous versions of this paper.

of new technical artifacts and tools for archiving and retrieving functional descriptions of existing artifacts (for a general discussion on this topic see (Carrara, Garbacz, & Vermaas, 2011), for semantics, formalizations, and logic of the term see (Borgo, Carrara, Garbacz, & Vermaas, 2009a, 2009b, 2010, 2011a, 2011b)).

Now, consider this methodological fact: a theory pertaining to a certain area of inquiry must be able to account not only for the success stories, but also, let us say, for the "failures". Thus, for example, a theory of possibility must give an account for impossibility; a theory of truth must give an account for falsehood; etc. The same is for a theory – including formalization, semantics and logic – of *(technical) function*: a good theory of *function* must include an account for *malfunction*.[2]

Following (Carrara & Jespersen, 2011, 2013) in this paper I sketch some semantical preliminaries to a *formalization* and a *logic of malfunction*. The paper resumes most of the ideas presented in the above quoted two papers.

The paper is divided in 6 sections and a conclusion. In section 2 I observe that 'malfunctioning' is syntactically a *modifier*: 'malfunctioning' stays for a *modifier* rather than a property. Then, the principal modifiers, *subsective, privative, intersective*, and *modal* are briefly summarized and discussed in section 3. I restrict my analysis of *malfunctioning* to three modifiers: *subsective, privative*, and *intersective modifier* (section 4) and I show that just the first twos are interesting for the notion at issue. In section 5 I propose to combine *malfunctioning* as *subsective* and *privative*, with two main views on *functions*: the *design view* and the *use view* of function. On the basis of the possible combinations a characterization of *malfunctioning* as a *subsective* or as a *privative* is given (in section 6).

2 Malfunction as a property modifier

Consider the following sentence:

(1) *a* is a malfunctioning *F*

where 'a' denotes an arbitrary technical artifact and 'F' a certain artifact property compatible with the modifier denoted by 'malfunctioning'. An instance of the schema in (1) would be:

---

[2] For a general introduction to the topic in engineering see (Del Frate, Franssen, & Vermaas, 2011), (Del Frate, 2012), and, in particular, (Del Frate, 2014).

Logic of Malfunction

(2) This is a malfunctioning corkscrew,

or

(3) *a* is a malfunctioning corkscrew.

How to analyse 'malfunctioning' in (2) and (3)? *Prima facie*, a simple analysis of the expression 'malfunctioning' is to say that the the term stays for a *property*. And one can go on to specify that the denotation of the term is a set of malfunction objects. And the denotation of 'corkscrew' is is another set of objects: corkscrews. The result is that (3) is analysable in terms of a conjunctive sentence like:

(4) *a* is malfunctioning and *a* is a corkscrew.

But, observe that, such a conjunctive analysis of (3) is wrong. Indeed, (3) does not factor out into the conjunction (4). The above analysis, to be right, would require that *a* would be malfunctioning, *pure and simple*, without qualifying with respect to what *a* would be malfunctioning. But what we say in colloquial discourse is, for instance, that *a* is a large drum kit – i.e., large for a drum kit, or large as far as drum kits go – *b* is a giant midget, *c* is a midget giant, and *d* is a malfunctioning amplifier. This goes to show that 'malfunctioning' is syntactically a *modifier*.

To expand the point and argue why (4) is not a correct analysis of (3), consider the following sentence, completely analogous to (3):

(5) *b* is a big table.

If we use the same analysis as above, (5) would be analysable in terms of a conjunctive sentence like:

(6) *b* is big and *b* is a table.

As for (4), what (6) requires is that there is something big *per se*. But, observe that everything is big relatively to a scale such that one can distinguish between, for example, tables that are *big*, or *medium-sized*, or *small*. Moreover, suppose that we accept (6) as the right analysis of (5) and consider the inference:

$$\frac{b \text{ is a big table}}{b \text{ is big}}$$

It is an invalid transition from a weaker proposition to a stronger one: In the above inference the premise attributes relative bigness to *b* (bigness, as far as tables go); the conclusion attributes absolute bigness to *b*. It is a fallacy, known as *secundum quid ad simpliciter*. In the same vein, one can say that in (3):

(3) *a* is a malfunctioning corkscrew.

Nothing malfunctions, *pure and simple,* but only with respect to a certain technical-artifact property. (3) will be incomplete without a specification of a property for a certain object *a*. The correct response to the claim "*a* is malfunctioning" is "*a* is a malfunctioning *what*?". So, the analysis of (3) in terms of (4) is wrong.

Which alternatives do we have? An alternative here analysed is to follow syntax and to construe the adjectives in (3) and (5) – 'malfunctioning' and 'big' – as denoting *modifiers*, rather than properties. In this way it is no longer an option to infer from (4), by *conjunction elimination* ($\wedge$ E), that *a* is a corkscrew.

The proposal (see Carrara & Jespersen, 2011, 2013) is to read the property of *being a malfunctioning F* as formed by means of a functional application – symbolized by '$[XY]$', X a function, and Y a functional argument of the modifier *Malfunctioning* (abbreviated *Malf*) to the property F. Functional application provides also the logic of predication, such that when predicating the property [*Malf F*] of *a*, the result of applying a property to an individual is a truth-value.

Consider, now, the following argument.

$$\frac{[[Malf\ F]\ a]}{Fa}$$

Is it correct? The answer to the question depends *first of all* on the *status* of the modifier. In the next section I briefly analyse the principal ones.

3 A taxonomy of modifiers

The principal modifiers here considered are: *subsective, privative, intersective*, and *modal*.[3] I briefly summarize and discuss them.

---

[3] See (Geach, 1956). See also (Gamut, 1991, sec. 6.3.11).

Logic of Malfunction

Subsective (M_s).

This is an example of *subsective modifier*:

(6) If a is a square box then a is a box.

Deductively, a *subsective modifier* behaves in the following way:

$$\frac{[[M_s\ F]\ a]}{Fa}$$

Observe that – if one forms extensions from intensional entities with $\|\ \|$, in our case *properties* – this rule is valid thanks to upward monotonicity:

Necessarily, $\| M_s F \| \subseteq \| F \|$.

Privative (M_p). [4]

This is an example of *privative modifier*:

(7) If a is a forged banknote then a is not a banknote.

Deductively, the *privative modifier* behaves in the following way.

$$\frac{[[M_p\ F]\ a]}{\neg Fa}$$

And it is valid because:

Necessarily, $\| M_p F \| \cap \| F \| = \emptyset$.

or equivalently:

Necessarily, $\| M_p F \| \subset \|\text{non-}F\|$.

That is, $\| M_p F \|$ is a proper subset of the complement of $\| F \|$.

Intersective (M_i).

An example of *intersective modifier* is:

---
[4]For privative modification in particular, see (Jespersen & Primiero, 2010)

(8) If a is a square box then a is a square and a is a box.

Deductively, *intersective modifier* behaves in the following way:

$$\frac{[[M_i\ F]\ a]}{M^*\ a \wedge Fa}$$

This rule of inference is valid because, necessarily, an element of the set $\|\ MF\ \|$ is an element of the intersection of $\|\ M^*\|$ and $\|\ F\ \|$:

Necessarily, $\|\ M_iF\ \|=\|\ M^*\| \cap \|\ F\ \|$

Notice that M is a property modifier, while M^* is a property. Therefore, M cannot be detached from $[M_iF]$ and predicated of a on pain of attempting to make a modifier behave as though it were a property: as said before a mistake.

How to obtain the desidered inference? In order to obtain the result one should apply a *rule of left subsectivity*. The rule works by quantifying away the original modified property, replacing it by a variable, and the original modifier in its turn modifies the variable. To exemplify the rule, let *Large* be the modifier and *Large** the property formed from it by *left subsectivity*. Then the following inference is valid:

$$\frac{[[Large\ cat]\ a]}{[Large^*\ a] \wedge [cat\ a]}$$

In general, whichever kind of modifier M may be, the rule of *left subsectivity* validates the inference (schema):

$$\frac{[[MF]\ a]}{M^*\ a}$$

The idea behind *left subsectivity* is straightforward. While it is not trivial that somebody is a wise man, say, it follows trivially from this – provided we allow ourselves to quantify over properties – that *there is a property with respect to which that person is wise*. And the property *wise** is the property of being wise with respect to some property (in this case the property of *being a man*). Again, consider this other example: if a is a typical hammer then there is a property with respect to which a is *typical**; the conclusion is not that a is typical, period, which would be nonsensical, just like inferring from b being a main station that b is main, period. The correct conclusions are that there is a property p such that a is a (typical p) and that there is a property q such that b is a (main q). The relevant derivation pertaining to *intersective modification*, using the rule of *left subsectivity* is then the following one:

$$\dfrac{\dfrac{[[M_i F]\, a]}{M * a}\;\text{left}-\text{subsectivity} \qquad \dfrac{[[M_i F]a]}{Fa}\;\text{subsection}}{M * a \wedge Fa}\wedge\text{I}$$

Modal (M_m). [5]

The exact logical properties of the modal modifier are still disputed in the literature (for a detailed analysis and discussion see (Carrara & Jespersen, 2013)). In particular, it remains an open question how to provide a positive *definition* of modal modifiers. The unique feature that *modal modifiers* possess is that they oscillate between *being subsective* and *being privative*. So, for example, if a is an alleged terrorist, then it is possible that a is a terrorist but it is also possible that a is not a terrorist. The premise is not strong enough to either cancel out or rule in either possibility. For any given instance of $[[M_m\ F]\ b]$, it is an open question whether M_m is *subsective* or *privative*, so it is an open question if the result of the application is Fb or $\neg Fb$. If one construes *empirical possibility* as existential quantification over worlds w_i and times t_i the *rule defining modal modification* has the following form:

$$\dfrac{[[M_m\ F]_{w,t} b]}{\exists w't'\ [F_{w't'}\ b] \wedge \exists w''t''\neg\ [F_{w''t''}\ b]}$$

Consider the above example of the alleged terrorist: the open question is whether the $<w,t>$ at which a is an alleged terrorist is identical to $<w',t'>$, where a is a terrorist, or to $<w'',t''>$, where a is not a terrorist. The set-theoretic counterpart of modal modification is the *union* of two disjoint sets. In the example mentioned before, the relevant union is the union of the set of terrorists at $<w',t'>$ and the set of non-terrorists at $<w'',t''>$. Observe that it is trivial that an alleged terrorist, just like any other individual, is a member of that union, but it is not trivial which of its two subsets an alleged terrorist, or any other individual, belongs to. For the above reasons, I am not going to make further use of the modal modifier below.

---

[5] On this see (Jespersen & Primiero, 2013); for an introduction to the problem see (Williamson, 2000) or section 1.3 of (Williamson, 2013).

4 Modifiers and malfunctioning

Let us exclude the modal modifier and restrict our analysis to the three remaining kinds: *subsective, privative, and intersective* modifier.

Take a malfunctioning corkscrew: Is it *subsective, privative* or *intersective*?

In general, as seen before, *subsective* and *privative* modifiers have the following characteristic logical behaviour:

$$\frac{[[Subsective\ F]\ a]}{Fa}$$

and

$$\frac{[[Privative\ F]\ a]}{\neg Fa}$$

Take our case: On one side, on the *subsective* conception, a malfunctioning F is still an F; on the other in *privative* conception, a malfunctioning F is *not* an F. Therefore, the following inference characterises the *subsective reading* of *Malf* (*Malf$_s$ F*).

$$\frac{[[Malf_s\ F]\ a]}{Fa}$$

Instead, the following inference characterises the *privative reading* (*Malf$_p$ F*).

$$\frac{[[Malf_p\ F]\ a]}{\neg Fa}$$

Notice that it is crucial that the central sentence be (1) and not, e.g.,

(9) This F is a malfunctioning F.

The sentence:

(1) a is a malfunctioning F

is neutral between the *subsective* and the *privative* conception: indeed, it does not identify the subject of predication via the property F. Instead, (9) prejudges in favour of the *subsective reading*: if the predicate 'F' truly applies to a malfunctioning F then it is a foregone conclusion that when being a malfunctioning F the artifact in question is an F.

What about malfunctioning F as an *intersective* modifier? This property is literally derivative: a needs to be a $[Malf_s \ F]$ or a $[Malf_p \ F]$ first before a can become malfunctioning$_s$*, i.e. $[Malf_s \ p]$, or malfunctioning$_p$*, $[Malf_p \ p]$. That can also observed if we consider, for example, the above relevant derivation pertaining to intersective *modification*, where we use the subsection rule.

In general, if *Malf* is the modifier, then the inference, in the *subsective case* is:

$$\frac{[[Malf_s \ F] \ a]}{[Malf^* a] \wedge Fa}$$

and, in the *privative case*, it is:

$$\frac{[[Malf_p \ F] \ a]}{[Malf^* \ a] \wedge \neg \ Fa.}$$

The first conjunct is obtained via *left subsectivity*; the respective second conjuncts follow from the respective conceptions of *Malf*.

In the end, we can concentrate just on two modifiers: *subsective* and *privative*.

5 Function and malfunction

In order to evaluate the logical *status* of *malfunction* let us consider the following question: Which *notion of function* is compatible with, or appropriate for, which notion of *malfunction*? There are two main views on *functions*: the *design view* and the *use view* of function.

In the first one, the *design view*, the technical functions of an artifact are the capacities or goals for which agents designed the artifact (on this position see, for example, Cummins, 1975). In *the design view* of function one holds that what defines an artifact a as an F is that a is a token of the designed artifact kind F. In details, an object a is judged to belong to the artifact kind F if a has the same function originally intended as that of *Fs*.

According to such a view *a* will be classified as belonging to the kind *clock* if it has been construed for having the same function the inventor of clocks has originally assigned to it, i.e. the function of measuring time. It is neither necessary nor sufficient to be an artifact that happens to function as a device for measuring time.

By contrast, in the *use view*, the technical functions of an artifact are the capacities or goals for which agents *use* the artifact. This view holds that what defines an artifact *a* as an *F* is that *a* is being, or could be, *used* as an *F*, irrespective of whether *a* was designed as an *F* in the first place. Its usability, or use, anchors it to a particular artifact kind, but in such a way as to make it vulnerable to failure to function in the way characteristic of the relevant artifact kind. In the *use view* conception an object *a* is judged to belong to kind *F* if its current function is that of *Fs*. If this hypothesis were correct we would have, for instance, that an object originally created as a bird house, but currently used as a doll house, will be categorized as belonging to the kind doll house (or toy, depending on the level of function which is taken to be relevant). This conception of function has been proposed by Neander (1991) and McLaughlin (2001). Note, *passim*, that if one takes the designer/creator of an artifact as one of its users, the use view may well subsume the designer/creator intentions account. It is a position Dennett, for example, considers when he says that: "The inventor is just another user, only circumstantially and defeasibly privileged in his knowledge of the functions and uses of his device. If others can find better uses for it, his intentions, clearheaded or muddled, are of mere historical interest." (Dennett, 1990, p. 186)

Consider the relationships between the two notions of malfunction, *subsective* and *privative* and the two notions of function *design* and *use*. The combination of the *subsective conception* of *Malf* with the *design view* is restrictive with respect to the property *F*: Indeed it is demanding for an artifact to qualify as an *F*; however it is permissive with respect to [*Malf F*]. No extra 'effort' is required of a malfunctioning *F* to remain an *F*. We write '*Des*' for the abbreviation of 'the design view' and '*Use*' for the abbreviation for 'the use view'. $Malf_s$, $Malf_p$ are as above. The first combination is then the following:

(1) *Des* + $Malf_s$. An artifact *a* was designed as an *F*, which is why *a* remains an *F*, even when *a* is a malfunctioning *F*.

The other interesting combination is between the *privative conception* of *Malf* and the *use view*. It is permissive with respect to *F*: Also what the design view would deem successful improper use of a as an F makes a an F, but it is restrictive with respect to [Malf F], the current incapacity to be used, or usable, as an F disqualifies a as an F:

(2) *Use* + *Malf*$_p$. An artifact a is a malfunctioning F, hence a was previously an F and no longer is one.

What about the other two combinations? One of them is logically impossible, because inconsistent:

(3) *Use* + *Malf*$_s$. This is the logically impossible combination. If a is a malfunctioning F then a is an F; and if a is a malfunctioning F then a fails to function as an F and is, therefore, not an F. So, if a is a malfunctioning F then a both is, and is not, an F.

Finally, to complete the logical combinations we have that:

(4) *Des* + *Malf*$_p$. An artifact a was designed as an F, but is not an F, because no malfunctioning F is an F.

Notice that, the *privative conception* of *Malf* is both compatible with the *design view* and the *use view*. This two-way compatibility goes to show that the privative view of *Malf* does not logically presuppose the *use view* of function at the expense of the *design view*. A difference, however, between combining *Malf*$_p$ with the *design view* or the *use view* is that on the latter design is not sufficient to preserve a's status as an F. According to this combination, a designed F that fails to function as an F is a malfunctioning F and, therefore, *not an F*. But, it is far from clear why it would be philosophically rewarding to go with (4). For, as far as *Malf* goes, the notion of design is rendered idle.

(1) and (2) are then the two most interesting combinations for our purpose here. In the next section we just focus on them.

6 Malfunction as a *subsective* and *privative* modifier

Let us first consider the subsective modifier and its inference:

$$\frac{[[Malf_s\ F]\ a]}{Fa}$$

A broken corkscrew, which fails to uncork bottles is still a corkscrew. Malfunctioning corkscrews are corkscrews, because they were *designed* as corkscrews. Their incapacity to perform the function they were designed to perform is an accidental feature and as such does not affect their *status*. The *status* of a corkscrew is anchored to its design, not to its structure or to the capacities that the structure enables. So, even if the artifact's structure cannot substain the function, still the function is at work.

Consider, now, the *privative view* and the relative inference:

$$\frac{[[Malf_p\ F]\ a]}{\neg Fa}$$

This view only tells us, strictly speaking, what a *malfunctioning F* is not, namely an *F*. The view maintains that a malfunctioning corkscrew is not a corkscrew that is a malfunctioning uncorking device; it is a non-corkscrew. Differently from the subsective view, the privatist holds that the incapacity to function as a corkscrew is an essential feature of a malfunctioning corkscrew. It affects its *status* by depriving it of being a corkscrew.

Observe, that this view rests content with not differentiating between those *non-F*'s that are *malfunctioning F*'s and those *non-F*'s that are not. Both kinds of *non-F* are located in the complement of any sets of *F*'s, but, of course, there is a very important difference between a malfunctioning corkscrew and a nuclear power plant.

The general problem with *privation* is that it tells us what something is not – for example that a forged banknote is not a banknote – and not also what something is. But, from the fact that *a* is a *malfunctioning F* it can be inferred that *a* used to be an *F*: a currently *malfunctioning F* has a past as an *F*. So one proposal is to introduce a meaning postulate, on behalf of the privative theory, that something that presently fails to function as an *F* previously did function as an *F*. Following the above suggestion, one can argue that the privative conception is more accurately rendered in the following way, where Ex is a privative modifier as referred to in a predicate like 'is an ex-Comunist':

$[Malf\ F]\ a] \to [Ex\ F]\ a \to \neg Fa.$

The property of being a *non-F* subsumes the property of being an *ex-F*, which in turn subsumes the property of being a *malfunctioning F*.

7 Conclusion

What is required as a preliminary to the construction of a rigorous theoretical framework within which to raise, discuss and possibly solve questions bearing on malfunction? More specifically, which are the basic elements for a logic of *malfunction*?

First, in the paper I observed that nothing malfunctions *simpliciter*. It is only with respect to a property (be it a particular property or the value of a variable ranging over properties) that it is meaningful to say that something malfunctions.

Secondly, I showed that there are two conceptions of (technical) malfunction, couched in terms of whether the property modifier *malfunctioning* is *subsective* or *privative*. The *subsective* conception of malfunction is restrictive with respect to F and permissive with respect to $[Malf\ F]$, whereas the *privative* conception of malfunction is restrictive with respect to $[Malf\ F]$ and permissive with respect to F.

This result is obtained combining *malfunction* with two main conceptions of *function*: the *design view*, where the technical functions of an artifact are the capacities or goals for which agents designed the artifact and the *use view*, where the technical functions of an artifact are the capacities or goals for which agents *use* the artifact. The notion of *design* is required to prevent an F failing to function as an F from ceasing to be an F. The notion of *use* is required to enable an F failing to function as an F to cease being an F.

References

Borgo, S., Carrara, M., Garbacz, P., & Vermaas, P. E. (2009a). A Formal Ontological Perspective on the Behaviors and Functions of Technical Artifacts. *Artificial Intelligence for Engineering Design, Analysis and Manufacturing, 23*(2), 3–21.

Borgo, S., Carrara, M., Garbacz, P., & Vermaas, P. E. (2009b). Towards the Ontological Representation of Functional Basis in DOLCE. In M. Okada & B. Smith (Eds.), *Interdisciplinary Ontology: Proceedings of the Second Interdisciplinary Ontology Meeting* (Vol. 2, pp. 3–16). Tokyo: Keio University Press.

Borgo, S., Carrara, M., Garbacz, P., & Vermaas, P. E. (2010). Formalizations of Functions within the DOLCE Ontology. In I. Horváth,

F. Mandorli, & Z. Ruzák (Eds.), *Proceedings of the Eighth International Symposium on Tools and Methods of Competitive Engineering, TMCE 2010, April 12-16, Ancona, Italy, Vol. 1, Tools and Methods of Competitive Engineering* (pp. 113–126). Delft University of Technology.

Borgo, S., Carrara, M., Garbacz, P., & Vermaas, P. E. (2011a). A Formalization of Functions as Operations on Flows. *Journal of Computing and Information Science in Engineering*, *11*(3), 031007-1–031007-14.

Borgo, S., Carrara, M., Garbacz, P., & Vermaas, P. E. (2011b). Two Ontology-driven Formalisations of Functions and Their Comparison. *Journal of Engineering Design*, *22*(11-12), 733–764.

Carrara, M., Garbacz, P., & Vermaas, P. E. (2011). If Engineering Function Is a Family Resemblance Concept: Assessing Three Formalization Strategies. *Applied Ontology*, *6*(2), 141–163.

Carrara, M., & Jespersen, B. (2011). Two Conceptions of Technical Malfunction. *Theoria*, *77*(2), 117–138.

Carrara, M., & Jespersen, B. (2013). A New Logic of Technical Malfunction. *Studia Logica*, *101*(3), 547–581.

Cummins, R. (1975). Functional Analysis. *Journal of Philosophy*, *72*, 741–765.

Del Frate, L. (2012). Preliminaries to a Formal Ontology of Failure of Engineering Artifacts. In *FOIS, Formal Ontology in Information Systems* (pp. 117–130). Amsterdam: IOS Press.

Del Frate, L. (2014). *Failure: Analysis of an Engineering Concept*. Delft University of Technology and Eindhoven University of Technology.

Del Frate, L., Franssen, M., & Vermaas, P. E. (2011). Towards a Transdisciplinary Concept of Failure for Integrated Product Development. *International Journal of Product Development*, *14*(1), 72–95.

Dennett, D. C. (1990). The Interpretation of Texts, People, and Other Artifacts. *Philosophy and Phenomenological Research*, *50*, 177–194.

Gamut, L. T. F. (1991). *Logic, Language, and Meaning, Vol. II*. Chicago and London: University of Chicago Press.

Geach, P. (1956). Good and Evil. *Analysis*, *17*, 33–42.

Jespersen, B., & Primiero, G. (2010). Two Kinds of Procedural Semantics for Privative Modification. In *New Frontiers in Artificial Intelligence* (pp. 252–271). Berlin: Springer.

Jespersen, B., & Primiero, G. (2013). Alleged Assassins: Realist and Constructivist Semantics for Modal Modification. In *Logic, Language, and Computation* (pp. 94–114). Berlin: Springer.

McLaughlin, P. (2001). *What Functions Explain: Functional Explanation and Self-reproducing Systems.* Cambridge: Cambridge University Press.

Neander, K. (1991). The Teleological Notion of 'Function'. *Australasian Journal of Philosophy, 69*(4), 454–468.

Williamson, T. (2000). The Necessary Framework of Objects. *Topoi, 19*(2), 201–208.

Williamson, T. (2013). *Modal Logic as Metaphysics.* Oxford and New York: Oxford University Press.

Massimiliano Carrara
University of Padua
Italy
E-mail: `massimiliano.carrara@unipd.it`

The Totality and Its Complement

FILIPPO CASATI AND NAOYA FUJIKAWA

Abstract: This short paper delivers a mereological account of nothingness considered as the absence of every thing. The formal system presented here is based on the paraconsistent mereology proposed by Cotnoir and Weber (Forthcoming) and it shows that the totality (defined as the mereological sum of everything that is self-identical) has a complement namely the mereological sum of everything that is not self-identical. It presents also some features of nothingness itself.

Keywords: mereology, paraconsistency, totality, nothing(ness)

1 Introduction

How can nothingness be problematic if there is actually nothing problematic? And, moreover, how is it even possible to question nothingness if there is actually nothing to question at all? Despite all these worries, the problem of nothingness constitutes one of the most important metaphysical riddle in the history of both western and eastern metaphysics (Givone, 2006). Among all the different metaphysical accounts, the most famous one describes nothingness as the absence of all objects. For instance, Heidegger thinks that "nothingness is the removal of all entities" (Heidegger, 2012, p. 132). A similar view is presented by Sartre as well: "nothingness is the lack of beings" (Sartre, 1956, p. 23). In more contemporary terms, Priest writes: "[Absolutely nothingness] is the complement of everything: the absence of every thing" (Priest, 2014, p. 65). According to those intuitions, we obtain nothingness by subtracting all objects from the totality of all objects. Indeed, it is quite intuitive to see that nothingness, as the absence of all things, could be interpreted as no things at all, namely no-thing.

The aim of this paper is to give a mereological description of nothingness as the absence of all objects, that is, nothingness as what is left if all objects are subtracted from the totality. A natural way to do this is to take nothingness in this sense as *the complement of the totality*. In mereology, the complement of an object x is defined as the sum of things which is disjoint from x (that is, the sum of things which don't share any part with x) and the sum of x and its complement is the totality, that is the sum of all objects. The complement of x is thus naturally taken as the remainder of x, that is, the remainder of the subtraction of all x from the totality. The complement of the totality seems to be a good candidate of a formal representation of nothingness in the sense in question.

However, the existence of the complement of x is not unrestrictedly admitted in mereology. A standard and natural condition of the existence of the complement of x is as follows: it exists if and only if there is something which is not a part of x. If we accept this biconditional, then the complement of the totality exists if and only if there is something which is not a part of the totality.

As one way to take the notion of the complement of the totality seriously, in this paper we explore a system of mereology which supposes that something is not a part of the totality. Since this supposition contradicts the fact that everything is a part of the totality (if not, it is not the totality), a system of mereology based on the classical logic must exclude this supposition. Classical extensional mereology is such a system, and thus it doesn't have the complement of the totality. To deal with this contradiction, a system of mereology with the supposition in question has to use some paraconsistent logic as its underlying logic. In this way, paraconsistency is essentially involved in our consideration.

In the next section, we propose an axiomatic system of mereology with the supposition that something is not a part of the totality. Before presenting the system, we give some preliminary considerations about the totality and its complement in paraconsistent mereology.

First of all, the totality. In mereology, the totality is defined as the sum of everything, and it has everything as its part. A natural way to define the totality is to define it as the sum of self-identicals:

(1) $1_= =_{df} lub[x|x=x]$

where $lub[x|\phi]$ is the $\leq$-least-upper-bound of all objects which satisfy ϕ, where $\leq$ is the parthood relation. Since everything is self-identical, $\forall x(x \leq 1_=)$.

Let us consider the complements of the totalities in a paraconsistent context. Usually the complement of x, $\overline{x}$, is defined as $lub[y|\neg y \bullet x]$, where $x \bullet y := \exists z(z \leq x \wedge z \leq y)$. Then, the complement of $1_=$ is as (2).

(2) $\overline{1_=} = lub[y|\neg y \bullet 1_=]$

The Totality and Its Complement

At this point, we have to consider whether $\overline{1_=}$ exists or not. Cotnoir and Weber (Forthcoming, p. 16) argue against the existence of the complement of the totality as follows. Suppose that $\top$ is 'a no-place connective that is satisfied by everything' (Cotnoir & Weber, Forthcoming, p. 16). The totality is defined as the sum of $\top$s: $1_\top = lub[x|\top]$. Now it is natural to think the complement of $1_\top$, $\overline{1_\top}$, is identical to the sum of $\bot$s ($lub[x|\bot]$), where $\bot$ is the absurdity constant 'which nothing satisfies on pain of triviality' (Cotnoir & Weber, Forthcoming, p. 16), that is, $\bot \to A$ holds for any A. Therefore, if $\overline{1_\top}$ existed, triviality immediately would follow, which should be avoided even in paraconsistent setting.

We completely agree with them about the nonexistence of $\overline{1_\top}$. However, we have another candidate of the complement of the totality, that is, $\overline{1_=}$. Does the existence of $\overline{1_=}$ lead to triviality? In a paraconsistent mereology, apparently it doesn't. Suppose that we have $\overline{1_=}$. Let us define $0_{\neq}$ as follows.

(3) $0_{\neq} =_{df} lub[x|x \neq x]$

As we will see in the next section, $0_{\neq}$ is the complement of $1_=$, that is

(4) $0_{\neq} = \overline{1_=}$

Given the reflexivity of the parthood relation, $0_{\neq}$ has at least itself as its part. There is thus something which is a part of the complement of $1_=$ ($\exists z z \leq 0_{\neq}$). What follows from this is that there is something which is not self-identical. Since everything is self-identical, this leads to contradiction. But, of course, paraconsistency blocks the inference from contradiction to triviality.

There is one problem of this response to Weber and Cotnoir's objection. Note that both $1_=$ and $1_\top$ have everything as their parts, that is, $\forall x x \leq 1_=$ and $\forall x x \leq 1_\top$. From this it follows that that $1_= = 1_\top$, since the parthood relation is antisymmetric. Therefore, if, as is expected, $\overline{1_\top} = lub[x|\bot]$, $\overline{1_=} = lub[x|\bot]$. Thus the argument above works only if a system doesn't contain a pair of formulas, one of which serves as $\top$ and the other of which as $\bot$.

This is a handful problem and what we propose here is just to suppose that a system of paraconsistent mereology doesn't contain $\top$ and $\bot$.[1] However, note that it is not clear whether paraconsistent mereology itself is strong enough to deliver such constants. Weber and Cotnoir's argument

---

[1] Another way to deal with this problem is to deny that the parthood relation is antisymmetric (cf. Cotnoir, 2010), which is not pursued here.

also depends on a supposition – the supposition that a system has $\top$ and $\bot$. Moreover, there is no obvious candidate of formulas which serve as $\top$ or $\bot$. It is clear that self-identity and non-self-identity do not constitute such a pair in paraconsistent setting. Even $\forall x \forall y (x \leq y)$, which is analogous to $\forall x \forall y (x \in y)$ (an absurdity formula in paraconsistent naive set theory, from which any formula follows), does not serve as $\bot$ in paraconsistent mereology.[2] According to what we discussed until now, there are no special reasons forcing us to suppose there are such constants nor that there are not. In absence of constraining arguments, each of the two suppositions are equally unsupported. Thus, we decide to endorse the position that there is neither $\top$ nor $\bot$ being perfectly aware that we do not have a definitive argument as our opponents.

Another worry about the existence of the complement of the totality concerns the *monism*. Given the totality has everything as its part, the complement seems to be a part of everything, that is, a bottom element. In a classical context, it follows from the existence of a bottom element that everything is identical to the bottom, and it immediately follows from this that every object is identical to each other (Varzi, 2009).

$\overline{1_=}$ is naturally expected to be a bottom element, that is $\forall z (\overline{1_=} \leq z)$. As we will see in the next section, this actually holds in our system. However, there is a non-monistic model of a system of paraconsistent mereology which include such a bottom, that is a model which has $\overline{1_=}$ but validates $\neg \exists x \forall y (x = y)$. Cotnoir and Weber (Forthcoming) have already shown that paraconsistent mereology can have such a non-monistic model with the bottom. Figure 1 represents such a model. In Figure 1, the solid arrow represents the parthood relation and the dashed arrow the non-parthood relation. What is important is that 0 is a part of 1 but is not a part of 1 as well. Because of this, $1 \neq 0$ and thus the model validates $\forall x \exists y (x \neq y)$, even thought 0 is the bottom.

Note that, partly because of the problem of monism, such a bottom elements is explicitly eliminated from the standard mereology. However, some theorists currently try to incorporate the bottom element into mereology. There are two different kinds of heresies. The first one tries to incorporate the bottom developing non-standard theories which reject some axioms but still rely on a classical logic framework. For instance, Martin (1965) defined the null item as 'what is not a part of itself' rejecting the unrestricted Reflexivity Axiom. Bunt (1985), Meixner (1997), Hudson (2006), and Se-

---

[2]We thank Zach Weber for suggesting this.

The Totality and Its Complement

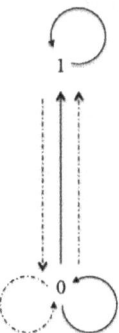

Figure 1: The model

gal (2014) abandoned the (quasi-) Supplementation Axiom. The second one tries to deal with the bottom presenting even higher non-standard theories which save the whole axiomatic structure of classic mereology but change the underlying logic into a non-classical (paraconsistent) one. We present a third heresy which changes both the axiomatic system and the underlying logic in the next section. Of course showing that the existence of $\overline{1_=}$ leads to neither triviality nor monism is not enough to guarantee its existence. Then, how to get the complement of $1_=$? As stated at the beginning of this section, the way which is explored in this paper is to suppose that something is not a part of the totality. Based on what we have considered, we should suppose that something is not a part of $1_=$.

Before moving into the next section, let us make a brief metaphysical comment on how we can see the relation between objecthood and nothingness from the view point of a paraconsistent mereology. If what is collected in the totality is an object, whatever is not collected in the totality is not an object. But, then, what makes an object an object? What is the criterion to decide if something is a thing or not? One way to state this criterion is to take the *objecthood* of an object to be the property of being self-identical. As such, the totality is the mereological sum of everything that is self-identical. Moreover, since the complement of the totality is the sum of everything which is not an object, the complement of the totality is the mereological sum of everything that is not self-identical.

Filippo Casati and Naoya Fujikawa

2 Formal implementation

2.1 Axioms

In this section, we propose an axiomatic system which has the existence of $\overline{1_=}$ as a theorem. Our proposal is based on the system PM in (Cotnoir & Weber, Forthcoming). The language of PM is the standard first-order language with identity and $\leq$, which represents the parthood relation. The notions of overlap, disjoint, $\leq$ least upper bound are defined as follows.

(5)   a. $x \bullet y =_{df} \exists z(z \leq x \wedge z \leq y)$

   b. $x \parallel y =_{df} \neg x \bullet y$

   c. $lub(x, \phi) =_{df} \forall y(\phi \to y \leq x) \wedge (\forall w \forall y(\phi \to y \leq w) \to x \leq w)$

PM consists of the following axioms.

PM0 Axioms of the weak relevant logic DKQ [3]

PM1 $\forall x(x \leq x)$

PM2 $\forall x \forall y(x \leq y \wedge y \leq x \leftrightarrow x = y)$

PM3 $\forall x \forall y \forall z(x \leq y \wedge y \leq z \to x \leq z)$

PM4 $\forall x(\exists y(y \not\leq x) \to \exists z(z \parallel x \wedge \forall y((y \parallel x \to y \leq z) \wedge (y \parallel z \to y \leq x))))$

PM5 $\exists x \varphi \to \exists z\, lub(z, \varphi)$

PM6 $\exists x \exists y(y \neq x) \to \neg \exists x \forall y(x \leq y)$

We omit PM6 from PM and add PM7, which says that something is not a part of the totality, to PM. The result is as follows.

PM0 Axioms of the weak relevant logic DKQ

PM1 $\forall x(x \leq x)$

PM2 $\forall x \forall y(x \leq y \wedge y \leq x \leftrightarrow x = y)$

---

[3] For the detail of DKQ, see the appendix of (Cotnoir & Weber, Forthcoming) and (Priest, 1987, sec. 18.3.). In DKQ, $\to$ is detachable and contrapositive.

The Totality and Its Complement

PM3 $\forall x \forall y \forall z (x \leq y \wedge y \leq z \rightarrow x \leq z)$

PM4 $\forall x (\exists y (y \not\leq x) \rightarrow \exists z (z \mathrel{\rotatebox[origin=c]{90}{$\models$}} x \wedge \forall y ((y \mathrel{\rotatebox[origin=c]{90}{$\models$}} x \rightarrow y \leq z) \wedge (y \mathrel{\rotatebox[origin=c]{90}{$\models$}} z \rightarrow y \leq x))))$

PM5 $\exists x \varphi \rightarrow \exists z\, lub(z, \varphi)$

PM7 $\exists x \exists y (x \not\leq y \wedge lub(y, x = x))$

In the next section, we see some results of this system.

2.2 Some results

Let us first define the notions of complement as follows.

(6) $\overline{x} =_{df} lub[y | y \mathrel{\rotatebox[origin=c]{90}{$\models$}} x]$

Followings holds.[4]

(7) a. $\forall x (x \leq lub[x | x = x])$

b. $\forall x (x \not\leq \overline{lub[x | x = x]})$

c. $\forall x (\overline{lub[x | x = x]} \leq x)$

d. $lub[x | x \neq x] = \overline{lub[x | x = x]}$

Note that $\overline{lub[x | x = x]}$ and $lub[x | x \neq x]$ are well-defined. The existence of $lub[x | x = x]$ is ensured by PM7 and PM4. The existence of $lub[x | x \neq x]$ is ensured by PM5 and PM7. Proofs are given in Appendix.

(7a) means that the sum of self-identicals is the totality, as we have expected. (7b) means that the complement of the totality is empty. (7c) means that the complement of the totality is the bottom. Since being empty and being the bottom are expected properties of nothingness, (7b) and (7c) support our claims that the complement of the totality is nothingness. (7d) shows that the sum of non-self-identicals and the complement of the totality are identical. (7d) is interesting because it connects our notion of nothingness as the complement of totality with Priest's notion of nothingness as the sum of non-self-identicals (the sum of the empty set defined as $\{x | x \neq x\}$) (Priest, Forthcoming).

However, nothingness is highly contradictory. Indeed, the negation of each of (7)s holds as well.

---

[4] $lub[x | \phi]$ is the $lub(x, \phi)$, that is, $lub[x | \phi] = t =_{df} \exists! x lub(x, \phi)$

(8) a. $\neg \forall x(x \leq lub[x|x=x])$

   b. $\neg \forall x(x \not\leq \overline{lub[x|x=x]})$

   c. $\neg \forall x(\overline{lub[x|x=x]} \leq x)$

   d. $lub[x|x \neq x] \neq \overline{lub[x|x=x]}$

Note that they reflect or are derived from *expected* features of nothingness as well. Since to get nothingness we suppose that something is not a part of the totality, the totality is the totality but not the totality at the same time. Moreover, the complement of the totality is going to be highly contradictory too. Indeed it is empty because the totality sums up all objects. However, since the totality is not the totality, then the complement of it is not going to be empty as well.

Summarizing these last four results we can say that (8a) means that the sum of self-identicals is not the totality because there is an element which is not a part of it. The (8b) means that the complement of the totality, that is, nothingness, is not empty because there is an element which is a part of it, namely a non-self-identical object. Finally, the (8c) means that the complement of the totality is not the bottom and (8d) shows that the sum of non-self-identicals and the complement of the totality are not identical.

3 Conclusion

This short note has tried to deliver a mereological account of nothingness considered as the absence of everything, of all objects. Our system, which is based on the paraconsistent mereology proposed by Cotnoir and Weber (Forthcoming), has shown that the totality (defined as the mereological sum of everything that is self-identical) has a complement. Such complement is the mereological sum of everything that is not self-identical. The paper has also presented some (maybe weird but, for sure, contradictory) features of nothingness.

4 Appendix

Proofs of (7) (7a) immediately follows from $\forall x(x=x)$ and $\forall x(x=x \to x \leq lub[x|x=x])$. Note that $\to$ is detachable. □

(7b) is shown as follows. Since $\forall x \forall y(x \leq \overline{y} \leftrightarrow \neg x \bullet y)$,[5] we have $y \leq$

---

[5](Cotnoir & Weber, Forthcoming, sec. 4.4)

The Totality and Its Complement

$\overline{lub[x|x=x]} \leftrightarrow \neg y \bullet lub[x|x=x]$. From (7a) and **PM1**, $y \bullet lub[x|x=x]$. Given that $\rightarrow$ is contrapositive, it follows that $y \not\leq \overline{lub[x|x=x]}$. □

To show (7c), first we show (9).

(9) $\forall x(x \barparallel \overline{lub[x|x=x]})$

Take any x and any y. From (7b), it follows that $y \not\leq x \vee y \not\leq \overline{lub[x|x=x]}$. Therefore, we have $\forall y(y \not\leq x \vee y \not\leq \overline{lub[x|x=x]})$, that is, $x \barparallel \overline{lub[x|x=x]}$. It follows that (9). (7c) is shown as follows. Take any x. First consider the case where $\exists y(y \not\leq x)$.[6] In this case, there is the complement of x, $\overline{x}$. Again since $\forall x \forall y(x \leq \overline{y} \leftrightarrow \neg x \bullet y)$, $\overline{lub[x|x=x]} \leq \overline{\overline{x}} \leftrightarrow \neg lub[x|x=x] \bullet \overline{x}$. From (9), we have the right-hand side of the biconditional, thus we have $\overline{lub[x|x=x]} \leq \overline{\overline{x}}$. Since $\overline{\overline{x}} = x$, $\overline{lub[x|x=x]} \leq x$ holds. Second consider the case where $\neg \exists y(y \not\leq x)$. This is equivalent to $\forall y(y \leq x)$, thus it immediately follows that $\overline{lub[x|x=x]} \leq x$. In both cases, $\overline{lub[x|x=x]} \leq x$ holds. Therefore we have (7c). □

Given (7c) and **PM2**, (7d) is proven if we show that $lub[x|x \neq x] \leq \overline{lub[x|x=x]}$ holds. Since $\forall y(y \barparallel lub[x|x=x] \rightarrow y \leq \overline{lub[x|x=x]})$ (from the definition of the complement), we have $lub[x|x \neq x] \barparallel lub[x|x=x] \rightarrow lub[x|x \neq x] \leq \overline{lub[x|x=x]}$. The antecedent of this conditional is shown as follows. $\exists y(y \leq lub[x|x \neq x]) \rightarrow \exists x(x \neq x)$[7] holds. Since $\forall x(x=x)$, we have $\neg \exists x(x \neq x)$. Given that $\rightarrow$ is contrapositive, from them, it follows that $\neg \exists y(y \leq lub[x|x \neq x])$, which is equivalent to $\forall y(y \not\leq lub[x|x \neq x])$. From this it follows that $\forall y(y \not\leq lub[x|x \neq x] \vee y \not\leq \overline{lub[x|x=x]})$, that is, $lub[x|x \neq x] \barparallel lub[x|x=x]$.[8] Therefore, $lub[x|x \neq x] \leq \overline{lub[x|x=x]}$.[9] □

Proofs of (8) (8a) immediately follows from PM7 which delivers $\exists x(x \not\leq lub[x|x=x])$. □

---

[6]Here we appeal to argument by cases, which is valid in DKQ. See (Cotnoir & Weber, Forthcoming, p. 23).

[7]Or, $\forall y(y \leq lub[x|x \neq x] \rightarrow \exists x(x \leq y \wedge x \neq x))$.

[8]Intuitively speaking: Since everything is self-identical, no object is non-self-identical. Therefore, no object is a part of the sum of non-self-identicals (given that if something is a part of the sum of non-self-identilcals, then something is non-self-identical). This means that everything is disjoint from the sum of non-self-identicals, and therefore it is disjoint from the sum of self-identicals.

[9]$\overline{lub[x|x=x]} \leq lub[x|x \neq x]$ immediately follows from (7c). This is also shown by the following facts. $\forall y(y \barparallel lub[x|x=x] \rightarrow y \leq lub[x|x \neq x])$ and $\overline{lub[x|x=x]} \barparallel lub[x|x=x]$.

(8b) is proved as follows. Given (8a) and $\forall y(y = y \to y \leq lub[x|x = x])$, $\exists x(x \neq x)$ follows. Therefore $\exists x(x \leq lub[x|x \neq x])$. From (7d) and the substitutibity of identicals, $\exists x(x \leq \overline{lub[x|x = x]})$ follows. □

(8c) is equivalent to $\exists x(\overline{lub[x|x = x]} \nleq x)$ and the latter immediately follows from (7b). □

Finally (8d). Given PM2 and (7b), $lub[x|x \neq x] \neq \overline{lub[x|x = x]}$ holds. □

References

Bunt, H. (1985). *Mass Terms and Model-theoretic Semantics*. Cambridge: Cambridge University Press.

Cotnoir, A. (2010). *Non-classical Mereologies and their Applications*. Unpublished doctoral dissertation, University of Connecticut.

Cotnoir, A., & Weber, Z. (Forthcoming). Inconsistent boundaries. *Synthese*.

Givone, S. (2006). *Storia del Nulla*. Segrate: Mondadori.

Heidegger, M. (2012). *Contributions to Philosophy*. Bloomington: Indiana University Press.

Hudson, H. (2006). Confining Composition. *Journal of Philosophy*, *103*, 631–651.

Martin, R. M. (1965). Of Time and the Null Individual. *Journal of Philosophy*, *62*, 723–736.

Meixner, U. (1997). *Axiomatic Formal Ontology*. Dordrecht: Kluwer.

Priest, G. (1987). *In Contradiction: A Study of the Transconsistent*. Leiden: Martinus Nijhoff.

Priest, G. (2014). *One: Being an Investigation into the Unity of Reality and of Its Parts, Including the Singular Object which is Nothingness*. Oxford: Oxford University Press.

Priest, G. (Forthcoming). Much Ado About Nothing. *Australian Journal of Logic*.

Sartre, J. P. (1956). *Being and Nothingness: An Essay on Phenomenological Ontology*. New York City: Philosophical Library.

Segal, A. (2014). Causal Essentialism and Mereological Monism. *Philosophical Studies*, *169*, 227–255.

Varzi, A. (2009). *Mereology*. http://plato.stanford.edu/entires/mereology.: Stanford Encyclopedia.

The Totality and Its Complement

Filippo Casati
University of St. Andrews
Scotland
E-mail: `fgec@st-andrews.ac.uk`

Naoya Fujikawa
Kyoto University
Japan
E-mail: `fjnaoya@gmail.com`

Conjunction in Paraconsistent Weak Kleene Logic

ROBERTO CIUNI

Abstract: In this paper I present Parconsistent Weak Kleene (PWK) and I prove that its conjunction is definable in the Logic of Paradox LP introduced by Graham Priest, thus establishing connections between the two formalisms. I also discuss two different possible readings of PWK-conjunction and propose a relational semantics for it, along the lines of the Routley-Meyer semantics for Relevant Logic.

Keywords: Weak Kleene Logic, paraconsistency, conjunction, fusion, Routley-Meyer semantics

1 Introduction

Halldén (1949) on *non-sense* and Prior (1957) on *contingent existents* independently discuss a logical framework that is based on the following three tenets: (a) some sentences cannot be assigned a classical truth value: they require a third, non-classical value; (b) the non-classical value propagates from one sentence A to any sentence B where A occurs as a subsentence; (c) *falsity avoidance* is what valid inferences aim at: they go from non-false premises to non-false conclusions. The so-called Weak Kleene Logic – or Bochvar Logic – introduced by Bochvar (1938) is built in accordance with tenets (a) and (b), but it assumes that *untruth avoidance* is what valid inferences aim at: valid reasoning goes from *true* premises to *true* conclusions. By contrast, if we apply tenet (c) and follow all the intuitions proposed both by Halldén (1949) and Prior (1957), we will obtain a paraconsistent counterpart of Weak Kleene Logic, which I call *Paraconsistent Weak Kleene* – PWK for short.

PWK adds to the paraconsistent logics based on three-valued matrices, the most famous of which is, today, the Logic of Paradox LP proposed by Priest (2006). At the same time, the conjunction from PWK lacks some important features of standard conjunctions – for instance, it fails Conjunction Simplification (CS). On the other hand, the connective shares features of a compatibility operator from Relevant Logic, namely the *fusion operator* – failure of CS itself, and the rule for introducing the connective.

This paper aims at establishing formal connections between PWK and LP. In particular, it proves that PWK-conjunction is definable via a complex formula from LP. The definition highlights a disjunctive feature encoded in PWK, which is in turn connected with failure of CS. Also, based on the definability result above, I provide a satisfaction-preserving translation from PWK to LP that explains failure of CS.

Beside providing these formal results, the paper confronts the issue of a possible reading of PWK-conjunction and proposes an alternative, relational semantics beside the usual many-valued one. This semantics goes along the lines of the Routley-Meyer semantics for Relevant Logic, although it does not pursue the formalization of a relevant conditional and focuses on notions that go beyond the standard ones that motivate Routley-Meyer semantics.

The paper proceeds as follows. The rest of this section clarifies some issues concerning the origins of PWK. Section 2 introduces PWK and its many-valued semantics, while familiarizing with some notable failures – including failure of CS – and other features of PWK-conjunction. Section 3 establishes the definability of PWK-conjunction in LP and provides a satisfaction-preserving translation from the former to the latter. Section 4 presents two possible readings of conjunction and sketches a Routley-Meyer semantics for PWK-conjunction. Section 5 briefly discusses the conceptual import of such semantics and connects that construction with the two possible readings from the previous section. Section 6 summarizes the content of the paper.

Discussing the philosophical interpretation that Halldén and Prior gave to PWK goes beyond the purposes of this paper, but a clarification is in order.

First, the name PWK does not appear either in (Halldén, 1949) or (Prior, 1957), but two facts make it clear that the logic discussed there is indeed PWK. First, Halldén (1949) or Prior (1957) use the language of first-order logic and a three-valued Kleene semantics to interpret the propositional connectives. Second, they accept (a)–(c), although they seem not to notice that this implies *paraconsistency*.

Conjunction in Paraconsistent Weak Kleene Logic

2 Paraconsistent Weak Kleene

The language $\mathcal{L}_1$ of PWK includes a countable set Atom = $\{p, q, r, s, \ldots\}$ of *atomic formulas* and the set $\text{Conn}_1 = \{\neg, \vee_1, \wedge_1, \supset_1, \equiv\}$ of connectives – subscripts will help in the comparison with LP below; no subscript is attached to negation and material biconditional, since since they are the same in both logics. The set Form_1 is determined by the usual recursive definition of a formula. I denote *arbitrary* formulas with latin capital letters A, B, ... I let $\{\mathbf{t}, \mathbf{f}, \mathbf{b}\}$ be the truth values of PWK, with $\mathbf{t}$ and $\mathbf{f}$ being (classical) truth and falsity, respectively, and $\mathbf{b}$ being the non-classical value. The truth conditions for arbitrary formulas in Form_1 are exemplified by the truth table concerning connectives in Conn_1 from Table 1:

	$\neg A$
t	f
b	b
f	t

$A \vee_1 B$	t	b	f
t	t	b	t
b	b	b	b
f	t	b	f

$A \vee_2 B$	t	b	f
t	t	t	f
b	t	b	b
f	f	b	f

$A \wedge_1 B$	t	b	f
t	t	b	f
b	b	b	b
f	f	b	f

$A \wedge_2 B$	t	b	f
t	t	b	f
b	b	b	f
f	f	f	f

$A \supset_1 B$	t	b	f
t	t	b	f
b	b	b	b
f	t	b	t

$A \supset_2 B$	t	b	f
t	t	b	f
b	t	b	b
f	t	t	t

$A \equiv B$	t	b	f
t	t	b	f
b	b	b	b
f	f	b	t

Table 1: Truth tables for PWK and LP

The tables for connectives in $\mathcal{L}_1$ make it clear that we could restrict

Conn$_1$ to $\neg$, $\wedge_1$ and set $A \vee_1 B := \neg(\neg A \wedge_1 \neg B)$, $A \supset_1 B := \neg A \vee_1 B$ and $A \equiv_1 B := (\neg A \vee_1 B) \wedge_1 (\neg B \vee_1 A)$. The notions of a *valuation* and *logical consequence* are standardly defined:

Definition 1 A PWK-*valuation* for Form$_1$ is the unique extension V_{PWK} : Form$_1 \to \{\mathbf{t}, \mathbf{b}, \mathbf{f}\}$ of a mapping Atom $\to \{\mathbf{t}, \mathbf{b}, \mathbf{f}\}$, that is induced by the truth tables for PWK of Table 1. I denote valuations of PWK by V_{PWK}, V_{PWK}, ... and let $\mathcal{V}_{\mathsf{PWK}}$ be the class of such valuations.

Definition 2 Let $\mathcal{D} = \{\mathbf{t}, \mathbf{b}\}$ be the set of *designated values* of PWK and let $\{A_1, \ldots, A_n\} \cup \{B\} \subseteq $ Form$_{\mathcal{L}_1}$. Then, B is an PWK-*consequence* of $\{A_1, \ldots, A_n\}$ (in symbols: $A_1, \ldots, A_n \vDash_{\mathsf{PWK}} B$) if, for all PWK-valuations $V_{\mathsf{PWK}} \in \mathcal{V}_{\mathsf{PWK}}$ for Form$_1$, if $V_{\mathsf{PWK}}(A) \in \mathcal{D}$ for all $A_i \in \{A_1, \ldots, A_n\}$, then $V_{\mathsf{PWK}}(B) \in \mathcal{D}$.

The extension of $\mathcal{D}$ makes it clear that Definition 2 is the formal expression of tenet (c) in Section 1. As usual, a formula $A \in $ Form$_1$ is valid if $\varnothing \vDash_{\mathsf{PWK}} A$, and satisfiability is defined accordingly. I skip the recursive definition of truth conditions in PWK for space limits; the tables for operators in Conn$_1$ from Table 1 suffices to understand how these go. The tables for $\neg, \equiv, \vee_1, \wedge_1, \supset_1$ follow the distinctive feature of PWK, that is the *Contamination Principle*:

CP For every valuation $V \in \mathcal{V}_{\mathsf{PWK}}$ and formula $A \in $ Form$_1$, if some atomic formula p in A takes value $\mathbf{b}$, then A takes value $\mathbf{b}$.

CP is in turn the formal expression of tenet (b) from Section 1.[1] As is clear from Table 1, a formula A in $\mathcal{L}_1$ takes the value it would take in classical logic if no atomic formula p in A has the non-classical value. The presence of two designated values is crucial, since it leads to the falsification of some classical cases of logical consequence. Here is some example:

1	$A, A \supset_1 B \nvDash_{\mathsf{PWK}} B$	MP
2	$\neg B, A \supset_1 B \nvDash_{\mathsf{PWK}} \neg A$	MT
3	$A \supset_1 B, B \supset_1 C \nvDash_{\mathsf{PWK}} A \supset_1 C$	TR $\supset_1$
4	$A \supset_1 (B \wedge_1 \neg B) \nvDash_{\mathsf{PWK}} \neg A$	RAA
5	$A \wedge_1 \neg A \nvDash_{\mathsf{PWK}} B$	ECQ

---

[1] Notice that $\neg$ and $\equiv$ behave in the same way in LP, which does not obey CP, however – as clear by the tables for $\vee_2, \wedge_2$ and $\supset_2$ from Table 1.

Conjunction in Paraconsistent Weak Kleene Logic

As for MP (Modus Ponens), suppose $V_{\mathsf{PWK}}(A) = \mathbf{b}$ and $V_{\mathsf{PWK}}(B) = \mathbf{f}$. This suffices to have the premises designated, but the conclusion undesignated. In order to falsify MT (Modus Tollens), just change the valuation above to $V_{\mathsf{PWK}}(B) = \mathbf{b}$ and $V_{\mathsf{PWK}}(A) = \mathbf{t}$. As for TR⊃1, suppose $V_{\mathsf{PWK}}(A) = \mathbf{t}$, $V_{\mathsf{PWK}}(B) = \mathbf{b}$ and $V(C) = \mathbf{f}$: we will have the premises designated and the conclusion false. Coming to RAA (Reductio ad Absurdum), $V(A)_{\mathsf{PWK}} = \mathbf{f}$ and $V(B) = \mathbf{b}$ falsifies the rule. Finally, any valuation $V_{\mathsf{PWK}} \in \mathcal{V}_{\mathsf{PWK}}$ such that $V_{\mathsf{PWK}}(A) = \mathbf{b}$ and $V_{\mathsf{PWK}}(B) = \mathbf{f}$ will falsify ECQ (Ex Contradictione Quodlibet), thus making PWK a paraconsistent logic. Failures 1–5 are standard in many paraconsistent logics. Let us now go to one of the main topics of the paper: conjunction in PWK. The semantics from Table 1 (for operators in Conn$_1$) suffices to see that *Conjunction Simplification* fails:

$$A \wedge_1 B \not\vDash_{\mathsf{PWK}} B \qquad \text{CS}$$

(Counterexample to CS: any valuation $V_{\mathsf{PWK}} \in \mathcal{V}_{\mathsf{PWK}}$ such that $V_{\mathsf{PWK}}(A) = \mathbf{b}$ and $V_{\mathsf{PWK}}(B) = 0$ is enough to falsify CS, since it will be such that $V_{\mathsf{PWK}}(A \wedge_1 B) = \mathbf{b}$). This marks a crucial difference with LP and other paraconsistent logics, where CS is a valid rule. On the other hand, the following rule links conjunction and disjunction:

$$A \wedge_1 B \vDash_{\mathsf{PWK}} A \vee_1 B \qquad \text{CD}$$

Since a conjunction is undesignated any time the disjunction of the conjuncts is – the latter indeed equates with A and B both being undesignated. Also, it is easy to see that if we have both $\vDash_{\mathsf{PWK}} A$ and $\vDash_{\mathsf{PWK}} B$, then we have $\vDash_{\mathsf{PWK}} A \wedge_1 B$. This secures that, when setting up a natural deduction for PWK – or similar proof-theoretic devices – we can assume the usual *Introduction Rule for Conjunction*. I do not define any proof theory here, but I feel free to say that $\wedge_1$ 'introduces as a standard conjunction' and talk about the 'introduction of $\wedge_1$'. Finally, we have that

Fact 1 It can be the case that $A_1, \ldots, A_n \vDash_{\mathsf{PWK}} B$ and $A_1 \wedge_1 A_2 \cdots \wedge_1 A_n \not\vDash B$.

For instance, suppose B is A_n. All valuations V_{PWK} such that, for all $i \in \{1, \ldots, n\}$, $V_{\mathsf{PWK}}(A_i), V(B) \in \mathcal{D}$ suffice to verify $A_1, \ldots, A_n \vDash_{\mathsf{PWK}} B$, while a valuation where $V_{\mathsf{PWK}}(A_n) = \mathbf{f}$ and $V_{\mathsf{PWK}}(A_j) = \mathbf{b}$ for all $j \in \{1, \ldots, n-1\}$ suffices to have $V(A_1 \wedge_1 A_2 \cdots \wedge_1 A_n) \in \mathcal{D}$ and $V(B) = \mathbf{f}$, which of course implies $A_1 \wedge_1 A_2 \cdots \wedge_1 A_n \not\vDash B$. This is possible because the set of PWK-valuations satisfying *all* formulas in $\{A_1, \ldots, A_n\}$ may not coincide with the set of PWK-valuations satisfying the conjunction of such formulas – the latter may also include valuations where *some* of $A_1, \ldots, A_n$ is undesignated, on condition that at least one of them has value **b**. Another interesting fact is the following:

Fact 2 It can be the case that $A \vDash_{\mathsf{PWK}} B$ and $A \wedge_1 C \nvDash B$.

Indeed, the set $\mathcal{V}_{\mathsf{PWK}}(A \wedge_1 C)$ of valuations satisfying $A \wedge_1 C$ needs not to be a subset of the set $\mathcal{V}_{\mathsf{PWK}}(A)$ of valuations satisfying A: given the behavior of $\wedge_1$, we can have a valuation $V_{\mathsf{PWK}} \in \mathcal{V}_{\mathsf{PWK}}$ such that $V_{\mathsf{PWK}}(A) = \mathbf{f}$, $V_{\mathsf{PWK}}(C) = \mathbf{b}$, and – accordingly – $V_{\mathsf{PWK}}(A \wedge_1 C) = \mathbf{b}$. In other words, $\mathcal{V}_{\mathsf{PWK}}(A \wedge_1 C) \nsubseteq \mathcal{V}_{\mathsf{PWK}}(A)$. Suppose now that $\mathcal{V}_{\mathsf{PWK}}(A) \subseteq \mathcal{V}_{\mathsf{PWK}}(B)$ where $\mathcal{V}_{\mathsf{PWK}}(B)$ is the set of the valuations satisfying B, and that (1) there is a valuation $V_{\mathsf{PWK}} \in \mathcal{V}_{\mathsf{PWK}}$ such that $V_{\mathsf{PWK}}(C \wedge_1 \neg C \wedge_1 A) = \mathbf{b}$ and (2) $V_{\mathsf{PWK}}(A) = V_{\mathsf{PWK}}(B) = \mathbf{f}$. This implies $\mathcal{V}_{\mathsf{PWK}}(A \wedge_1 C) \nsubseteq \mathcal{V}_{\mathsf{PWK}}(B)$. This suffices to prove the fact.

Failure of CS, Fact 1 and Fact 2 tell us straight out that $\wedge_1$ does not behave as the standard conjunction to which classical logic and its many-valued kins have made familiar to us. By contrast, failure of CS and Fact 2 are features $\wedge_1$ shares with the *fusion operator* (also known as *intensional conjunction*) from Relevant Logic – see (Mares, 2004, pp. 167–168) for a compact and yet insightful introduction to the notion. I defer the discussion of such common features to Section 4. For the time being, let us go to further results that will make our understanding of PWK-conjunction more complete.

More precisely, in the next section I prove the definability of PWK-conjunction in LP (Theorem 1 below). The definition will provide further insights on a possible reading for PWK-conjunction. Also, it will help explain the failure of CS in PWK (see Corollary 1 below, and related comment). After presenting these results, I will come back to the possible readings of PWK-conjunction.

3 Defining PWK-conjunction in LP

The language $\mathcal{L}_1$ of the Logic of Paradox LP obtains from $\mathcal{L}_2$ by replacing Conn_1 with the set $\mathsf{Conn}_2 = \{\neg, \vee_2, \wedge_2, \supset_2, \equiv\}$, and defining the set Form_2 accordingly. LP shares truth values *and* designated values with PWK, and the difference between the two lies in how the truth conditions assign values to arbitrary formulas. The truth tables involving the operators in Conn_2 from

Conjunction in Paraconsistent Weak Kleene Logic

Table 1 provides the truth conditions for arbitrary formulas in Form$_2$, and suffice to understand the difference with the truth conditions related to the connectives in PWK.

As is easy to check, also in LP we could start with $\neg$ and $\wedge_2$ as the only primitives and define the remaining ones exactly as in classical logic and in PWK.

The tables for $\vee_2, \wedge_2, \supset_2$ reflect the reading of **b** as *both true and false*: as one sees from the relevant truth tables in Table 1, an LP-disjunction has value **b** (both true and false) iff no disjunct is simply true and at least one disjunct is both true and false: this matches with the traditional falsity-conditions for a disjunction (being *false* if *both disjuncts are so*) and those for its truth (a disjunction is *true* iff *at least a disjunct is so*).[2] Similar considerations go for conjunction and the conditional in LP. The notions of a *valuation* ($V_{\mathsf{LP}} \in \mathcal{V}_{\mathsf{LP}}$) and *logical consequence* ($A_1, \ldots, A_n \vDash_{\mathsf{LP}} B$) are as those for PWK, *mutatis mutandis*. With this at hand, we can go to a comparison between PWK and LP.

First, it is easy to check that LP share failures 1–5 with PWK (see Beall, 2011; Priest, 2006): the same counterexamples we have seen in the case of PWK will do here. By contrast, it is easy to see that CS is valid in LP:

$$A \wedge_2 B \vDash_{\mathsf{LP}} B$$

Indeed, a conjunction is designated in LP iff *both conjuncts are designated*, as is clear from the relative truth table from Table 1. Interestingly, the non-simplifying conjunction $\wedge_1$ from PWK can be defined in terms of the simplifying conjunction $\wedge_2$ from LP. Let $A!$ be short for $A \wedge \neg A$. We have that:

Theorem 1 *Connective $\wedge_1$ of* PWK *is definable in* LP *as follows:*

$$A \wedge_1 B :=_{\mathsf{LP}} (A \wedge_2 B) \vee_2 (A! \wedge_2 \neg B) \vee_2 (\neg A \wedge_2 B!)$$

Proof by the table below, where $\xi(A, B)$ is short for $(A \wedge_2 B) \vee_2 (A! \wedge_2 \neg B) \vee_2 (\neg A \wedge_2 B!)$. The truth assignments in the last row to the right conforms to the ones for $\wedge_1$, as is clear by Table 1.

Table 2: Defining PWK's $\wedge_1$ in LP

---
[2] At the same time, if the two disjuncts have value **b** and **t** respectively, the disjunction has value **t**, once again in accordance with the conditions for a disjunction to be *true*.

A	B	$A \wedge_2 B$	$A! \wedge_2 \neg B$	$\neg A \wedge_2 B!$	$\xi(A,B)$	$A \wedge_1 B$
t	t	t	f	f	t	t
t	b	b	f	f	b	b
t	f	f	f	f	f	f
b	t	b	f	f	b	b
b	b	b	b	b	b	b
b	f	f	b	f	b	b
f	t	f	f	f	f	f
f	b	f	f	b	b	b
f	f	f	f	f	f	f

Theorem 1 suggests the following meaning for PWK-conjunction: '*either* all conjuncts are designated *or* at least one is non-classical *and* some other conjunct is *false*.'[3] An equivalent definition is $A \wedge_1 B := (A \wedge_2 B) \vee_2 A! \vee_2 B!$, as is easily checked. However, the definition I present above gives a more complete insights into the different *entries* of the truth tables for $\wedge_1$, since it refers directly to cases where one conjunct has value **b** and the other one has value **f**.

Theorem 1 is also interesting since it helps explain the failure of CS in PWK. The above definition of $\wedge_1$ in LP suffices to have a good hint at this, but in order to prove that a failure in PWK-consequence is explained by its connections with LP-consequence, the definability of (part of) $\mathcal{L}_1$ in $\mathcal{L}_2$ is not enough: we need a translation bridging the first with the second.

Once again, for every formula A, B from $\mathcal{L}_2$, we let $\xi(A, B)$ be the formula $(A \wedge_2 B) \vee_2 (A! \wedge_2 \neg B) \vee_2 (\neg A \wedge_2 B!)$. We let $\tau : \mathcal{L}_1 \to \mathcal{L}_2$ be a *translation function* from the language of PWK to that of LP, such that:

$$\begin{aligned}
\tau(p) &= p \\
\tau(\neg A) &= \neg(\tau(A)) \\
\tau(A \wedge_1 B) &= \xi(\tau(A), \tau(B)) \\
\tau(A \vee_1 B) &= \neg(\xi(\tau(\neg A), \tau(\neg B))) \\
\tau(A \supset_1 B) &= \neg(\xi(\tau(A), \tau(\neg B))) \\
\tau(A \equiv B) &= \xi(\neg(\xi(\tau(A), \tau(\neg B))), \neg(\xi(\tau(B), \tau(\neg A))))
\end{aligned}$$

Translation τ is *satisfaction preserving*, that is:

Proposition 1 *There exists a valuation* $V_{\mathsf{PWK}} \in \mathcal{V}_{\mathsf{PWK}}$ *such that* $V_{\mathsf{PWK}}(A) \in \mathcal{D}_{\mathsf{PWK}}$ *iff there exists a valuation* $V_{\mathsf{LP}} \in \mathcal{V}_{\mathsf{LP}}$ *such that* $V_{\mathsf{LP}}(\tau(A)) \in \mathcal{D}_{\mathsf{LP}}$.

---

[3] Given the reading of the non-classical value in LP, this equates with 'either all conjuncts are designated or at least one is a dialetheia'. As is well-known, in the paraconsistent tradition (especially that connected to paraconsistent truth theory) a *dialetheia* is a formula that is both true and false.

Conjunction in Paraconsistent Weak Kleene Logic

We prove Proposition 1 by structural induction on A. First, given a valuation $V_{\mathsf{PWK}} \in \mathcal{V}_{\mathsf{PWK}}$, we associate it with a valuation $V_{\mathsf{LP}} \in \mathcal{V}_{\mathsf{LP}}$ such that $V_{\mathsf{PWK}}(p) = V_{\mathsf{LP}}(p)$ for every atomic formula $p \in$ Atom – remember that both logics share the same set Atom. The case of atomic formulas is trivial, and so is the case of formulas of the form $\neg B$ or $A \equiv B$ given the definitions of negation and material biconditional in the two logics, and here I only present the cases of $\wedge_1$, $\vee_1$ and $\supset_1$.

- *Case $A = B \wedge_1 C$.* This implies that $V_{\mathsf{PWK}}(A), V_{\mathsf{PWK}}(B) \in \mathcal{D}_{\mathsf{LP}}$ or $V_{\mathsf{PWK}}(A) = \mathbf{b}$ and $V_{\mathsf{PWK}}(B) = \mathbf{f}$ (or vice versa). By induction hypothesis, this implies that $V_{\mathsf{LP}}(\tau(B)), V_{\mathsf{LP}}(\tau(C)) \in \mathcal{D}_{\mathsf{LP}}$ or $V_{\mathsf{LP}}(\tau(B)) = \mathbf{b}$ and $V_{\mathsf{LP}}(\tau(C)) = \mathbf{f}$ (or vice versa). But this equates with having $V_{\mathsf{LP}}((\tau(B) \wedge_2 \tau(C)) \vee_2 ((\tau(B) \wedge_2 \tau(\neg B)) \wedge_2 \neg \tau(C)) \vee_2 (\neg \tau(B) \wedge_2 (\tau(C) \wedge_2 \tau(C)))) \in \mathcal{D}_{\mathsf{LP}}$ – that is, $V_{\mathsf{LP}}(\xi(\tau(B), \tau(C))) \in \mathcal{D}_{\mathsf{LP}}$.

- *Case $A = B \vee_1 C$.* Consider that $A \vee_i B := \neg(\neg A \wedge_i \neg B)$ for $i \in \{1, 2\}$. The case thus reduces to $A = \neg(\neg B \wedge_1 \neg C)$. By induction hypothesis, this implies that $V_{\mathsf{LP}}(\tau(\neg B)), V_{\mathsf{LP}}(\tau(\neg C)) \neq \mathbf{t}$ or $V_{\mathsf{LP}}(\tau(\neg B)) \neq \mathbf{t}$ and $V_{\mathsf{LP}}(\tau(\neg C)) = \mathbf{t}$ (or vice versa). But this equates with having $V_{\mathsf{LP}}(\neg(\tau(B) \wedge_2 \tau(C)) \vee_2 \neg((\tau(B) \wedge_2 \tau(\neg B)) \wedge_2 \neg \tau(C)) \vee_2 \neg(\neg \tau(B) \wedge_2 (\tau(C) \wedge_2 \tau(C)))) \in \mathcal{D}_{\mathsf{LP}}$ – which simply means that $V_{\mathsf{LP}}(\neg \xi(\tau(\neg B), \tau(\neg C))) \in \mathcal{D}_{\mathsf{LP}}$.

- *Case $A = B \supset_1 C$.* Consider that $A \supset_i B := \neg(A \wedge_i \neg B)$ for $i \in \{1, 2\}$. The case thus reduces to $A = \neg(B \wedge_1 \neg C)$. The proof then goes as with the case above, with B replacing $\neg B$, thus holding $V_{\mathsf{LP}}(\neg \xi(\tau(B), \tau(\neg C))) \in \mathcal{D}_{\mathsf{LP}}$.

An immediate corollary of Proposition 1 is that PWK-consequence is preserved by LP-consequence, that is:

Corollary 1 *If $A_1, \ldots, A_n \vDash_{\mathsf{PWK}} B$, then $\tau(A_1), \ldots, \tau(A_n) \vDash_{\mathsf{LP}} \tau(B)$.*

(otherwise, there should be a $V_{\mathsf{LP}} \in \mathcal{V}_{\mathsf{LP}}$ such that $V_{\mathsf{LP}}(A_i) \in \mathcal{D}_{\mathsf{LP}}$ for $1 \leq i \leq n$ and $V_{\mathsf{LP}}(B) \notin \mathcal{D}_{\mathsf{LP}}$, while $V_{\mathsf{PWK}}(A_i) \in \mathcal{D}_{\mathsf{PWK}}$ for $1 \leq i \leq n$ and $V_{\mathsf{PWK}}(B) \in \mathcal{D}_{\mathsf{PWK}}$ for all $V_{\mathsf{PWK}} \in \mathcal{V}_{\mathsf{PWK}}$. But this would contradict Proposition 1.)

In turn, Corollary 1 explains failure of CS in PWK. Indeed, given $p, q \in$ Atom, we have $\xi(p, q) \nvDash_{\mathsf{LP}} q$. By $\tau(p) = p$ for all $p \in$ Atom and the contraposition of Corollary 1, we have $p \wedge_1 q \not\vDash_{\mathsf{PWK}} q$. A little adjustment adapts this to arbitrary formulas A and B.

Finally, notice that Theorem 1 and Proposition 1 together imply that PWK is a fragment of LP.

4 Possible readings of PWK-conjunction

Sections 2 and 3 have brought a number of insights on PWK-conjunction $\wedge_1$. On the one hand, $\wedge_1$ lacks some properties that are usually considered distinctive of conjunction, such as CS. While this does not imply that $\wedge_1$ cannot be legitimately considered a conjunction, it implies that it cannot be considered a 'standard conjunction', and that a conjunction-like notion $\wedge_1$ would express must still be brought to the fore of philosophical logic.

Part of the divergence of $\wedge_1$ from standard conjunction are explained by Theorem 1. In particular, the latter seems to motivate a given reading of the connective: $\wedge_1$ would be a particular form of disjunction (a 'disjunction in disguise,' let us say), with $A \wedge_1 B$ meaning 'either both A and B are designated, or one of them is non-classical and the other one is false.' Call this *'disguised disjunction reading.'*

The definability of $\wedge_1$ in terms of LP's connectives and the consequent reading of $\wedge_1$ seem to be quite conclusive. PWK-conjunction should be then read as a 'disjunction in disguise', that imposes the designatedness of both A and B when they have classical value, and requires that one of them is non-classical if the other is not-designated (that is, false).

However, there is an alternative reading that deserves some attention and that I now go to discuss. It is to such reading that I now turn. After discussing it, I will briefly compare the two readings.

The conjunction of PWK displays some interesting similarities with the *fusion* operator ∘ from Relevant Logic. Both ∘ and $\wedge_1$ do not simplify and yet introduce as a conjunction, and the fact that B follows from A does not imply that B follows from $A \circ C$ or $A \wedge_1 C$ – see Fact 2 for the latter. In Relevant Logic, ∘ expresses a notion of (informational) *compatibility*: $A \circ B$ holds (at a world w) only if A and B separately holds at two worlds w' and w'' the are compatible with w, respectively. In light of this interpretation, the two features just discussed seem in order: the compatibility of A and B does not implies that, say, B is true, and yet the validity of the two formula would be enough to prove that they are compatible.

The similarities with *fusion* makes it an attractive option to chart PWK-

Conjunction in Paraconsistent Weak Kleene Logic

conjunction as a compatibility operator – in the sense of the term that I have detailed earlier. Call this 'the *compatibility reading*.' In particular, the investigation of an intensional semantics for $\wedge_1$ seems the natural way to follow in order to give a proper formal counterpart to this reading (indeed, an intensional semantics is indeed at play when interpreting fusion).

I sketch a possible intensional reading of $\wedge_1$ below. Before doing this, let us familiarize with the differences between $\wedge_1$ and $\circ$. First, PWK-conjunction is *idempotent* – $A \wedge_1 A \vDash_{\mathsf{PWK}} A$ – while the fusion operator $\circ$ is not – A does not follow from $A \circ A$. Second, a PWK-conjunction of two formulas implies that at least one such formula is designated, as clear by CD, while compatibility operators do not display this feature – this comes with the fact that a 'disjunctive feature' is built up in $\wedge_1$. Finally, $\wedge_1$ differs from $\circ$ in that $\circ$ exchange with comma in the premises of an inference, while $\wedge_1$ does not – see Fact 2; this is actually something shared by the extensional conjunction of Relevant Logic.

Thus, an intensional semantics for $\wedge_1$ must preserve its similarities with $\circ$ and yet also respect the differences between the two. In what follows, I sketch an intensional semantics for $\wedge_1$ along the lines of the Routley-Meyer Semantics for Relevant Logic, and I will show that such semantics preserves all the above similarities and differences with the fusion operator. I will defer a more complete investigation to future work, where I plan to establish the equivalence of the semantics below and that from Section 2 with respect to $\mathcal{L}_1$. Also, I leave the discussion of the reading of the semantics below to Section 5.

4.1 Towards a Routley-Meyer semantics for $\wedge_1$

Let us start from the semantics by Routley and Meyer (1972). First, we endorse the reading of worlds $w, w', w'', \dots$ as *informational states*. This naturally comes with the intuition that there are incomplete worlds and inconsistent worlds beside complete ones, the latter being the only ones admitted by standard Kripke semantics – for a survey of the motivations, see the crystal-clear (Restall, 1999). Just to be clear: where $\vDash$ is the satisfaction relation (or 'assertion relation'), a *complete* world is a world w such that $w \vDash A$ or $w \vDash \neg A$ for *all* formulas A in the language; an *incomplete* world is a world w such that $w \nvDash A$ or $w \nvDash \neg A$ for *some* formulas A in the language.

Second, we assume the partial order $\leq$, with $w \leq w'$ reading '(the information from) w' extends (the information from) w.' We impose the usual *heredity condition*: if $w \models p$ and $w \leq w'$, then $w' \models p$ – it is easy to extend the conditions to literals.

Third, we endorse the semantics for negation from (Routley & Routley, 1972), which is also part of the Routley-Meyer semantics. This implies the definition of the Routley star $\star$ (being a function from world to world) and the imposition of the usual conditions on it – see (Restall, 1999) for this, such as $w^{\star\star} = w$.[4] As known since (Dunn, 1994), $w^\star$ is the strongest informational state that is compatible with w – $w^\star$ asserts anything w does not deny. We assume that $w \models \neg A$ iff $w^\star \not\models A$; as is well known, this implies that w is inconsistent relative to A iff $w^\star$ is incomplete relative to A.

Fourth and final, we assume the ternary relation R^3 between worlds under the particular reading that was mentioned by Dunn and Restall (2002): $R(w, w', w'')$ means that $w \leq w''^\star$ and $w' \leq w''^\star$ (given the connections between the Routley star and compatibility, this indeed implies that both w and w' are compatible with w'').

To the standard elements of Routley-Meyer semantics we add a distinguished world @, on which we impose that condition that @ $\models A$ or @ $\models \neg A$ for *all* formulas A in the language and @ $\models A$ *and* @ $\models \neg A$ for *some* formulas A – that is, @ is *inconsistent*, but not necessarily trivial.

We define logical consequence as preservation of satisfaction in every Routley-Meyer model at its distinguished world @: $A \models B$ iff for every Routley-Meyer model $\mathcal{M}$ with a distinguished world @, if $\mathcal{M}, @ \models A$, then $\mathcal{M}, @ \models B$. Validity is defined accordingly as @ $\models A$ for every model $\mathcal{M}$. The rationale of @ and the definitions of logical consequence and validity is to allow for contradictions to be satisfied and Excluded Middle to be valid – something we cannot have if we make incomplete worlds relevant for consequence and validity.

With this at hand, we can propose a definition of $w \models A \wedge_1 B$. Let us introduce the intended conditions one by one. One necessary condition is that:

(1) $\quad w' \models A$ or $w'' \not\models \neg B \quad$ for some worlds $w', w'' \in W$
$\qquad\qquad\qquad\qquad\qquad$ and such that $R^3(w', w'', w)$

Due to the semantics for negation, $w'' \not\models \neg B$ equates with $w''^\star \models B$; and by the definition of R^3 and $w \leq w''^\star$ iff $w'' \leq w^\star$, we have that condition (1) implies that $w^\star \models A$ or $w^\star \models B$. Take now @*. It is incomplete, since @

---
[4]The other usual conditions will not really play a role in sketching my proposal below, but I assume them for conformity with the original Routley-Meyer semantics.

is inconsistent. This given, condition (1) suffices to guarantee a reasonable version of CD: if $@ \vDash A \wedge_1 B$, then either $@ \vDash A$ or $@ \vDash B$.[5] Indeed, suppose $@ \vDash A \wedge_1 B$. By condition (1), we have $w \vDash A$ or $w' \nvDash \neg B$ for some $w, w' \in W$ such that $R(w, w', @)$. The latter implies $@^* \vDash A$ or $@^* \vDash B$. By this, the definition of Routley star, and the completeness/inconsistency condition on @, the above version of CD follows.

We now need appropriate conditions to avoid that w is complete relative to both A and B and satisfies just their negations:

(2) If $w^* \vDash \neg A$, then $w^* \nvDash \neg B$ and $w^* \nvDash B$
(3) If $w^* \vDash \neg B$, then $w^* \nvDash \neg A$ and $w^* \nvDash A$

The inconsistency condition on @ and the definition of logical consequence secure failure of CS, Fact 1 and Fact 2, beside idempotence and the introduction of $\wedge_1$.

Let us start with failure of CS. Suppose there is a model such that $@ \vDash A \wedge_1 B$. Suppose $@^* \vDash \neg A$. This is compatible with condition (1) above, under the proviso that $w''' \vDash B$. $@^* \vDash \neg A$ implies $@ \vDash \neg A$ and $@ \nvDash A$. We will thus have a model where $@ \nvDash A$, although $@ \vDash A \wedge_1 B$.

Notice that, in such cases, conditions (2) and (3) guarantee that, if @ satisfies $\neg A$, it is inconsistent relative to A – since $@^*$ is incomplete relative to the same formula. As is clear from Theorem 1 and Section 4, this is indispensable in modeling $\wedge_1$. Condition (2) also explains why CS holds for the special cases $A \wedge \neg A \vDash A$ and $A \wedge A \vDash A$ – the latter being idempotence of $\wedge_1$.

As for Fact 1, suppose that each of $\{A_1, \ldots, A_n\}$ is satisfied at @ and $A_1, \ldots, A_n \vDash B$. Suppose now that $A_1, \ldots, A_{n-1} \nvDash B$, to the effect that $A_n \vDash B$. Of course, there will be a model where $@ \vDash A_1 \wedge_1 A_2 \wedge_1 \cdots \wedge_n A_n$ and yet $@ \nvDash A_n$, due to failure of CS. This model is compatible with $@ \nvDash B$. A prove that the present relational semantics also guarantees Fact 2 goes along the same lines.

Finally, it is easy to show that $\wedge_1$ 'introduces like standard conjunction': if for for every model we have that $@ \vDash A$ and $@ \vDash B$, then for every model we will also have that for every model, $@ \vDash A \wedge_1 B$. In other worlds, the fact that both A and B are valid suffices to secure that $A \wedge_1 B$ is valid.

---

[5] We are keeping other connectives from PWK aside, except for negation, and so we cannot use $\vee_1$ and express CD. But the above is a reasonable approximation of it in the semantics.

5 Discussion

In the previous section I have provided a semantics for $\wedge_1$, without touching upon the issue of its interpretation. The conditions suggested for $\wedge_1$ fit the formal purpose we had, but what notion they capture has not yet been discussed.

Conditions (1)–(3) capture a complex notion of compatibility: they tell us that A and B are compatible in the sense that $\neg A$ and $\neg B$ are never satisfied together by a world that is compatible with @ or its twin @* without A and B also being satisfied together by @ – by contrast, A and B can be satisfied together without their negations sharing this feature.

Arguing that this notion can have some theoretical relevance goes beyond the purpose of this paper. An insight on its conceptual relevance, in my view, can emerge just from a theory of informational states and extensions of information. In particular, notions like the one above can be properly assessed by a formal theory that describe *exclusion, prevention, preclusion* between information state. The formal theory motivating Routley-Meyer semantics has not provided a systematic investigation on the relations among different properties of information in different yet related states. However, the extension of the relational semantics for Relevant Logic to a theory of relevant information and related notion is an interesting research direction, that I plan to pursue in future work.

Before closing, it is worth to sum up the import of Section 4 in the present discussion. There, we have mentioned two possible reading: the 'disguised disjunction reading' and the 'compatibility reading'. My aim here is not to assign primacy to any of them, but just to show that PWK-conjunction *can* be given a relational semantics – motivating the compatibility reading somehow – beside the many-valued semantics – that seems to be the perfect semantics for the disguised disjunction reading. Which one is the right semantics for $\wedge_1$ is an altogether different question.

6 Conclusion

In this paper I have presented the paraconsistent kin of Weak Kleene Logic PWK and proved that the conjunction $A \wedge_1 B$ from PWK is definable in LP as $(A \wedge_2 B) \vee_2 (A! \wedge_2 \neg B) \vee_2 (\neg A \wedge_2 B!)$ – where $B!$ is short for $B \wedge_2 \neg B$. Also, I have offered a satisfaction-preserving translation from PWK to LP that explains a distinctive feature of PWK: failure of Conjunction

Simplification.

These results have raised a natural question on the reading of PWK-conjunction. I have approached the issue by discussing two possible reading. One of them – the 'disguised disjunction reading' – perfectly fits the many-valued semantics that is usually defined for PWK. The other one – the 'compatibility reading' – seems to fit an intensional semantics better. I have sketched such a semantics, checked its adequacy w.r.t. the relevant features of PWK-conjunction, and briefly commented the meaning the this intensional semantics seem to attach to the connective.

The natural extension of this first investigation of an intensional semantics for PWK is the prove that the usual many-valued semantics for PWK and the relational semantics suggested here are equivalent with respect to the language $\mathcal{L}_1$ of PWK.

Also, a further direction is to provide a formal theory of the relations between different information at different states, and the resulting notions.

I plan to carry on with the research on these issues in some future work.

References

Beall, J. (2011). Multiple-conclusion LP and Default Classicality. *Review of Symbolic Logic*, 4(2), 326-336.

Bochvar, D. (1938). On a Three-valued Calculus and Its Application in the Analysis of the Paradoxes of the Extended Functional Calculus. *Matematicheskii Sbornik*, 287–308.

Dunn, M. (1994). Star and Perp: Two Treatments of Negation. In *Philosophical Perspectives*. Hoboken: Wiley-Blackwell.

Dunn, M., & Restall, G. (2002). Relevance Logic. In D. Gabbay & F. Guenthner (Eds.), (Vol. 6). Dordrecht: Kluwer Academic Publishers.

Halldén, S. (1949). *The Logic of Non-sense*. Unpublished doctoral dissertation, University of Uppsala.

Mares, E. (2004). *Relevant Logic. A Philosophical Interpretation*. Cambridge: Cambridge University Press.

Priest, G. (2006). *In Contradiction*. Oxford: Oxford University Press.

Prior, A. (1957). *Time and Modality*. Oxford: Clarendon University Press.

Restall, G. (1999). Negation in Relevant Logics (How I Stopped Worrying and Learned to Love the Routley Star). In D. Gabbay & H. Wansing (Eds.), *What Is Negation?* (pp. 53–76). Berlin: Springer.

Roberto Ciuni

Routley, R., & Meyer, R. (1972). Semantics for Entailment II – III. *Journal of Philosophical Logic*, *1*(1–2), 53–73 and 192–208.
Routley, R., & Routley, V. (1972). Semantics of First-degree Entailment. *Noûs*, *3*, 335–359.

Roberto Ciuni
Ruhr University Bochum
Germany
E-mail: `roberto.ciuni@rub.de`

The Intuitionistic Robinson Arithmetic(s)

MICHAL DANČÁK[1]

Abstract: Heyting arithmetic HA is a well established intuitionistic theory. As a counterpart to Peano arithmetic PA, it forms *the* theory of constructive arithmetic. Its induction-less variant, the intuitionistic Robinson arithmetic iQ, is not as well understood though. One of the problems with iQ is that there is no true agreement among logicians what iQ is. Looking at the literature one finds two distinct non-equivalent axiomatics of iQ. We'll look at the differences and compare these two axiomatics with respect to provability strength, models, and translations.

Keywords: Intuitionistic logic, Robinson arithmetic, non-classical theories

1 Introduction

The constructivism school puts some strict requirements on the concept of proof. If one claims there is an object with a certain property, one has to present this object and show it has the desired property. Merely proving that there must be such an object because it would be absurd otherwise does not suffice. One needs to *construct* the entity with a given property. This concept leads, on the formal side, to the intuitionistic logic. These ideas are of a strong interest in philosophy since the entirety of logic, mathematics, and other formal sciences rests on the notion of proof. That which is proven, is accepted; that which is refuted is discarded. Hypotheses rest in a limbo in-between. What we call proven or refuted is based solely on our formal syntactical system that stems from our semantics – our logic. The most popular arbiter for that which is provable is classical logic.

Classical logic tries to capture reasonable thinking. Everything either is or isn't, there is no middle ground. If something is not the case then it's opposite must be the case. A contradiction entails everything. A truth

---

[1]This output was created within the project Axiomatické teorie nad intuicionistickou logikou solved at Charles University in Prague from the Specific university research in 2014.

follows from anything, and so on. These sound like very obvious "truths" yet they aren't so obvious under a careful scrutiny. They lead to paradoxes that some other logics attempt to solve. Typical problems concern relevance, self-reference, sorites paradoxes, and so on.

Various logics can be interpreted in various ways. Fuzzy logics refine classical logic, modal logics attempt to capture the meaning behind necessity and possibility, epistemic logics behind knowledge or belief, doxastic logics deal with permission. Our view of intuitionistic logic will be that of information acquisition. We have some incomplete information to start with. As the time goes on, our information extends. So for example there are three possibilities regarding our information on p: either we know that p, or we know that $\neg p$, or we do not have information either way. If we have the information that p we will always have this information and similarly for $\neg p$. But if we are ignorant on the matter of p, we may later acquire an information that p truly holds, for instance. From that moment on we have the information that p is true and that the falsity of p is ruled out. But we keep in mind that we might have been given a different information from which we'd deduce that p does not hold.

2 Intuitionistic logic

We will be working in the classical predicate language with negation $\neg$ being defined by implication $\to$ and a contradiction $\bot$ in the usual way: $\neg \varphi \equiv \varphi \to \bot$. We will use the axiomatics of first-order intuitionistic logic as defined in (Moschovakis, 2015).

Definition 1 *The axioms of the first-order intuitionistic logic IQL are:*

1. $\varphi \to (\psi \to \varphi)$
2. $(\varphi \to \psi) \to ((\varphi \to (\psi \to \chi)) \to (\varphi \to \chi))$
3. $\varphi \to (\psi \to \varphi \wedge \psi)$
4a. $\varphi \wedge \psi \to \varphi$ 4b. $\varphi \wedge \psi \to \psi$
5a. $\varphi \to \varphi \vee \psi$ 5b. $\psi \to \varphi \vee \psi$
6. $(\varphi \to \chi) \to ((\psi \to \chi) \to (\varphi \vee \psi \to \chi))$
7. $(\varphi \to \psi) \to ((\varphi \to \neg \psi) \to \neg \varphi)$
8. $\neg \varphi \to (\varphi \to \psi)$
9. $\forall x \varphi(x) \to \varphi(t)$
10. $\varphi(t) \to \exists x \varphi(x)$,

where t is substitutable for x in φ. The deductive rules are:

The Intuitionistic Robinson Arithmetic(s)

MP	$\varphi, \varphi \to \psi \quad \vdash \quad \psi$
$\forall$-generalization	$\psi \to \varphi(x) \quad \vdash \quad \psi \to \forall x \varphi(x)$
$\exists$-generalization	$\varphi(x) \to \psi \quad \vdash \quad \exists x \varphi(x) \to \psi,$

where x is not free in ψ.

As in the propositional case, one cannot derive the usual classical tautologies of $\varphi \vee \neg \varphi$ or $\varphi \to \psi \equiv \neg \varphi \vee \psi$. Another important formula not provable in IQL is $\neg \forall x \varphi(x) \to \exists x \neg \varphi(x)$.

We'll be working with kripke style models for IQL. These models look like trees, i.e. graphs without circles, where every node has an inner structure in some language. The persistence condition extends to these languages and to domains of inner structures and realizations of functions and relations. Since our interest is focused on the language of arithmetic, we will make some simplifications in the definition.

Definition 2 *Let $\mathcal{K} = \langle W, R, \Vdash \rangle$ be a kripke model of a propositional intuitionistic logic (i.e. R is reflexive and transitive, $\Vdash$ is persistent and satisfies the usual forcing conditions). Let $\mathcal{D}$ be a function that assigns to each node $w \in W$ a non-empty domain and let $\mathcal{F}$ be a function that assigns to each node $w \in W$ realizations of the functions $0, S, +, \cdot,$ and the relation $=$. Call $\mathcal{M} = \langle \mathcal{K}, \mathcal{D}, \mathcal{F} \rangle$ a model of arithmetic if the following holds:*

$\mathcal{D}(w) \subseteq \mathcal{D}(v)$ for any wRv
$\mathcal{F}(w) \subseteq \mathcal{F}(v)$ for any wRv
$w \Vdash \forall x \varphi(x)$ iff $\forall v \, (wRv \Rightarrow \forall a \in \mathcal{D}(v) \, v \Vdash \varphi(a))$
$w \Vdash \exists x \varphi(x)$ iff $\exists a \in \mathcal{D}(w) \, w \Vdash \varphi(a)$

We will call such kripke models "global models" and its nodes "local models". Each local model has its own elements called numbers and its own realizations of all arithmetic functions and relation $=$. Both of these are persistent though; other local models accessible via R may contain new numbers but they always have to contain all the old ones. Since all the arithmetic functions are total, their realizations may differ only on these new numbers. Zero will always be the same zero, the successor of zero will also stay the same, and so on. The equality relation $=$ obviously need not be total although some previously distinct numbers may become equal in some accessible local model.[2] We will define an ordering on each local model in

---

[2] As mentioned, one can view this as an acquisition of a new information. For example, we may start with the model of natural numbers in the root and eventually discover new non-standard numbers. Going further down the R relation we may obtain a new information that these new numbers are actually renamed standard numbers so we can identify them with the standard numbers via the $=$ relation.

the usual way: $x \leq y \equiv \exists z\ z + x = y$ and $x < y \equiv \exists z\ S(z) + x = y$.

The problem with non-classical counterparts of classical theories is with ambiguousness of their axiomatics. All axioms of a given theory may be rewritten in infinitely many equivalent forms even if it means simply adding dummy variables, e.g. $\forall x, y\ (S(x) = S(y) \to x = y)$ is classically equal to $\forall x, y\ (\neg S(x) = S(y) \lor x = y)$ or to $\neg \exists x, y\ \neg(\neg x = y \to \neg \exists z\ S(x) = S(y))$. Some axiom forms are obviously "wrong". Dummy variables only convolute the formula. Barring that, why would some axiom be written in one form but not the other, i.e. why is this form: $\forall x, y\ (S(x) = S(y) \to x = y)$ of the axiom of arithmetic used while this form: $\forall x, y\ (\neg S(x) = S(y) \lor x = y)$ is not? Who is the arbiter? Unless the original author explicitly states what the "correct" axiomatic is, the set of axioms (and deductive rules) remains somewhat fuzzy.[3]

In this paper, we will focus on the intuitionistic variant of Robinson arithmetic (classical version was first introduced in Mostowski, Robinson, & Tarski, 1953), i.e. Heyting arithmetic without the induction schema. Using the scientific community as our arbiter for the axioms, we arrive at two distinct axiomatics.

3 Intuitionistic Robinson arithmetic

Define the following set of well-known axioms.

Q1	$\forall x, y$	$(S(x) = S(y) \to x = y)$
Q2	$\forall x$	$\neg(S(x) = 0)$
Q3$^\lor$	$\forall x$	$(x = 0 \lor \exists y\ (S(y) = x))$
Q3$^\to$	$\forall x$	$(x \neq 0 \to \exists y\ (S(y) = x))$
Q4	$\forall x$	$(x + 0 = x)$
Q5	$\forall x, y$	$(x + S(y) = S(x + y))$
Q6	$\forall x$	$(x \cdot 0 = 0)$
Q7	$\forall x, y$	$(x \cdot S(y) = x \cdot y + x)$
Ind–φ		$(\varphi(0) \land \forall x\ (\varphi(x) \to \varphi(S(x)))) \to \forall x\ \varphi(x)$

Axioms Q1, Q2, and both versions of Q3 describe the behavior of the successor function. Axiom Q1 says that S is an injective function, axiom Q2 says that 0 is not in the range of function S while axioms Q3$^\lor$ and Q3$^\to$ state that everything else is in the range of the successor function. Axioms

---
[3]This whole problem occurs only when studying non-classical logics that are not strictly stronger than classical logic. If one were to study, say, a set theory over modal logic, no problems would arise unless some form of substitution law is broken.

The Intuitionistic Robinson Arithmetic(s)

Q4 and Q5 recursively define addition and axioms Q6 and Q7 recursively define multiplication.

Definition 3 *The set of axioms Q1, Q2, Q3$^\vee$ and Q4 through Q7 composes the strong intuitionistic Robinson arithmetic* iQ$^\vee$. *The same axioms with Q3$^\vee$ replaced by Q3$^\rightarrow$ composes the weak intuitionistic Robinson arithmetic* iQ$^\rightarrow$. *Adding the induction schema* Ind$-\varphi$ *for* $\varphi \in \Gamma$ *to* iQ$^\rightarrow$ *yields the axiomatic system for* iΓ, *the logic of limited induction. If we let* Γ *be the set of all formulas FLA, we obtain Heyting arithmetic* HA.

The version of classical Q defined with the stronger Q3$^\vee$ may be found for instance in (Boolos, Burgess, & Jeffrey, 2007; Burgess, 2005) while the version with Q3$^\rightarrow$ appears in (Hájek & Pudlák, 1993; Mostowski et al., 1953; Švejdar, 2007) and others. One might be hesitant to call the disjunctive version iQ$^\vee$ constructive. The stronger Q3$^\vee$ states that everything is either zero or has a predecessor but gives us no hint, no construction that would point us in the direction of the correct disjunct. On the other hand iQ$^\vee$ is still just a theory in intuitionistic logic that cannot prove $\varphi \vee \neg\varphi$ for an arbitrary formula φ or any other formula that would already entail classical logic.

There arises a natural question. Are iQ$^\vee$ and iQ$^\rightarrow$ equivalent? And if they aren't, why aren't there two distinct Heyting arithmetics? The answer lies in the power of induction schema. First of all, note that Q3$^\vee$ implies Q3$^\rightarrow$ just from the IQL axioms. HA can prove the converse implication with the help of induction:

(i) $0 = 0$ axiom of IQL
(ii) $0 = 0 \vee \exists y(S(y) = 0)$ follows from (i) by IQL
(iii) $\forall x\, (S(x) = 0 \vee \exists y\, S(y) = S(x))$ IQL, set $y = x$

(ii) is an instantiation of Q3$^\vee$ with 0, (iii) is and instantiation of Q3$^\vee$ with $S(x)$. Since both are provable, use Ind$-$Q3$^\vee$ to conclude the proof of Q3$^\vee$. Note that the induction hypothesis is not needed at all – the assumption that Q3$^\vee$ holds when instanced with x is not used anywhere. We have done without the axiom Q3$^\rightarrow$ as well. Reformulating Q3$^\vee$ slightly (having y be bounded by x) allows us to prove this using only a bounded induction, i.e. a theory iΔ_0.

We have established that iQ$^\vee \vdash$ iQ$^\rightarrow$ and that it doesn't matter which version of the third axiom we use when defining HA. All that remains is to demonstrate that iQ$^\rightarrow \nvdash$ iQ$^\vee$, i.e. that there exists a model of iQ$^\rightarrow$ that is not a model of iQ$^\vee$. One such model follows:

Michal Dančák

$\mathcal{M}_1$

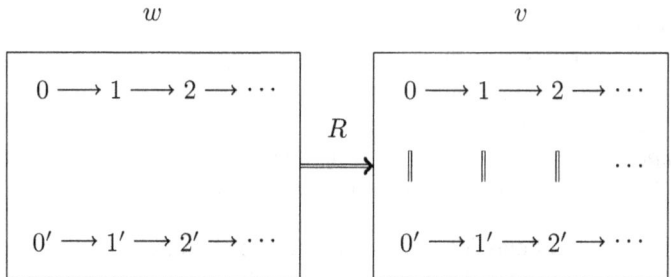

Addition and multiplication in both local models can be defined using the following table:

+	n	n'
m	$m+n$	$m+n$
m'	$(m+n)'$	$(m+n)'$

·	n	n'
m	$m\cdot n$	$(m\cdot n)'$
m'	$m\cdot n$	$(m\cdot n)'$

The model above contains two nodes w and v with wRv. Both local models have the same domain with two copies of natural numbers. In v these two copies are made identical with the use of = relation. The Q3$^\vee$ axiom doesn't hold in w: instantiate it with $0'$ for x. Q3$^\vee$ then claims that either $0' = 0$ (in w) or that there is a predecessor of $0'$ (in w). Neither of these two are true. On the other hand, Q3$^\rightarrow$ holds in both local models w and v. It holds in v trivially since it is a leaf that is essentially equivalent to the natural model $\mathbb{N}$ and therefore behaves like $\mathbb{N}$ in all respects. It holds in w because

$$w \Vdash \forall x\, (x \neq 0 \rightarrow \exists y\, (S(y) = x))$$
$$w \Vdash a \neq 0 \rightarrow \exists y\, (S(y) = a) \qquad \text{for all } a \in w$$

I.e. if a given a is not equal to 0 and in no local model accessible by R is equal to 0, then it must have some predecessor. The number $0'$ (and 0) does not satisfy the antecedent of the implication so the implication instance with $a = 0'$ (and $a = 0$) holds. All the other numbers have predecessors. The reader is free to verify that $\mathcal{M}_1$ satisfies all other axioms of Q3$^\rightarrow$. There are some other examples of formulas that are provable in the stronger iQ$^\vee$ while

The Intuitionistic Robinson Arithmetic(s)

not provable in the weaker iQ$^\rightarrow$. The model $\mathcal{M}_1$ given above also shows that iQ$^\rightarrow$ ⊬ $\forall x, y\, (x+y = 0 \to (x = 0 \land y = 0))$ (let $x = 0$ and $y = 0'$) and that iQ$^\rightarrow$ ⊬ $\forall x\, (x = 0 \lor x \neq 0)$ (let $x = 0'$). The second formula gives us an alternative axiomatic of iQ$^\vee$, since iQ$^\vee$ ⊣⊢ iQ$^\rightarrow$ + $\forall x\, (x = 0 \lor x \neq 0)$. The missing link can therefore be said to be the decidability of zero. Thus the philosophical question of what is the true intuitionistic Robinson arithmetic boils down to the question whether we accept the decidability of zero as an intuitionistically valid theorem.

The most glaring example of a formula on which iQ$^\vee$ and iQ$^\rightarrow$ disagree is $\forall x \exists y\, (x \neq y)$. The following model is due to Vítězslav Švejdar.

$\mathcal{M}_2$

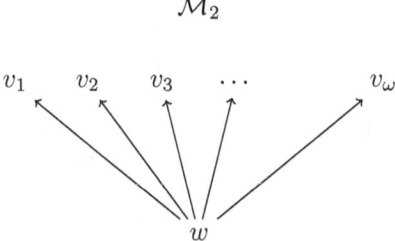

Let the domains of all local models be the same, namely a set of all polynomials over some fixed variable z with natural coeficients with principal coeficient being positive. So for example all natural numbers (polynomials of degree 0) will be in the domains, along with elements like $2z + 5$ or $4z^3 + z^2 + 7$. The domains therefore look like this:

$$
\begin{array}{cccc}
0 \to 1 \to 2 \to \cdots & & z^2 \to z^2 + 2 \to \cdots \\
 & & z^2 + 1 & \\
z \to z+2 \to \cdots & & z^2 + z \to z^2 + z + 2 \to \cdots \\
z+1 & & z^2 + z + 1 & \\
2z \to 2z+2 \to \cdots & & z^2 + 2z \to z^2 + 2z + 2 \to \cdots \\
2z+1 & & z^2 + 2z + 1 & \\
\vdots & & \vdots &
\end{array}
$$

In other words there are ω^ω many copies of natural numbers. The successor function is symbolized by the arrows and realized by $+1$, addition and multiplication is defined as usual on the polynomials. The local models differ only in the distribution of elements into equivalence classes. There are no equalities in w (apart from those enforced by the reflexivity of $=$). Two polynomials $p(z)$ and $q(z)$ of v_n are equal iff $p(n) = q(n)$. For example $5z^2 + 3z + 1$ and $z^3 + 2z^2 + z + 7$ are equal in v_3. Two polynomials $p(z)$ and $q(z)$ of v_ω are equal iff either both are of zero degree and denote the same number or if both are of non-zero degree. Then the element z of w is a counterexample for formula $\forall x \exists y \, (x \neq y)$. The argument is as follows. Pick an arbitrary witness y. If it is a polynomial of non-zero degree, x and y are equal in v_ω and thus $w \nVdash x \neq y$. If y is of zero degree, i.e. some natural number n, x and y are equal in v_n and once again $w \nVdash x \neq y$.

We have established that there are many differences between $\mathsf{iQ}^\vee$ and $\mathsf{iQ}^\rightarrow$. But our main question shouldn't be "which of these two logics is *the* intuitionistic Robinson arithmetic?" It should be "what do we want from our theory?" Do we want a strong theory that may not fully correspond with our notion of constructivism? Or do we prioritize a fully constructive theory with no axiom in the disjunctive form that is however too weak to prove anything?

4 Local models

What theories do models of $\mathsf{iQ}^\vee$ or $\mathsf{iQ}^\rightarrow$ have to locally satisfy? This is a natural question to ask whenever one studies kripke models of intuitionistic theories. Do the nodes need to locally satisfy the classical counterpart of the theory? The answer is not always simple as is apparent when one looks at Heyting arithmetic. Countable models of Peano arithmetic are not constructive by Tennenbaum's theorem (with the exception of the model of natural numbers) so we cannot study them directly. Different indirect methods give partial answers. It was shown that finite models of HA have to be locally PA (van Dalen, Mulder, Krabbe, & Visser, 1986) and in the general case they have to be at least $I\Delta_1$ (Marković, 1993). Wehmeier (1996) also showed that HA is complete with respect to the set of all kripke models, where every node is locally PA. The case of local models of intuitionistic Q is much simpler due to the absence of induction schema. We will first recursively define a useful notion of local and global formulas.

- Every atomic formula is local.

- If φ, ψ are local, then $\varphi \wedge \psi$ is local.

- If φ, ψ are local, then $\varphi \vee \psi$ is local.

- If φ is local, then $\exists x\, \varphi$ is local.

- Every local formula is global.

- If φ is local, then $\neg \varphi$ is global.

- If φ is local and ψ is global, then $\varphi \to \psi$ is global.

- If φ is global, then $\forall x\, \varphi$ is global.

The definitions of local and global formulas stem from the conditions for connectives and quantifiers in kripke models. $x = 0 \vee \exists y(S(y) = x)$ is an example of a local formula. For any node w it holds that the following conditions are equivalent:

$$w \Vdash x = 0 \vee \exists y(S(y) = x)$$
$$w \Vdash x = 0 \text{ or } w \Vdash \exists y(S(y) = x)$$
$$w \Vdash x = 0 \text{ or } \exists a \in \mathcal{D}(w)\, w \Vdash S(a) = x$$

Both formulas $x = 0$ and $S(a) = x$ are atomic and their forcing relation is therefore identical to the satisfaction relation. In other words when we ask whether a local formula is forced in a node, we ask whether some formulas hold in the same node locally. This property fails for global formulas. Knowing that w locally satisfies $x \neq 0$ for example is not enough to deduce that w forces $x \neq 0$ because there may be some accessible nodes that satisfy $x = 0$. However if we assume that *all* nodes of a given model satisfy $x \neq 0$, then we can conclude that *all* nodes force $x \neq 0$. These properties of local and global formulas are summarized in the following lemma.

Lemma 1 *Let φ be a local formula and $\mathcal{M}$ an arbitrary kripke model of IQL. Then*

$$\forall w \in \mathcal{M}\ (w \Vdash \varphi \textit{ iff } w \vDash \varphi).$$

Let φ be a global formula and $\mathcal{M}$ an arbitrary kripke model of IQL. Then

$$\forall w \in \mathcal{M}\, w \Vdash \varphi \textit{ iff } \forall w \in \mathcal{M}\, w \vDash \varphi.$$

Proof. The case of local formulas is trivial so let us consider the global case. Let φ be a global formula. We will proceed by induction.

Case φ local:

Apply the local case of the lemma.

Case $\varphi \equiv \psi \to \chi$ for ψ local and χ global:

Let $\forall w\ w \Vdash \psi \to \chi$. Then $\forall w \forall v\ (wRv \Rightarrow (v \nVdash \psi \text{ or } v \Vdash \chi))$. $v \nVdash \psi$ implies $v \nvDash \psi$ because ψ is local. $v \Vdash \chi$ implies $v \vDash \chi$ by the induction hypothesis. Therefore $\forall w \forall v\ (wRv \Rightarrow (v \nvDash \psi \text{ or } v \vDash \chi))$ and $\forall w \forall v\ (wRv \Rightarrow v \vDash \psi \to \chi)$. By the reflexivity of relation R we can say that $\forall w\ w \vDash \psi \to \chi$.

Let $w \vDash \psi \to \chi$ for all nodes w instead. Then $\forall w\ (w \nvDash \psi \text{ or } w \vDash \chi)$. Using the fact that ψ is local and by the induction hypothesis deduce that $\forall w\ (w \nVdash \psi \text{ or } w \Vdash \chi)$ and thus $\forall w\ (w \Vdash \psi \to \chi)$.

Other cases are similar. $\square$

Obviously not all formulas are global. For instance $x = 0 \lor x \neq 0$ or $x = 0 \to \neg\neg x = 0$ are not global. Note however that all axioms of $\text{iQ}^\lor$ are global, giving us the direct result that for any kripke structure $\mathcal{M}$ over IQL

$$\mathcal{M} \vDash \text{iQ}^\lor \text{ iff } \forall w \in \mathcal{M}\ w \vDash Q.$$

Therefore models of $\text{iQ}^\lor$ are characterized by Q: all nodes in a model of $\text{iQ}^\lor$ locally satisfy Q while a collection of classical models of Q partially ordered in such a way as to satisfy all persistance conditions yields a model of $\text{iQ}^\lor$. The question becomes more complicated in the case of $\text{iQ}^\to$ because axiom $Q3^\to$ is not global. It pairs a global premise of an implication with a local conclusion. Model $\mathcal{M}_1$ as shown in section 3 is an example of a model of $\text{iQ}^\to$ that does not locally satisfy all axioms of Q. Clearly all axioms of $\text{iQ}^\to$ with the exception of $Q3^\to$ still have to be locally satisfied but this does not give us the whole story. Taking only node w of model $\mathcal{M}_1$ does not result in a model of $\text{iQ}^\to$. Therefore Q without the third axiom does not give us the full characterization of $\text{iQ}^\to$. Moreover not every model of $\text{iQ}^\to$ needs to contain a local model of Q. It is a general fact of predicate intuitionistic logic that leaves in kripke models (nodes with no successors other than themselves) are classical, i.e. if w is a leaf, then for any formula φ it holds that $w \Vdash \varphi$ iff $w \vDash \varphi$, or equivalently $w \Vdash \varphi \lor \neg\varphi$. Leaves of models of $\text{iQ}^\to$ are therefore locally models of full Robinson arithmetic Q. Yet there exist models of $\text{iQ}^\to$ such that no node is locally a model of Q. One such model can be constructed as a generalization of model $\mathcal{M}_1$.

The Intuitionistic Robinson Arithmetic(s)

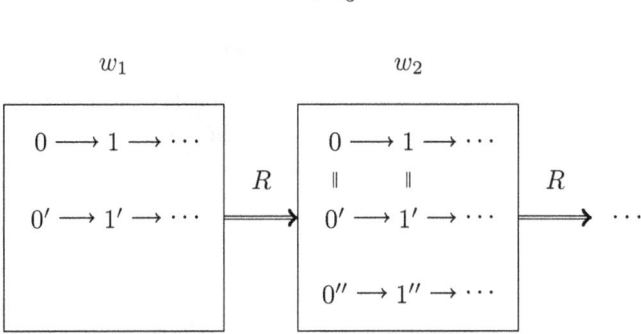

Node w_n has $n+1$ copies of natural numbers where copies 0 through n are made equal and copy $n+1$ is a fresh set of new members. Addition and multiplication can be defined similarly to $\mathcal{M}_1$. None of the nodes locally satisfy Q3$^\rightarrow$, since we keep introducing new non-zero elements without a predecessor. Yet the whole model globally satisfies Q3$^\rightarrow$, because all the new elements without a predecessor will turn out to be equal to 0 eventually.

Model $\mathcal{M}_1$ not only shows that theory iQ$^\rightarrow$ is not characterizable by Q, it also shows that iQ$^\rightarrow$ is not characterizable by any classical theory. If it was characterizable by some theory T then all nodes of $\mathcal{M}_1$ would have to locally satisfy T, i.e. $w \vDash T$ and $v \vDash T$. However taking only node w by itself does not yield us a kripke model of iQ$^\rightarrow$ even though $w \vDash T$ in such a kripke structure.

5 Translations and interpretations

A double-negation translation is any translation of formulas of classical logic into formulas of intuitionistic logic that adds two negations in front of some connectives and either in front or after some quantifiers and commutes with others. The purpose of these translations is to obtain a way to translate provability results between classical logic and intuitionistic logic. We will be using the Gödel-Gentzen translation ^{G} but the results that follow hold for all the other translations commonly used (see Ferreira & Oliva, 2011 for the remaining translations). φ^G is recursively defined as follows:

- $\varphi^G \equiv \neg\neg\varphi$ if φ is atomic
- $(\varphi \vee \psi)^G \equiv \neg\neg(\varphi^G \vee \psi^G)$

- $(\exists x \varphi)^G \equiv \neg\neg \exists x \varphi^G$

- ^{G} commutes with all other connectives and quantifiers

It is a known result that $\mathsf{PA} \vdash \varphi$ iff $\mathsf{HA} \vdash \varphi^G$ (and the same holds for other double-negative translations). Will the same result still be valid when we drop the induction schema? The answer to this question is yes, translations still work as usual: $\mathsf{Q} \vdash \varphi$ iff $\mathsf{iQ}^\vee \vdash \varphi^G$ iff $\mathsf{iQ}^\rightarrow \vdash \varphi^G$. In other words $\mathsf{iQ}^\vee$ and $\mathsf{iQ}^\rightarrow$ coincide on all formulas of the form φ^G. The proof is rather simple and we will not reproduce it here since it follows the same steps as Gödel's original proof (1986).

We have already established that $\mathsf{iQ}^\vee \vdash \mathsf{iQ}^\rightarrow$ while $\mathsf{iQ}^\rightarrow \nvdash \mathsf{iQ}^\vee$, i.e. that $\mathsf{iQ}^\vee$ is strictly stronger than $\mathsf{iQ}^\rightarrow$ with respect to the provability relation. What about the (relativized) interpretability relation? It is known that classical Q is a rather powerful theory when one considers interpretations: Q interprets $\mathsf{I}\Delta_0$, i.e. a theory with bounded induction.[4] The same fact holds for intuitionistic $\mathsf{iQ}^\vee$ as was shown by Sterken (2008). But can the weaker $\mathsf{iQ}^\rightarrow$ interpret $\mathsf{iQ}^\vee$? This is still an open question and an ongoing research.

6 Conclusion

We examined the two versions of intuitionistic Robinson arithmetics and compared their properties. We showed some prototypical formulas that distinguish these two logics and studied their locality properties. There is still a lot of research left to do however, the most important being the question whether $\mathsf{iQ}^\rightarrow$ can interpret $\mathsf{iQ}^\vee$. The current working hypothesis is that the method of shortening cuts which is used to show that some theory can be relatively interpreted in another theory will not work here since most (if not all) recursive models of $\mathsf{iQ}^\rightarrow$ that are not models of $\mathsf{iQ}^\vee$ can define natural numbers on a cut with the same formula. Without this method it would be difficult to find an interpretation of $\mathsf{iQ}^\vee$ in $\mathsf{iQ}^\rightarrow$ and in fact our current conjecture is that it is downright impossible and that $\mathsf{iQ}^\rightarrow$ cannot interpret $\mathsf{iQ}^\vee$.

So what is *the* intuitionistic Robinson arithmetic? There is no one good answer to this question. There are two suitable candidates (and plenty of poor ones) and while they differ in many aspects, none of them represents

---

[4]In fact Q interprets an even stronger theory $\mathsf{I}\Delta_0 + \Omega_1$. Ω_1 is an axiom that postulates a totality of a certain limited kind of an exponential function. Defining $|x|$ as the length of the binary representation of number x, $\Omega_1 \equiv \forall x \exists y\, 2^{|x|\cdot|x|} = y$.

the intuitionistic Q any better than the other. If we accept the decidability of zero as an intuitionistically valid theorem of Robinson arithmetic, we arrive at logic iQ$^\vee$. If not, we get iQ$^\rightarrow$. Both logics deserve further research however.

References

Boolos, G. S., Burgess, J. P., & Jeffrey, R. C. (2007). *Computability and Logic*. New Jersey: Princeton University Press.

Burgess, J. P. (2005). *Fixing Frege*. New Jersey: Princeton University Press.

Ferreira, G., & Oliva, P. (2011). On Various Negative Translations. In S. van Bakel, S. Berardi, & U. Berger (Eds.), *Classical Logic and Computation 2010* (pp. 21–33).

Gödel, K. (1986). On Intuitionistic Arithmetic and Number Theory. In S. Feferman, J. W. D. Jr., S. C. Kleene, G. H. Moore, R. M. Solovay, & J. van Heijenoort (Eds.), *Kurt Gödel: Collected Works: Volume I* (pp. 287–296). Oxford: Oxford University Press. (Translated from the original "Zur Intuitionistischen Arithmetik und Zahlentheorie")

Hájek, P., & Pudlák, P. (1993). *Metamathematics of First Order Arithmetic*. Berlin: Springer.

Marković, Z. (1993). On the Structure of Kripke Models of Heyting Arithmetic. *Mathematical Logic Quarterly*, *39*, 531–538.

Moschovakis, J. (2015). Intuitionistic Logic. In E. N. Zalta (Ed.), *The Stanford Encyclopedia of Philosophy* (Spring 2015 ed.). http://plato.stanford.edu/archives/spr2015/entries/logic-intuitionistic/.

Mostowski, A., Robinson, R. M., & Tarski, A. (1953). *Undecidable Theories*. Amsterdam: North Holland.

Sterken, R. (2008). *Concatenation as a Basis for Q and the Intuitionistic Variant of Nelson's Classic Result*. Unpublished master's thesis, Universiteit van Amsterdam.

van Dalen, D., Mulder, H., Krabbe, E. C. W., & Visser, A. (1986). Finite Kripke Models of HA Are Locally PA. *Notre Dame Journal of Formal Logic*, *27*(4), 528–532.

Švejdar, V. (2007). An Interpretation of Robinson Arithmetic in its Grzegorczyk's Weaker Variant. *Fundamenta Informaticae*, *81*, 347–354.

Michal Dančák

Wehmeier, K. F. (1996). Classical and Intuitionistic Models of Arithmetic. *Notre Dame Journal of Formal Logic*, *37*(3), 452–461.

Michal Dančák
Charles University in Prague
The Czech Republic
E-mail: mdancak@seznam.cz

How to Validly Quantify into Hyperintensional Nonpropositional Attitudes

MARIE DUŽÍ[1]

Abstract: In this paper I demonstrate how to validly quantify into hyperintensional contexts involving non-propositional attitudes like *seeking, solving, calculating*, and *wanting to become* in their respective *de dicto* and *de re* variants. The main results are the following. First, it is always valid to quantify *into* hyperintensional attitude contexts and *over* hyperintensional entities. Second, factive attitudes (e.g. *finding the site of Troy, having solved the equation sin(x) = 0*) validate, furthermore, quantifying *over* intensional or extensional entities, respectively, and so do non-factive attitudes, both empirical and non-empirical ones, provided the respective constituent of the attitude complement presents the sort of object that is to be quantified over. I focus mainly on mathematical attitudes, which are non-controversially hyperintensional. Concerning empirical attitudes, I analyse those that are hyperintensional, because their intensional variant would yield a contradiction, like seeking a yeti without seeking an abominable snowman. As a result I formulate and prove four rules for quantifying into hyperintensional attitudes *de dicto* and two rules for quantifying into attitudes *de re*, in their active and passive form, respectively.

Keywords: Quantifying in, procedural semantics, substitution method, Transparent Intensional Logic, non-propositional attitudes

1 Introduction

A cogent philosophical reason for studying *non-propositional attitudes* is that non-propositional attitude verbs (like 'to design', 'to solve', 'to construct') are part and parcel of both a common vernacular and more technical and scientific vocabularies, so a semantic analysis of them should not

---

[1] This research has been supported by the internal grant agency of VSB-TU Ostrava, project No. SP2014/157, "Knowledge modelling, process simulation and design" and by the Grant Agency of the Czech Republic, project No. 15-13277S, Hyperintensional logic for natural language analysis.

be missing from any theory designed to analyze a wide array of expressions. Surprisingly, though, non-propositional attitudes have received much less attention than 'that-clause' attitudes in analytic philosophy of language. Since non-propositional attitudes are no less important than their propositional cousins, making up half the sphere of attitudes, the situation ought to be rectified by treating them at length. In this paper I study how to validly apply existential quantification into non-propositional attitude contexts. Existential quantification into attitude contexts is known as *quantifying-in*.

Propositional attitudes divide into relations (-in-intension) to propositions (sets of possible worlds) and to hyperpropositions, i.e. modes of presentation of propositions. The same bifurcation applies to non-propositional attitudes, which also divide into relations-in-intension to intensions and to hyperintensions, i.e. modes of presentation of the intensions of possible-world semantics. Non-propositional attitudes come in many different types. First, the complement of an empirical attitude can be an intension of any type; for instance an individual office or role like *Miss Universe 2014*, or a property of individuals like *being a Platonist*. Second, the complements of mathematical attitudes are invariably hyperintensional. It can be a mathematical mode of presentation of a truth-value, of a number, of a set of real numbers, etc. Moreover, both empirical and mathematical attitudes come in two variants, to wit, *de dicto* and *de re*.

In this paper I deal with hyperintensional attitudes. I demonstrate how to validly quantify into hyperintensional contexts involving non-propositional attitudes like *seeking, solving, calculating*, and *wanting to become* in their respective *de dicto* and *de re* variants. I investigate quantifying-in by investigating the logic of instances of the following inferential schema:

$$\frac{a \text{ has an attitude } \textit{Att} \text{ whose complement is } X(b)}{\exists x \, (a \text{ has } \textit{Att} \text{ toward } X(x))}$$

The conclusion extracts a component x from X and quantifies over it. But we need to be careful not to draw illicit inferences. For instance, if the premise is that a is seeking an abominable snowman then the conclusion ought *not* be that there is an *individual*, i.e. an extensional entity, x such that a is seeking x. For if there are no abominable snowmen, the inference will have crossed the line from logic into magic, conjuring up an individual out of thin air. Yet the conclusion can be, for instance, that there is an x such that a is seeking x (i.e. a is seeking *something* of a particular type), or that there is an x such that x is abominable and a is seeking x (i.e. a is seeking *something abominable*).

Quantifying into Hyperintensional Attitudes

For a non-empirical example, if the premise is that *a* is calculating the last decimal of the expansion of π then we cannot validly infer that there is a *number y* such that *a* is calculating *y*, because there is no last decimal of π. And even if *per impossibile* the expansion of π had a unique last decimal, the conclusion that *a* is calculating a number would not be correct, because it is nonsense to calculate a number. What we can validly derive is, for instance, that *a* is calculating *something* of a particular type, or that there is a number such that *a* is calculating the last decimal of its expansion.

Examples such as these seem to suggest that if we quantify *over* a hyperintensional context, the respective quantificational ranges of *x*, *y* cannot be those of extensional entities such as individuals or numbers. Rather, the complement *X* of a hyperintensional attitude belongs to hyperintensions, hence we quantify over a hyperintension. Moreover, if we quantify *into* any position within *X*, then we need to carefully specify the type of the entity to be quantified over and prove the validity of such an inference. In order to offer a principled answer, I apply Transparent Intensional Logic, which is a typed, extensional logic of hyperintensions that preserves compositionality of meaning, referential transparency and extensional principles like substitutivity of identicals and existential quantification also in hyperintensional contexts. In TIL we explicate hyperintensions as abstract, structured *procedures* that are rigorously defined as so-called *constructions*. Accordingly, the attitudes I am going to focus on are relations (-in-intension) between an agent and a construction.

The main results are the following. First, it is always valid to quantify *into* hyperintensional attitude contexts and over hyperintensional entities. Second, factive attitudes (e.g. *finding the site of Troy, having solved the equation sin(x) = 0*) validate, furthermore, quantifying *over* intensions or extensions, respectively, and so do non-factive attitudes, both empirical and non-empirical ones, provided the respective constituent of the attitude complement presents the sort of object that is to be quantified over.

Applying logical operations to hyperintensional contexts is no trivial matter. The technical complications we are confronted with are rooted in *displayed* constructions, which contrast with *executed* constructions. For instance, if *a* calculates the cosine of π then *a* is related to the displayed mathematical construction $Cos(\pi)$ and not what the construction constructs when executed. Thus we cannot derive a conclusion of the form $\exists x$ [*Calculate a Cos(x)*], because the variable *x* occurs inside the hyperintensional context of the displayed construction *Cos(x)*. Yet any argument of the form "*a* calculates the cosine of π, hence *a* calculates the cosine of something" is obviously valid. But the quantifier has to reach across the attitude operator and grab an occurrence of a variable inside the scope of the operator. The problem is similar to trying to quantify into quoted contexts. Still the context in which a construction is displayed is more permissive than a "displayed" context, and in TIL we have a way out (or perhaps rather, a way in). In order to validly infer the conclusion, we need to *pre-process* the construction $Cos(\pi)$ in order to put the position occupied by π within reach of the quantifier. To this end we apply the *substitution method* that makes use of the functions *Sub* and *Tr* defined below.

The rest of the paper is organised as follows. Section 2 sets out the foundations of my background theory TIL. In Section 3 I formulate and prove four rules for quantifying into hyperintensional attitudes *de dicto* and two rules for quantifying into attitudes *de re*, in their active and passive forms, respectively. Rule R_1 is a straightforward rule of quantifying *over* hyperintensional contexts and *over hyperintensions*. R_2 is a rule for quantifying *into* hyperintensional contexts and *over* hyperintensions. R_2', a variant of R_2, and R_3 are rules for quantifying *into* hyperintensional contexts and *over the products* of hyperintensions. R_2' is applicable only at an additional assumption, namely that the respective hyperintension succeeds in presenting an object to be quantified over. R_3 is applicable in case that the respective hyperintension is Trivialisation of the object to be quantified over. R_4 is a rule for factive attitudes like *finding* or *having solved*. Due to the inherent factivity of the attitude, this rule is applicable without an additional explicit assumption. Finally, R_5 and R_6 are rules for quantifying into attitudes *de re* in their passive and active forms, respectively. They quantify over the extensions that are the values of hyperintensions occurring *de re*. These rules are applicable due to the validity of the principle of existential presupposition that all *de re* attitudes come with. Application of each rule is illustrated by examples. Concluding remarks are in Section 4.

2 Logical foundations

What makes TIL suitable for the job of quantifying into hyperintensional contexts is the fact that the theory construes the semantic properties of the sense and denotation relations as remaining invariant across different sorts of linguistic contexts and that its ramified type theory enables quantification of any order.[2] The syntax of TIL is Church's (higher-order) typed λ-calculus[3] with a procedural (as opposed to denotational) semantics, according to which a linguistic sense is an abstract procedure detailing how to arrive at an object of a particular logical type. TIL *constructions* are such procedures.

Moreover, TIL is an extensional logic of hyperintensions and of intensions. By extensional logic I mean such a logic that validates extensional principles in all kinds of context, whether extensional, intensional or hyperintensional. In fact, testing whether existential quantification comes out valid provides a clear method for testing various theories of non-extensional contexts.[4] In TIL we see no problem in principle with applying existential quantification to non-extensional contexts. Yet we do see a string of logical and semantic challenges that call for formally worked-out, principled solutions. This paper is devoted to providing the solutions required.

Technically speaking, some constructions are modes of presentation of functions, including 0-place functions such as individuals and truth-values, and the rest are modes of presentation of other constructions. Thus, with constructions of constructions, constructions of functions, functions, and functional values in our stratified ontology, we need to keep track of the traffic between multiple logical strata. The ramified type hierarchy does just that. What is important, in this paper, about this traffic is, first of all, that constructions may themselves figure as functional arguments or values.

Definition 1 (types of order 1) *Let B be a base, where a base is a collection of pair-wise disjoint, non-empty sets. Then:*

(i) Every member of B is an elementary type of order 1 over B.
(ii) Let α and $\beta_1,\ldots,\beta_m$ ($m > 0$) be types of order 1 over B. Then the collection $(\alpha\ \beta_1\ \ldots\ \beta_m)$ of all m-ary partial mappings from $\beta_1 \times \ldots \times \beta_m$ into α is a functional type of order 1 over B.
(iii) Nothing else is a type of order 1 over B.

For the purposes of natural-language analysis, we are currently assuming

---

[2]*Indexicals* being the only exception: while the sense of an indexical remains constant, its denotation varies in keeping with its contextual embedding. See (Duží, Jespersen, & Materna, 2010, §3.4).

[3]See (Church, 1956).

[4]Kaplan (1990, p. 14) and Bealer (1982, p. 13) mention quantifying-in as a challenge and a 'classical puzzle' (Bealer) which 'pure semantics' and (hyper-) intensional logic must address adequately.

Marie Duží

the following base of *ground types*, each of which is part of the ontological commitments of TIL:

- o: the set of truth-values $\{\mathbf{T}, \mathbf{F}\}$;
- ι: the set of individuals (constant universe of discourse);
- τ: the set of real numbers (doubling as temporal continuum);
- ω: the set of logically possible worlds (logical space).

Certain constructions, *qua* objects of attitudes, figure as functional arguments of the binary relations-in-intension that, relative to a world-time pair of evaluation, take an agent and an attitude relatum to a truth-value. Moreover, since constructions are being quantified into, we consequently need constructions of one order higher than those being quantified over. Thus we define

Definition 2 (construction)

(i) Let valuation v assign object o to variable x. Then x v-constructs *the object* o.

(ii) Let X be any object whatsoever (i.e. an extension, an intension, or a construction). Then ^0X is the Trivialization *of X, which constructs X without any change of X.*

(iii) Let X v-construct *a function f of type* $(\alpha\beta_1 \ldots \beta_m)$ *and let* $Y_1, \ldots, Y_m$ v-construct entities $B_1, \ldots, B_m$ of types $\beta_1, \ldots, \beta_m$, respectively. Then the Composition $[X\ Y_1 \ldots Y_m]$ v-constructs *the value (an entity, if any, of type α) of f on the tuple argument* $\langle B_1, \ldots, B_m \rangle$. Otherwise the Composition $[X\ Y_1 \ldots Y_m]$ does not v-construct *anything and so is v-improper.*

(iv) Let $x_1, \ldots, x_m$ be pair-wise distinct variables v-constructing *entities of types* $\beta_1, \ldots, \beta_m$ and let Y be *a* construction v-constructing *an α-entity*. Then the Closure $[\lambda x_1 \ldots x_m\ Y]$ v-constructs *the following function f of type* $(\alpha\beta_1 \ldots \beta_m)$. Let $v(B_1/x_1, \ldots, B_m/x_m)$ be a valuation identical with v at least up to assigning objects $B_1/\beta_1, \ldots, B_m/\beta_m$ to variables $x_1, \ldots, x_m$. If Y is $v(B_1/x_1, \ldots, B_m/x_m)$-improper (see (iii)), then f is undefined on $\langle B_1, \ldots, B_m \rangle$. Otherwise the value of function f on $\langle B_1, \ldots, B_m \rangle$ is the α-entity that is $v(B_1/x_1, \ldots, B_m/x_m)$-constructed by Y.

(v) Let X v-construct *an object o. Then the* Single Execution ^1X v-constructs o. Let X be either a non-construction *or a v-improper* construction. Then ^1X is v-improper.

(vi) Let X v-construct a construction Y and let Y v-construct object Z (possibly itself a construction). Then the Double Execution ^2X v-constructs Z. Otherwise ^2X is v-improper.
(vii) Nothing else is a construction.

Here are some explicative remarks. A *variable* constructs an object by having that object as its value dependent on a valuation function v arranging variables and objects in a sequence. *Trivialization* is TIL's objectual counterpart of a non-descriptive constant term, which simply provides a particular object. *Composition* is the procedure of functional application, rather than the functional value (if any) resulting from application. *Closure* is the procedure of forming a function by lambda abstraction, rather than the resulting function. Variables and Trivializations are the *atomic* constructions of TIL, Composition, Closure and Single/Double Execution are the *molecular* constructions. An *atomic* construction is a structured whole with but one constituent part, namely the construction itself. A *molecular* construction is a structured whole with more parts than just itself.

The definition of the typed universe of TIL amounts to a definition of the ramified hierarchy of types which divides into three parts; firstly, simple types of order 1, which were already defined by Definition 1; secondly, constructions of order n; thirdly, types of order $n+1$.

Definition 3 (ramified hierarchy of types)
T_1 (types of order 1). *See Definition 1.*
C_n (constructions of order n)
  (i) Let x be a variable ranging over a type of order n over B. Then x is a construction of order n over B.
  (ii) Let X be a member of a type of order n over B. Then ^0X, ^1X, ^2X are constructions of order n over B.
  (iii) Let $X, X_1, \ldots, X_m$ ($m > 0$) be constructions of order n over B. Then $[X X_1 \ldots X_m]$ is a construction of order n over B.
  (iv) Let $x_1, \ldots, x_m, X$ ($m > 0$) be constructions of order n over B. Then $[\lambda x_1 \ldots x_m X]$ is a construction of order n over B.
  (v) Nothing is a construction of order n over B unless it so follows from C_n (i)-(iv).
T_{n+1} (types of order $n+1$) *Let $*_n$ be the collection of all constructions of order n over B. Then*
  (i) $*_n$ and every type of order n are types of order $n+1$ over B.
  (ii) If $m > 0$ and $\alpha, \beta_1, \ldots, \beta_m$ are types of order $n+1$ over B, then $(\alpha \beta_1 \ldots \beta_m)$ (see T_1 (ii)) is a type of order $n+1$ over B.

(iii) Nothing else is a type of order $n+1$ over B.

Logical objects like *truth-functions* and *quantifiers* are extensional: $\wedge$ (conjunction), $\vee$ (disjunction) and $\supset$ (implication) are of type (ooo) and $\neg$ (negation) of type (oo). The *quantifiers* $\forall^{\alpha}, \exists^{\alpha}$ are type-theoretically polymorphous, total functions of type $(o(o\alpha))$, for an arbitrary type α, defined as follows. The *universal quantifier* $\forall^{\alpha}$ is a function that associates a class A of α-elements with **T** if A contains all elements of the type α, otherwise with **F**. The *existential quantifier* $\exists^{\alpha}$ is a function that associates a class A of α-elements with **T** if A is a non-empty class, otherwise with **F**.

Below all type indications will be provided outside the formulae in order not to clutter the notation. Furthermore, 'X/α' means that an object X is (a member) of type α. '$X \to_v \alpha$' means that the type of the object (if any) v-constructed by X is α. We write '$X \to \alpha$' if what is v-constructed does not depend on a valuation v. This holds throughout: the variables $w \to_v \omega$ and $t \to_v \tau$. If $C \to_v \alpha_{\tau\omega}$ then the frequently used Composition $[[Cw]t]$, which is the intensional descent (a.k.a. extensionalization) of the α-intension v-constructed by C, will be encoded as 'C_{wt}'. When using constructions of truth-functions, I often omit Trivialization and use infix notation to conform to standard notation in the interest of better readability. Also when using constructions of identities of α-entities, $=_{\alpha}/(o\alpha\alpha)$, Trivialization and the type subscript are often omitted, and infix notion is used when no confusion arises. Moreover, the outermost brackets of Closure will be occasionally omitted.

Possible-world intensions are functions with the domain ω of possible worlds. With both hyperintensions and possible-world intensions in its ontology, TIL has no trouble assigning either hyperintensions or intensions to variables as their values. However, the technical challenge of untying an x occurring in an attitude hyperintensional context $(Att^* \ldots x \ldots)$ from $Att^* \to (o\iota *_n)_{\tau\omega}$ and $\exists$-binding it requires two non-standard devices. We need the function *Sub* (for 'substitution'), and occasionally we also need the function *Tr* (for 'Trivialization').

Definition 4 (Sub^n, Tr$^{\alpha}$) *Let $C_1/*_{n+1} \to *_n$, $C_2/*_{n+1} \to *_n$, and $C_3/*_{n+1} \to *_n$ v-construct constructions D_1, D_2, and D_3, respectively. Then the Composition $[^0Sub^n\ C_1\ C_2\ C_3]$ v-constructs the construction D that results from D_3 by collisionless substitution of D_1 for all occurrences of D_2 in D_3. The function $Tr^{\alpha}/(*_n\alpha)$ returns as its value the Trivialization of its α-argument.*

Example 1 Let variable $y \to_v \tau$. Then $[^0Tr^\tau y]\ v(\pi/y)$-constructs $^0\pi$. The Composition $[^0Sub\ ^1[^0Tr^\tau y]\ ^0x\ ^0[^0Cot\ x]]\ v(\pi/y)$-constructs the Composition $[^0Cot\ ^0\pi]$.

Remark 1 *Note that there is a substantial difference between the construction Trivialization and the function* Tr^α. *Whereas* ^0y *constructs just the variable* y *regardless of valuation, y being ^{0}-bound in* ^0y, $[^0Tr^\alpha y]$ v-*constructs the Trivialization of the object v-constructed by y. Hence y occurs free in* $[^0Tr^\alpha y]$. *Below I will omit the superscripts n and α and write simply 'Sub' and 'Tr' whenever no confusion can arise.*

3 Rules for quantifying in

In (Duží & Jespersen, 2010) we specified a general rule for quantifying into hyperintensional contexts. However, this rule is rather too general, and we need some more specific rules. The goal of this paper is to fill this gap. Below I specify four rules for quantifying into hyperintensional contexts of non-propositional attitudes *de dicto* and two rules for hyperintensional attitudes *de re*.

Let $Att^* \to (o\iota*_n)_{\tau\omega}$ be an arbitrary construction of a hyperintensional attitude relation; $a \to_v \iota$ a construction of an individual; $\exists^*/(o(o*_n))$ the existential quantifier over constructions; $C/*_n$ a construction of an attitude complement; $c \to_v *_n$ a variable ranging over constructions. Rule R_1 is a straightforward rule for quantifying *over* the complements of hyperintensional attitudes:

$$(R_1)\ \frac{[Att^*_{wt}\ a\ ^0C]}{[^0\exists^* \lambda c\ [Att^*_{wt}\ a\ c]]}$$

Proof. Let $Empty^*/(o*_n)$ be an empty class of constructions. Then
1. $[Att^*_{wt}\ a\ ^0C]$ ∅
2. $[\lambda c\ [Att^*_{wt}\ a\ c]^0C]$ 1, λ-abstraction
3. $\neg[^0Empty^* \lambda c\ [Att^*_{wt}\ a\ c]]$ 2, Definition 2 (iii,iv)
4. $[^0\exists^* \lambda c\ [Att^*_{wt}\ a\ c]]$ 3, EG

□

Example 2

$$\frac{\text{Tilman is seeking the last decimal of } \pi}{\text{Tilman is seeking something}}$$

Marie Duží

$$\frac{\lambda w \lambda t\ [^0Seek^*_{wt}\ ^0Tilman\ ^0[^0Last\_Dec\ ^0\pi]]}{\lambda w \lambda t\ [^0\exists^* \lambda c\ [^0Seek^*_{wt}\ ^0Tilman\ c]]}$$

Types. $Seek^*/(o\iota *_n)_{\tau\omega}$; $Tilman/\iota$; $Last\_Dec/(\nu\tau)$: a function assigning to a real number of type τ its last decimal natural number of type ν; π/τ.

Rule (R_2) is the rule for quantifying *into* the complements of hyperintensional attitudes, *over* hyperintensions. Let $C(X)$ be the construction of the attitude complement with a constituent X and let $c, d \rightarrow_v *_n$. Since $C(X)$ and thus also X occur *displayed* in the hyperintensional context, we must apply the substitution method in order to use the variable c free for logical manipulation:

$$(R_2)\ \frac{[Att^*_{wt}\ a\ ^0C(X)]}{[^0\exists^* \lambda c\ [Att^*_{wt}\ a\ [^0Sub\ c\ ^0d\ ^0C(d)]]]}$$

Proof. Let $=_* /(o *_n *_n)$ be the identity of constructions, and let $Empty^*/(o*_n)$ be an empty class of constructions. Then

1. $[Att^*_{wt}\ a\ ^0C(X)]$ ∅
2. $[^0=_*\ [^0Sub\ ^0X\ ^0d\ ^0C(d)]\ ^0C(X/d)]$ 1, Definition of *Sub*
3. $[Att^*_{wt}\ a\ [^0Sub\ ^0X\ ^0d\ ^0C(d)]]$ 1, 2, Leibniz
4. $[\lambda c\ [Att^*_{wt}\ a\ [^0Sub\ c\ ^0d\ ^0C(d)]]\ ^0X]$ 3, λ-abstraction
5. $\neg [^0Empty^* \lambda c\ [Att^*_{wt}\ a\ [^0Sub\ c\ ^0d\ ^0C(d)]]]$ Definition 2 (iii,iv)
6. $[^0\exists^* \lambda c\ [Att^*_{wt}\ a\ [^0Sub\ c\ ^0d\ ^0C(d)]]]$ 5, EG

□

Example 3

$$\frac{\text{Tilman computes the value of } sine \text{ at } tg(\frac{\pi}{2})}{\text{There is a construction } c \text{ such that Tilman computes the value of sine at } c}$$

$$\frac{\lambda w \lambda t [^0Compute_{wt}\ ^0Tilman\ ^0[\lambda x\ [^0Sin\ x]\ [^0Tg\ [^0/\ ^0\pi\ ^02]]]]}{\lambda w \lambda t [^0\exists^* \lambda c [^0Compute_{wt}\ ^0Tilman\ [^0Sub\ c\ ^0d\ ^0[\lambda x [^0Sin\ x]\ d]]]]}$$

Additional types. $Compute/(o\iota *_n)_{\tau\omega}$; $x \rightarrow_v \tau$; $Sin, Tg/(\tau\tau)$; $\pi, 2/\tau$.

In this case we cannot derive that there is a construction of a *number* such that Tilman computes the value of sine at this number, because there is no such number that would be the value of tangent at $\frac{\pi}{2}$. Yet though Tilman's activity is futile he can be attempting at executing the improper construction

$[\lambda x \ [^0Sin \ x] \ [^0Tg \ [^0/ \ ^0\pi \ ^02]]]$. Hence there is such a construction c that Tilman computes the value of sine at c.

If the attitude complement construction is proper we can use a variant (R'_2) of the rule (R_2) that quantifies over the *product* of a construction c. Let $Proper/(o*_n)$ be the class of constructions that are not v-improper for any valuation v; $X \to \alpha$; $c \to_v *_n$; $^2c, x \to_v \alpha$; $= /(o\alpha\alpha)$; $\exists/(o(o\alpha))$. Then

$$(R'_2) \frac{[Att^*_{wt} a \ ^0C(X)] \land [^0Proper \ ^0X]}{[^0\exists^* \lambda c \ ^0\exists \lambda x \ [[^0= \ x \ ^2c] \land [Att^*_{wt} a \ [^0Sub \ c \ ^0d \ ^0C(d)]]]]}$$

Proof.
1. $[Att^*_{wt} \ a \ ^0C(X)] \land [^0Proper \ ^0X]$ ∅
2. $[Att^*_{wt} \ a \ ^0C(X)]$ 1, ∧E
3. $[^0=_* \ [^0Sub \ ^0X \ ^0d \ ^0C(d)] \ ^0C(X/d)]$ 2, Definition of *Sub*
4. $[Att^*_{wt} \ a \ [^0Sub \ ^0X \ ^0d \ ^0C(d)]]$ 2, 3, Leibniz
5. $[^0Proper \ ^0X]$ 1, ∧E
6. $[\lambda x \ [^0= \ x \ X] \ X]$ 5, Def. of *Proper*, λ-abstr.
7. $[[\lambda x \ [^0= \ x \ X] \ X] \land$
   $[Att^*_{wt} \ a \ [^0Sub \ ^0X \ ^0d \ ^0C(d)]]]$ 6, 4 ∧I
8. $[\lambda c \ [[\lambda x \ [^0= \ x \ ^2c] \ X] \land$
   $[Att^*_{wt} \ a \ [^0Sub \ c \ ^0d \ ^0C(d)]]] \ ^0X]$ 7, λ-abstr., $X = \ ^{20}X$
9. $\neg [^0Empty^* \lambda c \ [[\lambda x \ [^0= \ x \ ^2c] \ X] \land$
   $[Att^*_{wt} \ a \ [^0Sub \ c \ ^0d \ ^0C(d)]]]]$ 8, Definition 2 (iii,iv)
10. $[^0\exists^* \lambda c \ ^0\exists \lambda x \ [^0= \ x \ ^2c] \land$
    $\land [Att^*_{wt} \ a \ [^0Sub \ c \ ^0d \ ^0C(d)]]]$ 9, EG

□

Example 4

$$\frac{\text{Tilman computes the value of } sine \text{ at } tangens \text{ of } \pi}{\begin{array}{c} \text{There is a construction of a number such that} \\ \text{Tilman computes the value of sine at this number} \end{array}}$$

$$\frac{\lambda w \lambda t \ [^0Compute_{wt} \ ^0Tilman \ ^0[\lambda y \ [^0Sin \ y] \ [^0Tg \ ^0\pi]]]}{\begin{array}{c} \lambda w \lambda t \ [^0\exists^* \lambda c \ ^0\exists \lambda x \ [x = \ ^2c] \land \\ [^0Compute_{wt} \ ^0Tilman \ [^0Sub \ c \ ^0d \ ^0[\lambda y [^0Sin \ y] \ d]]]] \end{array}}$$

Marie Duží

For an empirical example assume that 'abominable snowman' and 'yeti' denote one and the same property. Moreover, I am stipulating that 'is a yeti' and 'is an abominable snowman' are a pair not of synonymous but merely equivalent predicates. Their respective meanings are co-intensional, but not procedurally isomorphic (co-hyperintensional). The rationale for this stipulation is that the latter predicate has a molecular structure thanks to the application of the property modifier denoted by 'abominable' to the property denoted by 'snowman', whereas the former predicate is atomic. Hence if Tilman is seeking an abominable snowman without seeking a yeti, his attitude must be hyperintensional. Otherwise such a search would be impossible.

Example 5

$$\frac{\text{Tilman is seeking an abominable snowman and not seeking a yeti}}{\substack{\text{There is a property such that} \\ \text{Tilman is seeking an instance of it via some construction } c \\ \text{and not seeking an instance of it via some construction } d}}$$

$$\frac{\lambda w \lambda t \, [[^0Seek^*_{wt} \, ^0Tilman \, ^0[^0Abominable \, ^0Snowman]] \wedge \neg [^0Seek^*_{wt} \, ^0Tilman \, ^{00}Yeti]]}{\lambda w \lambda t \, [^0\exists^* \lambda c \, ^0\exists^* \lambda d \, ^0\exists \lambda p \, [[p = \, ^2c] \wedge [p = \, ^2d] \wedge [^0Seek^*_{wt} \, ^0Tilman \, c] \wedge \neg [^0Seek^*_{wt} \, ^0Tilman \, d]]]}$$

Types. $Seek/(o\iota *_n)_{\tau\omega}$; $Abominable/((o\iota)_{\tau\omega}(o\iota)_{\tau\omega})$: property modifier; $Yeti, Snowman/(o\iota)_{\tau\omega}$; $c, d \rightarrow_v *_n$; $^2c, {}^2d \rightarrow_v (o\iota)_{\tau\omega}$; $p, q \rightarrow_v (o\iota)_{\tau\omega}$.

Note that the premise demands the double Trivialization $^{00}Yeti$, because Tilman seeks what $^{00}Yeti$ constructs, namely ^0Yeti.

If the constituent X is the *Trivialization* of an α-object b, thus guaranteeing that the construction is proper, then we can access this hyperintensional context to quantify over this α-object. To this end we have another special rule R_3 of quantifying *into* a hyperintensional context and over *intensions* or *extensions*. This rule makes use of the function Tr:

$$(R_3) \; \frac{[Att^*_{wt} a \, ^0C(^0b/y)]}{[^0\exists \lambda x \, [Att^*_{wt} a \, [^0Sub \, [^0Tr \, x] \, ^0y \, ^0C(y)]]]}$$

Proof. According to Definition 4 the Composition $[^0Sub\ [^0Tr\ x]\ ^0y\ ^0C(y)]$ $v(b/x)$-constructs the construction $C(^0b/y)$ in which the occurrences of y have been replaced by ^0b. Therefore the following proof-steps are truth-preserving:

1. $[Att^*_{wt}\ a\ ^0C(^0b/y))]$ ∅
2. $[^0=_*\ [^0Sub\ [^0Tr\ x]\ ^0y\ ^0C(y)]\ ^0C(^0b/y)]$ 1, Definition 4
3. $[Att^*_{wt}\ a\ [^0Sub\ [^0Tr\ x]\ ^0y\ ^0C(y)]]$ 1, 2, Leibniz
4. $[\lambda x\ [Att^*_{wt}\ a\ [^0Sub\ [^0Tr\ x]\ ^0y\ ^0C(y)]]\ ^0b]$ 3, λ-abstr.
5. $\neg[^0Empty\ \lambda x\ [Att^*_{wt}\ a\ [^0Sub\ [^0Tr\ x]\ ^0y\ ^0C(y)]]]$ Def. 2 (iii)
6. $[^0\exists \lambda x\ [Att^*_{wt}\ a\ [^0Sub\ [^0Tr\ x]\ ^0y\ ^0C(y)]]]$ 5, EG

□

Example 6

$$\frac{\text{Tilman is seeking an abominable snowman}}{\text{Tilman is seeking something abominable}}$$

$$\frac{\lambda w \lambda t\ [^0Seek^*_{wt}\ ^0Tilman\ ^0[^0Abominable\ ^0Snowman]]}{\lambda w \lambda t\ [^0\exists \lambda p\ [^0Seek^*_{wt}\ ^0Tilman\ [^0Sub\ [^0Tr\ p]\ ^0q\ ^0[^0Abominable\ q]]]]}$$

Example 7   For a mathematical example, we have this one:

$$\frac{\text{Tilman is seeking the last decimal of } \pi}{\text{There is a number such that Tilman is seeking its last decimal}}$$

$$\frac{\lambda w \lambda t\ [^0Seek^*_{wt}\ ^0Tilman\ ^0[^0Last\_Dec\ ^0\pi]]}{\lambda w \lambda t\ [^0\exists \lambda x\ [^0Seek^*_{wt}\ ^0Tilman\ [^0Sub\ [^0Tr\ x]\ ^0y\ ^0[^0Last\_Dec\ y]]]]}$$

Types: $x, y/*_1 \rightarrow_v \tau; \exists/(o(o\tau))$.

Next I specify a rule for *factive* attitudes like *finding* or *having solved*. Factive attitudes are defined in terms of the notion of *requisite*.[5] A requisite is an analytically necessary relation-in-extension between two intensions such that, necessarily, any object that instantiates one intension also instantiates the other (though not necessarily the other way around). For instance, if the property of being catholic is a requisite of the office of the Bishop of Rome (i.e. the office of Pope) then, necessarily, whoever occupies the papal office must also be a catholic.

---
[5] See (Duží et al., 2010, §4.1).

Marie Duží

Let $Att^*_f/(o\iota *_n)_{\tau\omega}$ be a factive objectual attitude. The type of *requisite* relevant for factive hyperintensional attitudes is $Req/(o(o\iota\alpha *_n)_{\tau\omega}(o\iota *_n)_{\tau\omega})$ and is defined as follows. Let $Ident^\alpha/(o\iota\alpha *_n)_{\tau\omega}$ be a relation between an individual, an α-object and a construction such that the individual succeeds in identifying the α-object as the product of this construction, $x \to_v \iota$, $c \to_v *_n$, $^2c \to_v \alpha$, then:

$$[^0Req\ ^0Ident^\alpha\ Att^*_f] = \forall w \forall t \forall x\, c\, [[Att^*_{fwt}\ x\ c] \supset [^0Ident^\alpha_{wt}\ x\ ^2c\ c]]$$

$Ident^\alpha$ is a requisite of Att^*_f iff, necessarily for all individuals x and all constructions c, it holds that if x has an attitude Att^*_f to c then x has identified the α-product ^2c of construction c. Hence we *define* factive hyperintensional attitudes as those for which the above requisite relation holds.

The rule R_4 for factive objectual attitudes is this:

$$(R_4)\ \frac{[Att^*_{fwt}\ a\ ^0C]}{[^0\exists \lambda x\,[[x = C] \wedge [^0Ident_{wt}\ a\ x\ ^0C]]]}$$

Additional type: $C/*_n \to_v \alpha$.

Proof.

1. $[Att^*_{wt}\ a\ ^0C]$ ∅
2. $[^0Req\ ^0Ident^\alpha\ ^0Att^*_f]$ Definition of factivity
3. $\forall w \forall t \forall x\, c\, [[Att^*_{fwt}\ x\ c] \supset$
   $[^0Ident^\alpha_{wt}\ x\ ^2c\ c]]$ Definition of *Req*.
4. $[[Att^*_{fwt}\ a\ ^0C] \supset [^0Ident^\alpha_{wt}\ a\ ^{20}C\ ^0C]]$ 3, ∀E, a/x, $^0C/c$
5. $[^0Ident^\alpha_{wt}\ a\ ^{20}C\ ^0C]$ 1, 4, MPP
6. $[^0Ident^\alpha_{wt}\ a\ C\ ^0C]$ 5, Def. 2 (vi): $^{20}C = C$
7. $[^0Proper\ ^0C]$ 6, Definition 2 (iii)
8. $[^0=_\alpha\ x\ C]$ 7, Definition 2
9. $[^0Ident^\alpha_{wt}\ a\ x\ ^0C]$ 6, 8, Leibniz
10. $[[^0=_\alpha\ x\ C] \wedge [^0Ident^\alpha_{wt}\ a\ x\ ^0C]]$ 8, 9, ∧I
11. $\neg[^0Empty\ \lambda x\,[[x = C] \wedge$
    $[^0Ident^\alpha_{wt}\ a\ x\ ^0C]]]$ 10, λ-abstraction
12. $[^0\exists \lambda x\,[[^0=_\alpha\ x\ C] \wedge [^0Ident^\alpha_{wt}\ a\ x\ ^0C]]]$ 11, EG

□

The steps (6), (7) and (8) above call for elucidation. In the Composition $[^0Ident^\alpha_{wt}\ a\ C\ ^0C]$ the construction C occurs both displayed and executed. The first occurrence of C is executed while the second is displayed. Hence the first occurrence of C is a constituent of the Composition $[^0Ident_{wt}^\alpha \ a \ C \ ^0C]$. Since this Composition v-constructs **T**, according to Definition 2, (iii) this constituent must be proper. Thus we have step (7): $[^0Proper \ ^0C]$. And since C is v-proper, it v-constructs an α-object, which in turn justifies step (8): $[^0=_\alpha \ x \ C]$.

Example 8 If Tilman is seeking the millionth digit of the decimal expansion of π, he may succeed in his effort and identify that number. Thus we have another argument that illustrates the application of (R_4):

$$\frac{\text{Tilman finds the millionth digit of the decimal expansion of } \pi}{\text{There is a number such that Tilman identifies it as the millionth digit of the decimal expansion of } \pi}$$

$$\frac{\lambda w \lambda t \ [^0Find_{wt} \ ^0Tilman \ ^0[^0Mill\_Dec \ ^0\pi]]}{\lambda w \lambda t \ [^0\exists \lambda x \ [^0Ident_{wt} \ ^0Tilman \ x \ ^0[^0Mill\_Dec \ y]]]}$$

Types: $Find/(o\iota*_n)_{\tau\omega}$; ν: the type of naturals; $x \to_v \nu$; $Mill\_Dec/(\nu\tau)$: the function that associates a real number with its millionth decimal digit; $Ident/(o\iota\nu*_n)_{\tau\omega}$.

(R_1) through (R_4) are rules for quantifying into hyperintensional attitudes *de dicto*. Next I define a pair of rules for quantifying into hyperintensional contexts *de re*, both in a passive and an active variant. These rules are variants of the rules introduced in (Duží & Jespersen, 2012):

$$(R_5 \ act) \ \frac{[Att_{wt}^* \ a \ [^0Sub \ [^0Tr \ X] \ ^0y \ ^0C(y)]]}{[^0\exists \lambda x \ [Att_{wt}^* \ a \ [^0Sub \ [^0Tr \ x] \ ^0y \ ^0C(y)]]]}$$

$$(R_5 \ pas) \ \frac{[\lambda x \ [Att_{wt}^* \ a \ [^0Sub \ [^0Tr \ x] \ ^0y \ ^0C(y)] \ X]]}{[^0\exists \lambda x \ [Att_{wt}^* \ a \ [^0Sub \ [^0Tr \ x] \ ^0y \ ^0C(y)]]]}$$

Types: $X/*_n \to \alpha$; $x, y/*_n \to_v \alpha$; $\exists/(o(o\alpha))$; y free in $C(y)/*_n$.

Proof. (of R_5 act)

1.	$[Att_{wt}^* \ a \ [^0Sub \ [^0Tr \ X] \ ^0y \ ^0C(y)]]$	∅
2.	$[\lambda x \ [Att_{wt}^* \ a \ [^0Sub \ [^0Tr \ x] \ ^0y \ ^0C(y)] \ X]$	1, λ-abstr.
3.	$\neg[^0Empty \ \lambda x \ [Att_{wt}^* \ a \ [^0Sub \ [^0Tr \ x] \ ^0y \ ^0C(y)]]]$	2, Def. 2 (iii)
4.	$[^0\exists \lambda x \ [Att_{wt}^* \ a \ [^0Sub \ [^0Tr \ x] \ ^0y \ ^0C(y)]]]$	3, EG

□

Proof. (of R_5 *pas*) Since X occurs extensionally in the assumption, the proof is trivial:

1. $[\lambda x\ [Att^*_{wt}\ a\ [^0Sub\ [^0Tr\ x]\ ^0y\ ^0C(y)]\ X]]$ $\varnothing$
2. $\neg[^0Empty\ \lambda x\ [Att^*_{wt}\ a\ [^0Sub\ [^0Tr\ x]\ ^0y\ ^0C(y)]]]$ 1, Def. 2 (iii)
4. $[^0\exists \lambda x\ [Att^*_{wt}\ a\ [^0Sub\ [^0Tr\ x]\ ^0y\ ^0C(y)]]]$ 2, EG

$\square$

Example 9 In mathematics a *de re* attitude is characterized by an *extensional* occurrence of a constituent that v-constructs the value of the denoted function. This value is then anaphorically referred to in the attitude complement.

$$\frac{\text{The last decimal of } \pi \text{ is being calculated by Tilman}}{\text{There is a number such that Tilman calculates it as the last decimal of } \pi}$$

The argument is valid, though drastically unsound: the premise *presupposes* the existence of the last decimal of π. And if *per impossibile* a number n were the last decimal of π then Tilman would be obtaining n as a result of his calculation of the last decimal of π. Hence, existential generalization simply serves to make explicit an implicit ontological presupposition incurred in the premise. In this case it is the presupposed existence of the value of the function denoted by 'the last decimal of' at the argument π. In order for $[^0Last\_Dec\ ^0\pi]$ to be a constituent occurring extensionally, we must again deploy the functions *Sub* and *Tr*.

$$\frac{\lambda w \lambda t\ [^0Calculate_{wt}\ ^0Tilman\ [^0Sub\ [^0Tr\ [^0Last\_Dec\ ^0\pi]]\ ^0y\ ^0[[^0Last\_Dec\ ^0\pi] = y]]]}{\lambda w \lambda t\ [^0\exists \lambda x\ [^0Calculate_{wt}\ ^0Tilman\ [^0Sub\ [^0Tr\ x]\ ^0y\ ^0[[^0Last\_Dec\ ^0\pi] = y]]]]}$$

If we analyzed the premise similarly as in Example 2, that is, in terms of the Closure

$$\lambda w \lambda t\ [^0Calculate_{wt}\ ^0Tilman\ ^0[^0Last\_Dec\ ^0\pi]]$$

the existence of a last decimal of π would neither be presupposed, nor would it follow. The Composition $[^0Last\_Dec\ ^0\pi]$ occurs displayed, that is hyperintensionally here, but in the *de re* case it must occur executed extensionally.

4 Conclusion

Above I demonstrated how to validly quantify into various sorts of hyperintensional contexts involving objectual (i.e. non-propositional) attitude complements. This cluster of logical results was obtained in virtue of an extensional logic of hyperintensions, namely Transparent Intensional Logic, which comes with a context-invariant, top-down semantic theory that yields universal transparency to the exclusion of opaque contexts. An extensional logic of hyperintensions is permissive or inclusive in the sense of validating all the rules of extensional logic, including substitution of identicals and existential generalization, also in hyperintensional contexts, but must at the same time be restrictive or exclusive with regard to what counts as admissible operands.

I have proved two major things. First, it is always valid to quantify *into* hyperintensional attitude contexts and *over* hyperintensional entities. Second, factive empirical attitudes (e.g. *having solved an equation*) validate, furthermore, quantifying over intensions and extensions, and so do non-factive attitudes, both empirical and non-empirical (e.g. *calculating the millionth decimal of the expansion of* π), provided the respective constituent of the attitude complement presents the sort of object that is to be quantified over. Moreover, in case of hyperintensional attitudes *de re* it is always valid to quantify over the product of the extensionally occurring construction.

References

Bealer, G. (1982). *Quality and Concept*. Oxford: Claredon Press.

Church, A. (1956). *Introduction to Mathematical Logic*. Princeton: Princeton University Press.

Duží, M., & Jespersen, B. (2010). Transparent Quantification into Hyperintensional Contexts. In M. Peliš & V. Punčochář (Eds.), *The Logica Yearbook 2010* (pp. 81–98). London: College Publications.

Duží, M., & Jespersen, B. (2012). Transparent Quantification into Hyperintensional Contexts de re. *Logique & Analyse*, 220, 513–554.

Duží, M., Jespersen, B., & Materna, P. (2010). *Procedural Semantics for Hyperintensional Logic. Foundations and Applications of Trasnsparent Intensional Logic*. Berlin: Springer.

Kaplan, D. (1990). Dthat. *Syntax and Semantics*, 9, 221–243.

Marie Duží

Marie Duží
Technical University Ostrava, Department of Computer Science FEI
The Czech Republic
E-mail: `marie.duzi@gmail.com`

Carnap's Tolerance and Friedman's Revenge

NOAH FRIEDMAN-BIGLIN[1]

Abstract: In this paper, I defend Rudolf Carnap's Principle of Tolerance from an accusation, due to Michael Friedman, that it is self-defeating by prejudicing any debate towards the logically stronger theory. In particular, Friedman attempts to show that Carnap's reconstruction of the debate between classicists and intuitionists over the foundations of mathematics in his book *The Logical Syntax of Language*, is biased towards the classical standpoint since the metalanguage he constructs to adjudicate between the rival positions is fully classical. I argue that this criticism is mistaken on two counts: (1) it fails to fully appreciate the freedom with regard to the construction of linguistic frameworks that Carnap intended his Principle to embody, and (2) Friedman's objection underestimates the extent to which the evaluation of a framework is task-relative. I conclude that Tolerance is not self-undermining in the way that Friedman claims it is. While this is a restricted conclusion – and is not a vindication of Carnap's views on logic and mathematics *tout court* – it nonetheless suggests that his tolerant perspective has been dismissed too quickly, even by his supporters.

Keywords: Rudolf Carnap, Michael Friedman, Principle of Tolerance, logical pluralism, object-language/metalanguage distinction

Perhaps the most well-known feature of Rudolf Carnap's book *The Logical Syntax of Language* (*LSL*) is his 'Principle of Tolerance'.[2] It codifies the thought that any theory can be considered as an account of any phenomenon, as long as it is precisely specified. This is a strikingly permissive view, and the question that concerns us here is whether its permissiveness is also its undoing. Arguments to this effect have dogged the Principle since Carnap first proposed it, but in this paper I will focus on one recent exam-

---

[1] For feedback on earlier versions, I would like to thank Peter Milne, Fenner Tanswell, the Arché WIP group, and audiences at LOGICA 2014, BAHPS 2014, and ECAP8.

[2] In what follows, I will refer to the Principle of Tolerance by its full name or as either 'the Principle', or 'Tolerance'.

ple.³ In his (2001) work, Michael Friedman argues that Tolerance cannot do the philosophical work Carnap intends it to because it biases any choice between rival positions in favor of the logically stronger view. In particular, Friedman attempts to show that Carnap's reconstruction of the debate over the foundations of mathematics in *LSL* is biased towards the classical standpoint since the metalanguage he constructs to adjudicate between the rival positions is fully classical; the Carnapian intuitionist would therefore be forced to affirm in the metalanguage what they deny at the object level. To address this problem, we must first be clear about how Carnap meant his Principle to be understood. I begin by detailing Friedman's concerns with the usual – and more narrow – understanding of Tolerance in section 1. In section 2, I outline a new interpretation of Tolerance which I call the 'wide' reading. On this understanding, there are absolutely no constraints whatsoever on what can be proposed as a linguistic framework. Because of the permissiveness of this reading, the majority of the section is occupied by an argument for treating the wide reading as the intended one. I conclude the paper by putting the wide interpretation to work, so to speak, and use it in section 3 to show how Friedman's concerns miss their mark.

1   Metalevel Tolerance and Friedman's revenge

In his 2001 paper "Tolerance and Analyticity in Carnap's Philosophy of Mathematics", Michael Friedman raises a problem for the way that he, and many others, read Carnap's Principle of Tolerance. On his interpretation of the Principle, there are no constraints on the proposing of a linguistic framework; we are just as free to accept a language in which our background logic is constructivist as we are to accept one where it is classical. But when we consider how we might *choose between* various proposed languages, according to Friedman, problems emerge. Carnap tells us that, where before we might give philosophical arguments that one language correctly captures the nature of a concept (as, for example, logical consequence), when we attempt to work in the spirit of the Principle, we see that these debates are vexed – Carnap says these debates are "wearisome" and concern "pseudo-problems" (Carnap, 1937, p. *xiv – xv*). What we ought do instead is to determine the consequences of adopting the proposed languages; once we

---

³Other philosophers who have given arguments to similar effect include: Michael Potter (2002), Hilary Putnam (1983), Thomas Ricketts (1994), and Eino Kalia as reported in (Carus, 2007).

Carnap's Tolerance and Friedman's Revenge

know what the consequences of each proposal are, we are in a position to make a choice. Carnap tells us that this choice will be made on the basis of the *purely pragmatic* features of the languages in question. That is, we should evaluate the languages on the basis of their relative simplicity, economy, expressiveness etc. However, things are not so easy when we get down to the details of what 'determining the consequences of adopting a proposed language' amounts to, and it is precisely here that Friedman thinks that Tolerance falls down. In this section, I take up Friedman's worry, and attempt to make clear its structure.

Suppose that one wants to follow Carnap's advice in determining which of these two logico-mathematical systems are best for formalizing the language of science: intuitionist or classical. Presumably, in order to establish an answer to this question, we will have to construct a suitable syntax-language which can express all the necessary concepts for *both* languages; that is, we will need a metalanguage which is capable of describing both languages so that we can adequately investigate the consequences of adopting them in this metalanguage. Moreover, this metalanguage ought be neutral with respect the differences between the two languages, otherwise we prejudge the issue. However, as Friedman notes,

> In giving a metatheoretical description of [the language of classical mathematics], we therefore need a metalanguage even stronger than the language of classical mathematics itself (containing, in effect, classical mathematics plus a truth-definition for classical mathematics). And we need this strong metalanguage, not to prove the consistency of the classical linguistic framework in question, but to simply describe and define this framework in the first place so that questions about the consequences of adopting it (including the question of consistency) can then be systematically investigated. (Friedman, 2001, pp. 242–243)

That is, any metalanguage that is going to be adequate for describing the language of classical mathematics will have to include not only the resources to define concepts like "analytic-in-$\mathcal{L}$" and "consequence-in-$\mathcal{L}$" (where $\mathcal{L}$ is the language of classical mathematics), but also the totality of classical mathematics itself. Moreover, since it is to contain all of classical mathematics, it will also contain the resources used by classical mathematics. So, the metalanguage we construct to judge between the intuitionist proposal and the classical one – which, recall, is supposed to be neutral between the proposed languages – is committed to resources that the intuitionist would reject, namely those used in classical mathematics. In Friedman's words:

> In order to apply the principle of tolerance, we must view [the] choice [between intuitionistic and classical languages] as a purely pragmatic decision about "linguistic forms" having no ontological implications about "facts" or "objects" in the world. [...] Accordingly, we must view the logico-mathematical rules in question, in both linguistic frameworks, as sets of purely analytic sentences. Given Carnap's own explication of the distinctions between logical and descriptive terms, analytic and synthetic sentences, however, we must have already adopted the classical logico-mathematical rules in the metalanguage. Thus, to understand the choice between classical and intuitionistic logico-mathematical rules in accordance with the principle of tolerance, we must have already built the former logico-mathematical rules into our background syntactic metaframework. We must have already biased the choice against the intuitionist in the very way we set up the problem. (Friedman, 2001, pp. 242–243)[4]

In other words, in order to treat the logico-mathematical rules of the two candidate object-languages as analytic (in their respective languages), we must be able to draw the distinction between analytic and synthetic sentences in those languages. This distinction is drawn in the metalanguage, and so the relevant metalanguage must therefore have certain resources. In this particular case, as was noted above, it must have the resources of classical mathematics (otherwise it will be unable to draw the required distinction for the classical object-langauge). But, if it does, then it would seem that, again, the Carnapian intuitionist would be forced to affirm in the metalanguage the very rules she took to be in dispute at the object level. Because of the inability to neutrally frame the decision between competing frameworks, much less provide a place to stand from which we might neutrally assess the choice between them, Friedman concludes that the Principle of Tolerance is self-undermining. That is, any attempt to actually *use* the Principle, understood in the way that Friedman does, must fail to be neutral in the way it must be to do the work that Carnap envisions for it.

---

[4]For the purposes of this paper, I set aside questions which Friedman alludes to about Tolerance and ontology. Interested readers are commended to the discussions found in (Hirsch, 2009) and (Eklund, 2009).

2 Tolerances: wide and narrow

As we saw just above, if we understand the Principle in the way that Friedman does, the Carnapian project looks doomed. In this section, I distinguish two readings of the Principle of Tolerance. The first, which I argue has been the more common of the two, I call the 'narrow' reading. According to it, when Carnap says that there are "no morals in logic", he does not mean that all bets are off, so to speak (Carnap, 1937, p. 52). Rather, he means that anyone is free to propose any formal system for our *first-order* theorizing.[5] The various proposed languages are then compared by examining the consequences to which each leads. A determination about which to adopt is reached on the basis of the pragmatic merits of the proposed languages. In this section, I will show that despite the popularity of this narrow reading it is nonetheless mistaken. My strategy will be to focus on the textual evidence from *LSL*, and I will conclude from this evidence that Carnap is committed instead to an alternative understanding of the Principle whereby the voluntarism and pragmatic orientation that Tolerance expresses extends to the full extent of the linguistic hierarchy. I call this alternative understanding the 'wide' reading, and will detail it below.

Carnap is often bold in his writing and nowhere is this tendency more apparent than in his discussions of the Principle of Tolerance. While its full formulation, given in section §17 of *LSL*, is dramatic, it is eclipsed entirely by what he says in the introduction:

> For language, in its mathematical form, can be constructed according to the preferences of any [point of view]; so that no question of justification arises at all, but only the question of the syntactical consequences to which one or other of the choices leads, *including the question of non-contradiction*. (Carnap, 1937, p. *xv*, my emphasis)

Here, Carnap denies that there are any constraints on language construction outright; he even gives up consistency as a necessary condition on a language in order for it to be considered. Moreover, he also denies that one

---

[5] In *LSL*, Carnap is primarily concerned with languages for mathematics, but there is nothing special about this example; in his introduction, Carnap says "The standpoint we have suggested [...] relates not only to mathematics, but to all questions of logic" (Carnap, 1937, p. *xv*), and moreover, in the first section of the book proper, he says, "The method of syntax which will be developed in the following pages will not only prove useful in the logical analysis of scientific theories – it will also help in the *logical analysis of the word-languages*" (Carnap, 1937, p. 8, original emphasis).

must justify one's choices. The only factor that is relevant to the question of which language to adopt, Carnap says, are the consequences which accepting a candidate language leads:

> The fact that no attempts have been made to venture still further from the classical forms is perhaps due to the widely held opinion that any such deviations must be justified – that is, that the new language-form must be proved to be 'correct' and to constitute a faithful rendering of 'the true logic'.
>
> To eliminate this standpoint, together with the pseudo-problems and wearisome controversies which arise as a result of it, is one of the chief tasks of this book. In it, the view will be maintained that we have in every respect complete liberty with regard to the forms of language; that both the forms of the construction for sentences and the rules of transformation (the latter are usually designated as "postulates" and "rules of inference") may be chosen quite arbitrarily. (Carnap, 1937, pp. *xiv–xv*)

So, according to Carnap, there is no need for a proof of a proposed language's adequacy. That is, we can suggest any language we wish for a task – "we have *complete* liberty in *every* respect" – and neither our proposed languages nor our eventual choice need be provably 'correct', whatever the salient standard of correctness might be. Finally, we look at the Principle itself:

> *In logic there are no morals.* Everyone is at liberty to build up his own logic, i.e. his own form of language, as he wishes. All that is required of him is that, if he wishes to discuss it, he must state his methods clearly, and give syntactical rules instead of philosophical arguments. (Carnap, 1937, p. 52, original emphasis)

Though it comes a fair way into the book, at least in comparison to the quotations from the introduction, the phrasing here is nonetheless of a piece with the tone of those early passages. What is key from our present perspective is, again, the emphasis on liberty and freedom from restrictions on which languages can be proposed. Even the last clause – "all that is required of him is that, *if he wishes to discuss it*, he must state his methods clearly, and give syntactical rules instead of philosophical arguments" – is phrased in terms of a task. It is only in the case that one wants to discuss one's form of logic that this condition applies; even the need to give syntactical rules is not an absolute. However, since we must be cautious against reading too much into the rhetoric of the introduction and the early parts of the book, and, moreover, since everything we have said so far is consistent with both the wide and narrow understandings of the Principle, we now turn to the details of Carnap's behavior in *LSL*. In order to get clear on the intended interpretaion of Tolerance we will examine two episodes from the book: firstly his treatment of intuitionism, and secondly his consistency proof for Language II.[6] These examples will serve to highlight the use that he makes of the Principle, and in each case we will examine a slightly different aspect of Carnap putting Tolerance to work.

2.1 Putting Tolerance to work part 1: intuitionism

Though Carnap's discussion of intuitionism is short – it comes in section §16 of the book, and runs for only a few pages – it is highly significant. He begins by expressing frustration that no one has given a formal treatment of the intuitionistic view, and moreover that some of the intuitionists see the task of giving a formalism as unnecessary.[7] Because of this lack of a formal treatment, he claims, it is impossible to make sense of the claims over the nature of logic that intuitionists make:

> Once the fact is realized that all pros and cons of the Intuitionist discussions are concerned with the forms of a calculus, questions will no longer be put in the form: "What *is* this or that like?" but instead we shall ask: "How *do we wish to arrange* this or that in the language to be constructed?" or, from the theoretical standpoint: "What consequences will ensue if we construct a language in this or that way?" (Carnap, 1937, pp. 46–47, original emphasis)

Of course, this "realization" just is the Principle of Tolerance applied to case of the debate between intuitionists and classical logicians. By the lights of

---

[6] In sections 2.1 and 2.3 below, I assume familiarity with both Language I and Language II from *LSL*. Carnap gives the basic setup of I in Part I of *LSL* and the setup for II in Part III; a useful commentary on these frameworks can be found in the introduction to (Wagner, 2009).

[7] Carnap does note Heyting's book as an interesting first attempt, though does not have more to say about it than that at this stage of the book (Carnap, 1937, p. 46). When he takes the issue up again, however, he complains that Heyting's formalization is inadequate because the distinction between object-language and syntax-language is not drawn (Carnap, 1937, pp. 249–250).

the Principle, since the intuitionists have not yet produced a formal treatment of their view, or at least not one that Carnap deems adequate for discussion, then for the purposes of determining which language should form the basis of our mathematical reasoning, or which logical principles can be accepted in our mathematical practice, we are free to make precise their claims in any way we think fit. Carnap puts the point this way:

> It is in order to exclude [indirect proofs which lead] to an unlimited, non-constructive existential sentence that Brouwer renounces the so-called *Law of the Excluded Middle*. The language-form of I, however, shows that the same result can be achieved by other methods – namely, by means of the exclusion of the unlimited operators. [...] Thus Language I fulfills the fundamental conditions of Intuitionism in a simpler way than the form of language suggested by Brouwer (and partially carried out by Heyting). (Carnap, 1937, p. 48, original emphasis)

Since there are many ways we might give a precisely specified linguistic framework which meets the "fundamental conditions" of some philosophical view – in this case intuitionism – there must be a way to decide between these proposals. The way to adjudicate between them indicated by Carnap in the quotation, just as the Principle says it should be, is by comparing their pragmatic features. In this case, he thinks that Language I achieves the same aims as Heyting's proposed formalization of Brouwer's philosophical view, but that I does so in a simpler way. On that basis, and only on that basis, it is to be preferred.

Returning to the question of the proper interpretation of the Principle of Tolerance, what stands out about Carnap's treatment of intuitionism in *LSL* is its early placement in the book. Though it comes before the formal statement of the Principle, it serves to set the stage for that statement.[8] His discussion is, in that way, a case study in how to *act* in accordance with Tolerance. That is, instead of engaging with debates over the nature of negation, or of whether quantification is restricted, instead he gives the rules for a framework that he claims achieves the same aims. Anyone who disagrees is invited to do the same, and the choice between Carnap's Language I and any alternative framework will be made on the basis of the consequences of their adoption. Another part of *LSL* that helps make clear the way in which

---

[8] In the fist edition, the statement of the Principle comes in the very next section. For the second edition, Carnap inserted a short part, section §16a, on Wittgenstein's theory of identity.

Carnap's Tolerance and Friedman's Revenge

Carnap understood the Principle is his purported consistency proof, which we examine in section 2.3 below. However, before moving to the question of consistency, we pause briefly to discuss what it is, exactly, that Carnap is tolerant of.

2.2 The limits of Carnapian Tolerance

Carnap's reconstruction of the intuitionist position is a drastic departure from anything that Brouwer would have accepted. To begin with, Carnap's focus on formalizations of a language is antithetical to Brouwer's perspective.[9] Carnap is perfectly aware of this situation, and, moreover, shows no hesitation in ignoring Brouwer's view:

> We hold that the problems dealt with by Intuitionism can be exactly formulated only by means of the construction of a calculus, and that all the non-formal discussions are to be regarded merely as more or less vague preliminaries to such a construction. (Carnap, 1937, p. 46)

This dismissal of all non-formal discussion before the construction of a linguistic framework is at the heart of the Principle. It is what Carnap means when he says that one must "[...] give syntactical rules instead of philosophical arguments" (Carnap, 1937, p. 52). In other words, what the Principle enjoins us to be tolerant of is precisely formulated languages which are proposed for adoption, and *not* any philosophical justifications for those proposals. In the absence of a proposed formal language, these philosophical considerations will be completely superfluous because it will not be clear which position they support; that is, since there are many different logics that could be thought to be the formal precisification of Brouwers' claims, and so those philosophical claims do not settle the issue of which of the formal presentations we should pick. Carnap shows this implicitly by producing a linguistic framework, namely Language I, which he claims captures the spirit of intuitionism, despite its obvious departures from the philosophical claims that Brouwer makes.[10] With this in mind, we now turn to the second case study in Carnap's use of the Principle.

---

[9] See (Mancosu, 1998, p. 2).

[10] For example, the Law of the Excluded middle is valid in I, while rejected by Brouwer. See (Carnap, 1937, p. 48 and p. 34 Theorem 13.2).

2.3 Putting Tolerance to work part 2: consistency

Carnap addresses the question of consistency several times in *LSL*. The first two times are in the introduction, where he dismisses the worry that allowing a language to represent its own syntax will result in contradictions along with the demand that a proposed language be consistent in order to be considered for adoption. Despite this early dismissal, Carnap nonetheless offers a proof of the consistency of Language II in section §34*i*, and remarks on the issue again in section §59, "General Syntax". I will largely constrain my discussion to the earlier section, §34*i*, because the two sections are very similar.

The proof that Carnap gives for Language II is somewhat laborious, and we will not be concerned to cover the details. More interesting for our current investigation are Carnap's comments after the proof on the relationship between his result and Hilbert's program. After a few remarks to the effect that his term 'definite syntactical concepts' is approximately equivalent to Hilbert's 'proof with finite means', he says,

> Whether with such a restriction [to the use of only definite syntactical concepts in a consistency proof for classical mathematics], or anything like it, Hilbert's aim can be achieved at all, must be regarded as at best very doubtful in view of Gödel's researches on the subject (see §36). [...] The proof we have just given of the non-contradictoriness of Language II, in which classical mathematics is included, by no means represents a solution to Hilbert's problem. Our proof is essentially dependent upon the use of such syntactical terms as 'analytic', which are indefinite to a high degree, and which, in addition, go beyond the resources at the disposal of Language II. (Carnap, 1937, p. 129)

This passage is critical to the wide understanding of Tolerance. One of the tasks that Carnap set himself in *LSL* was to show that all of classical mathematics can be formalized, and that the language constructed to do it does not contain any contradictions. The latter challenge is completed by the proof, at least in a somewhat attenuated sense. Strictly speaking, what Carnap shows, as he comments somewhat later on, is that *"everything mathematical can be formalized, but mathematics cannot be exhausted by one system*; it requires an infinite series of ever richer languages" (Carnap, 1937, p. 222, original emphasis). So, the first challenge is impossible –

there is no single language which can serve for the formalization of *all* of mathematics. We might think that because of this fact, the consistency proof for II only shows the consistency of that part of mathematics which can be formalized in II. But, in order to complete the proof, Carnap has had to make use of resources that go beyond those available in II, namely the concepts 'analytic-in-II' and 'consequence-in-II' which are both indefinite in II. In the proof, these concepts are used in the metalanguage where they may very well be definite.[11] This not only shows, as was remarked above, that the consistency of mathematics cannot be demonstrated because there is no single language which can capture all of mathematics, it also shows that any purported proof of a language's consistency is only ever a proof of *relative* consistency. That is, we can only ever show that, by the lights of one language, some other language is consistent; in the case of Carnap's proof, we show that II is consistent if the metalanguage for II is.

What attracts our attention now, however, is his frank admission that the proof offers no absolute certainty, as well as his total lack of apparent concern over the effect that this has on the status of the proof. Carnap says:

> Our proof is essentially dependent upon the use of [...] syntactical terms [...] which are indefinite to a high degree, and which, in addition, go beyond the resources at the disposal of Language II. Hence, the significance of the presented proof of non-contradictoriness must not be over-estimated. Even if [our proof] contains no formal errors, it gives us no absolute certainty that contradictions in the object-language II cannot arise. For, since the proof is carried out in a syntax-language which has richer resources than Language II, we are in no wise guaranteed against the appearance of contradictions in this syntax-language, and thus in our proof. (Carnap, 1937, p. 129)

This is all said without further comment. By his silence, Carnap seems to suggest that, were we to have concerns about whether or not the proof is good, then we are free to construct a third language, capable of serving as a metalanguage for the metalanguage for II, and then to carry out a similar proof of the consistency of II's metalanguage in it. This new proof will have the same epistemic character as our original consistency proof, namely that

---

[11] Of course, the metalanguage for II (which we may as well call III) will have corresponding concepts which are indefinite in *it*. We know, for example, that the concept 'analytic-in-III' will be indefinite in III for analogous reasons. This fact is related (indeed is nearly equivalent) to Tarski's Theorem on the indefinability of truth. See (Procházka, 2006) for a discussion.

it does not guarantee us against paradox in the language in which we conduct the proof. What must lay behind Carnap's rather *laissez faire* attitude towards consistency proofs is, I claim, wide Tolerance. That is, since we are just as free to construct metalanguages as object-languages, we can always find a place to stand from which we can investigate any question that might interest us.

The Principle of Tolerance, on either the wide or the narrow interpretation, already licenses Carnap's dismissal of foundational concerns. That is, the complaint that our proof does not guarantee absolute consistency, but only consistency relative to some other theory, has no bite for him because leaving aside such demands for absolutes is just what the Principle enjoins us to do. However, what we see in the quotation is not simply Carnap's anti-foundationalism, but the thorough-going nature of his understanding of Tolerance. Recall that, for Carnap, statements about which language to choose can only be evaluated relative to a stated goal. This too is consistent with either the wide or narrow interpretation of the Principle. So, according to either interpretation, if our goal is to show that our object-language is consistent, then we need to construct a syntax-language with the appropriate resources (as, for example, being able to express the concept "analytic-in-I"). This point generalizes, and choice of task will have consequences for the way in which we should construct the syntax-language. Carnap puts the point directly himself in section §45:

> Our attitude towards the question of indefinite terms conforms to the principle of tolerance; in constructing a language we can either exclude such terms (as we have done in Language I) or admit them (as in Language II). It is a matter to be decided by convention. If we admit indefinite terms, then strict attention must be paid to the distinction between them and the definite terms; especially when it is a question of resolubility. Now this holds equally for the terms of syntax. [...] Some important terms of the syntax of transformations are, however, indefinite (in general) [...] (Carnap, 1937, pp. 165)[12]

So far, everything that Carnap has says is consistent with both the wide and the narrow interpretation. However, he continues on to say:

> If we wish to introduce these [indefinite] terms also, we must

---

[12]Friedman also notes this passage at (Friedman, 2001, p. 227).

use an indefinite syntax-language (such as Language II). (Carnap, 1937, pp. 165–166)

So, we now see that just as with the object-language, there are no morals at the metalinguistic level either.[13] That is, we are free to construct our syntax-language in any way we see fit, with the recognition that some ways of doing so may fare 'better' than others for particular tasks. This claim, namely that Carnap not only intended for Tolerance to apply at the object level, but at the level of syntax-languages as well, just is the distinction between the wide and narrow interpretations. So, while either interpretation can accommodate the task-relativity of Tolerance at the object level, it is wide Tolerance that enjoins us to follow the Principle all the way up the linguistic hierarchy.

3 Wide Tolerance and the boundless ocean

I begin with an obvious, but as I will show ultimately ineffective, route for Carnap to escape Friedman's argument. Friedman suggests the Carnap might be available to adopt a weaker, intuitionistically acceptable metalanguage (as, for example, one that excludes unlimited universal quantification), instead of the full-blooded classical one he in fact adopts in *LSL*. In his 2007 paper, Ricketts takes a similar line against Friedman's concerns. He says:

> There is nothing inconsistent or untoward in an advocate of weak logic for the language of science using a strong meta-language both to set forth her favored language and to compare it with other proposed languages for science.[14] Nevertheless, a tolerant advocate of a weaker logic *may* balk at the use of a strong meta-language. [...] The refusal of our advocate of a weaker logic to use a strong meta-language does not, however, close off the prospect of metamathematical comparisons of calculi. In such circumstances, the discussants will have to restrict themselves to those descriptions of the languages under consideration that are available in a meta-language they all share. (Ricketts, 2007, p. 219. Original emphasis.)

---

[13] Ricketts makes a similar point in his (Ricketts, 2007, p. 219).
[14] Ricketts' "languages for science" are what I have called "object-languages" in this paper; what is meant by both expressions is the languages used for our first-order theorizing.

Ultimately, and correctly in my view, Friedman finds this proposal lacking. While it would allow for the exploration of the consequences of adopting the two competing frameworks without prejudice to one over the other, what he says it will not do is allow for a "[...] sharp contrast between *merely* pragmatic questions of 'linguistic form' having no ontological import, on the one side, and genuine theoretical claims, on the other" (Friedman, 2001, p. 244, original emphasis). To see why this is the case, it is helpful to recall the reason we adopted the full strength of classical mathematics in the metalanguage in the first place, namely to be able to make use of indefinite notions like analyticity. These notions are what serve to mark the distinction between the genuine theoretical questions, and those which are settled by linguistic conventions. So, if we adopt a metalanguage which is too weak to characterize such notions in all the languages under consideration, then while we might get the intuitionist back into the game, so to speak, we may lose the ability to demarcate the boundaries of the field of play. So it would seem that the obvious maneuver – taking the strongest mutually agreeable language as metalanguage – will not be a solution in general. But there is another solution available, and it is to this that we now turn.

Friedman thinks that the tension his argument reveals in Carnap's Tolerance is ultimately fatal to the program in *LSL*.[15] However, I contend that the tension is merely apparent, and stems from a misconstrual of Tolerance as requiring the narrow reading. In particular, he has neglected to note a shift in the salient task that is critical to seeing that Tolerance is in no way undermined by the use of intuitionistically unacceptable resources in the metalanguage; however, this shift is only noticeable once one thinks that Tolerance extends to metalanguages as well.[16] Recall that for Carnap languages are evaluated relative to a particular task. To give two examples, in the case that Carnap is concerned with in *LSL* where our concern is to give a logical foundation for classical mathematics, then we ought pick a language which embraces non-constructive proof techniques. Conversely, were we to be concerned to ensure that we can decide for each numerical predicate whether it applies to a given number or not, then we ought pick a language which is constructed along the lines of Carnap's Language I.

---

[15](Friedman, 2001, p. 244).

[16]The arguments I make in this section have some similarity to arguments that Ricketts makes in his (Ricketts, 2007), especially in section III of that essay. There are differences, however. In particular, my focus here is on how the two facets of Tolerance interact to defuse certain objections; Ricketts simply states that these objections have no bite.

Carnap's Tolerance and Friedman's Revenge

Returning to the case at hand – that is to deciding between classical and intuitionist mathematics – what we will need to do is to construct a language that allows us to investigate the consequences of adopting each proposed language without prejudging the issue. The metalanguage so constructed need not be the same one in which we might, for example, investigate whether these same object-languages are consistent; after all, the task at hand has changed. Tolerance runs all the way up the linguistic hierarchy on the wide reading, and therefore so does task relativity. Friedman's argument assumes that there needs to be a single metalanguage, chosen once and for all; for him, it not only serves as a place to stand when considering questions about a particular language, it is also supposed to simultaneously be a perspective from which we adjudicate all disputes over the form that a language should take. But, this insistence on a single metalanguage for these disparate tasks is precisely the kind of absolutism that Carnap sought to combat with Tolerance.

There is another way in which Friedman's argument does not hit its mark. As he points out, if we want to prove facts about our object-languages (such as, for example, their consistency or completeness), we must do so in some language or other. This language must have certain resources in order for the proofs to go through, as we have noted several times. We then use the results of this investigation of these languages in order to come to a determination of which of the two to adopt as the language of our first-order theorizing. Friedman claims that this prejudices the decision of which object-language to adopt because we have made essential use of resources in the metalanguage that proponents of weaker logics do not accept as valid. It is this last step that is problematic from Carnap's perspective; it relies on a notion of validity *simpliciter* which he rejects. On Carnap's picture, inferences are only valid (or invalid) relative to a particular language, and so the fact that one treats an inference as valid in one language does not entail that one must treat it as valid in every language. Though this point is obvious in the case of different object-languages, as the case at hand of classical and intuitionistic logics illustrates, it is somewhat more subtle when examining the case of an object-language and its metalanguage. However, Carnap's view is that metalanguages are constructed in just the same way as object-languages are. This means that just as with object-languages, the validity of inferences in metalanguages are language relative. So, accepting certain inferences for the task of investigating the consequences of adopting a language does not thereby commit one to the unlimited validity of those inferences. That is, the intuitionistically inclined Carnapian can still entertain the notion that inferences like double negation elimination and unrestricted universal quantification are invalid in our mathematical reasoning quite independently of accepting them for investigating the consequences of rejecting those inferences on our mathematical practice. In this way, the notion of validity in the metalanguage floats free from the notion of validity in the object-language, and conversely.

In this paper, I have developed a reading of Carnap's Tolerance which I dubbed the 'wide' reading. According to it, as I argued, there are no constraints whatsoever on the construction of languages, and therefore no constraints on what languages we can consider as potential languages to adopt for formalizing some practice. Moreover, the wide reading ensures that the task-relativity of our evaluation of languages applies at the level of metalanguages in just the same way it does at the level of object-languages. I showed that this wide reading is the most plausible way to understand Carnap's intended meaning for the Principle of Tolerance by closely examining his uses of the Principle in *LSL*. Finally, I turned attention to an argument by Michael Friedman, namely that Tolerance is self-undermining due to the prejudicial nature of the metalinguistic considerations necessary to assess proposed linguistic frameworks. I considered a possible route of escape suggested by Ricketts, namely to adopt a weaker but mutually agreeable metalanguage from which to settle disputes, but concluded that it left the Tolerance-inclined logician in a difficult place: unable to define some of the languages that one might want. However, I offered an alternative suggestion, one that does not face this problem. It relied on the wide interpretation of Tolerance, and made essential use of the task-relativity of language selection.

References

Carnap, R. (1937). *The Logical Syntax of Language*. London: Kegan Paul, Trench, Trubner & Co.

Carus, A. (2007). *Carnap and Twentieth-Century Thought: Explication as Enlightenment*. Cambridge: Cambridge University Press.

Eklund, M. (2009). Carnap and Ontological Pluralism. In D. Chalmers, D. Manly, & R. Wasserman (Eds.), *Metametaphysics: New Essays in the Foundations of Ontology* (pp. 130–156). Oxford: Oxford University Press.

Friedman, M. (2001). Tolerance and Analyticity in Carnap's Philosophy of Mathematics. In J. Floyd & S. Shieh (Eds.), *Future Pasts: The Analytic Tradition in Twentieth-Century Philosophy* (pp. 223–255). Oxford: Oxford University Press.

Hirsch, E. (2009). Ontology and Alternative Languages. In D. Chalmers, D. Manly, & R. Wasserman (Eds.), *Metametaphysics: New Essays in the Foundations of Ontology* (pp. 231–259). Oxford: Oxford University Press.

Mancosu, P. (1998). *From Brouwer to Hilbert: The Debate on the Foundations of Mathematics in the 1920s*. Oxford: Oxford University Press.

Potter, M. (2002). *Reason's Nearest Kin: Philosophies of Arithmetic from Kant to Carnap*. Oxford: Oxford University Press.

Procházka, K. (2006). Consequence and Semantics in Carnap's *Syntax*. *Miscellanea Logica*, 6, 77–113.

Putnam, H. (1983). Philosophers and Human Understanding. In *Realism and Reason: Philosophical Papers, Volume 3* (pp. 184–204). Cambridge: Cambridge University Press.

Ricketts, T. (1994). Carnap's Principle of Tolerance, Empiricism, and Conventionalism. In P. Clark & B. Hale (Eds.), *Reading Putnam* (pp. 176–200). Oxford: Blackwell Publishing.

Ricketts, T. (2007). Tolerance and Logicism: Logical Syntax and the Philosophy of Mathematics. In M. Friedman & R. Creath (Eds.), *The Cambridge Companion to Carnap* (pp. 200–225). Cambridge: Cambridge University Press.

Wagner, P. (2009). *Carnap's Logical Syntax of Language*. Basingstoke: Palgrave Macmillan.

Noah Friedman-Biglin
The Arché Philosophical Research Centre,
and the Saint Andrews – Stirling Graduate Programme in Philosophy
Scotland, United Kingdom
E-mail: nfriedma@gmail.com

Fifty (More or Less) Shades of Logical Consequence

LUIS ESTRADA-GONZÁLEZ[1]

Abstract: In this paper I argue against the view that logical consequence is a truth-preserving relation and, in general, that it is a reflexive and transitive relation. I expound a number of non-reflexive or non-transitive notions of logical consequence and argue that they satisfy the requirements to be considered part of the so-called "core tradition" of logic, and that they thus give rise to logics bona fide.

Keywords: logical consequence, reflexivity, transitivity, non-Tarskian logics, Universal Logic

1 Introduction

The earliest abstract axiomatization of logical consequence proposed by Tarski (1928) can be reconstructed as follows.[2] Let $\mathbb{F}$ be a non-empty collection of, say, formulas. For any formulas φ and ψ and sets of formulas Γ and Δ, $\Vdash$ is a (Tarskian) relation of logical consequence if and only if the following hold:

(T1) $\varphi \Vdash \varphi$ (Reflexivity)
(T2)   If $\Delta \Vdash \varphi$ then $\Gamma, \Delta \Vdash \varphi$ (Monotonicity)
(T3)   If $\Gamma, \varphi \Vdash \psi$ and $\Delta \Vdash \varphi$ then $\Gamma, \Delta \Vdash \psi$ (Transitivity)

---

[1] This paper was written under the support from both the European Union through the European Social Fund (Mobilitas grant no. MJD310) and the CONACyT project CCB 2011 166502 "Aspectos filosóficos de la modalidad". I want to thank Daniel Cohnitz, Indrek Lõbus, Carlos Romero, the audiences of LOGICA 2014 at Hejnice (especially Dorothy Edgington and Graham Priest), the Seminario de Investigadores IIF-UNAM (especially Atocha Aliseda-Llera, Axel Arturo Barceló-Aspeitia, Delia Belleri, Matthieu Beirlaen, Maite Ezcurdia, Cristian Alejandro Gutiérrez-Ramírez, Matthieu Fontaine, Pau Luque, Moisés Macías-Bustos, and Ricardo Mena) for useful discussion, comments, and suggestions on previous stages of this paper that helped a lot to improve it. Charlie Donahue and Wiebe van der Hoek deserve special mention for their encouragement and useful advices.

[2] In his subsequent works he added other constraints, but the literature sanctions as Tarskian these three.

Luis Estrada-González

I take for granted here that the logicality of relations of consequence which do not meet Monotonicity has been secured, and thus that non-monotonic logics are really logics. Thus, another option is requiring of logical consequence just to be *generalized Tarskian*, i.e. that logical consequence is at least reflexive and transitive (as in Avron, 1991, Beall & Restall, 2006 or Paoli, in press). It is important to mention, but not intending to be exhaustive as to possible stances, the one in which logical consequence is required just to be *weakly Tarskian*, that is, that logical consequence has to meet at least some – probably weaker than (T1)–(T3) – forms of reflexivity, transitivity and monotonicity (cf. Aliseda-Llera, 2006, p. 90, Gabbay, 1994).

The question I am concerned with here is whether the relation of logical consequence has to be regarded as truth-preservation (from premises to conclusions, or *forwards truth-preservation*), and hence as reflexive and transitive. This question is important because there is always the worry that every new development in logic, non-reflexive and non-transitive logics in this case, results into something that is not really logic. For instance, Quine (1970, p. 70) famously held that some many-valued logics were but "uninterpreted theory, abstract algebra". Dummett expressed the same stance towards some work in formal semantics (for example see Dummett, 1998).

Beall and Restall have provided an explicit argument for the reflexivity and transitivity of logical consequence and deem as not genuinely logical the systems where those properties fail. This is interesting because their pluralism aims to be a good framework in which to understand contemporary logic (see Beall & Restall, 2006). However, in defending that logical consequence is forwards truth-preserving, and hence reflexive and transitive, they rule some activity in contemporary logic out, not only that coming from mathematics and computer science which could be suspect of philosophical naivety, but also including some philosophically motivated systems,[3] and even probably Aristotle's own founding non-reflexive notion of deduction.

The claim I defend is that logical consequence has not to be identified with truth-preservation and that logical consequence is not necessarily reflexive or transitive either. Therefore, my main target here is the idea that logical consequence must be at least generalized Tarskian, although I advance ideas for the idea that logical consequence needs not to satisfy any property, and hence against the demands for it being weakly Tarskian.

---

[3] Only in the non-transitive side one can find a relevant logic for so-called 'epistemic gain' in (Tennant, 1994), a logic for unrestricted set comprehension in (Weir, 1998), a theory for an unrestricted truth predicate in (Ripley, 2012), or a logic for vague expressions in (Zardini, 2008).

Fifty Shades of Logical Consequence

The plan of the paper is as follows. For the sake of definiteness, in the second section I reconstruct the argument against non-reflexive and non-transitive logics ("Non-transitive or non-reflexive systems of 'entailment' may well model interesting phenomena, but they are not accounts of *logical consequence*") put forward by Beall and Restall. I chose it because it is relatively simple yet makes clear the kind of philosophical background operating when deciding what to accept as a logic, in this case what they call the "core tradition" of logic, namely that logical consequence is a necessary, formal and normative truth-preserving relation used to evaluate arguments we actually employ. In the third section I expound a proposal akin to mine but which I find wanting, though, namely Béziau's historical-practical-inductive-analogical argument to show that logical consequence needs not to satisfy any principles, and a fortiori, that it needs not be reflexive or transitive (see Béziau, 1994, 2001, 2006, 2010). In the fourth section I present two non-Tarskian notions of logical consequence already known in the literature and show that, formally, they are not very different from the generalized Tarskian notion, but rather easy generalizations from this. In the fifth section I argue that non-Tarskian notions of consequence satisfy the requirements to be considered part of the core tradition of logic, that is, they can be used to analyze inferential relations in arguments actually employed and are neccesary, formal, and normative relations, so they give rise to logics bona fide albeit they are not truth-preserving, transitive, or reflexive. Finally, in the sixth section I introduce more abstract notions of logical consequence and argue that they still have the credentials to be considered part of such core tradition of logic.

A disclaimer is in order. The discussion of logical consequence in this paper will be cashed entirely in terms closer to model theory. This is done for simplicity. Sometimes the so-called "substructural logics" are thought to be "non-Tarskian logics", for deducibility or derivability is assimilated to consequence, (forms of) Transitivity is (are) assimilated to (forms of) Cut or (forms of) Monotonicity is (are) assimilated to (forms of) Weakening, but this is not so straightforward: There is a way, namely an "external" point of view in which logical consequence relates whole sequents rather than formulas in the sequentes, in which these substructural logics invariably induce Tarskian logics (see for example Béziau, 2001, section 5; Mossakowski, Diaconescu, & Tarlecki, 2009, p. 97; Paoli, 2002, p. 24; Wójcicki, 1988, p. 117). Given this fact and the complications introduced by the external and internal views of proof-theories, I do not discuss Gentzenian approaches here.

2 An argument for generalized Tarskian logical consequence

Based on a more or less widespread understanding of logic, Beall and Restall (2006) have put forward an argument to exclude from the realm of "proper" logic what they take to be merely so-called "logics".[4] They say:

– "Logic, whatever it is, must be a tool useful for the analysis of the inferential relationships between premises and conclusions expressed in arguments we actually employ. If a discipline does not manage this much, it cannot be *logic* in its traditional sense." (Beall & Restall, 2006, p. 8, italics in the original)

– "(...) any settling of the relation of logical consequence must be a necessary, normative, and formal relation on propositions." (Beall & Restall, 2006, p. 29)

– "We hold that deductive validity is a matter of the preservation of truth in all cases (...) This analysis of validity owes a great deal to the tradition (...) It is also connected intimately with the constraints on consequence that we have already seen (...) So, the analysis of logical consequence as preservation of truth in all cases goes some way to explaining how a relation of logical consequence is necessary, normative, and formal." (Beall & Restall, 2006, p. 23–24)

– "Non-transitive or non-reflexive systems of 'entailment' may well model interesting phenomena, but they are not accounts of *logical consequence.* One must draw the line somewhere and, pending further argument, we (defeasibly) draw it where we have. We require transitivity and reflexivity in logical consequence." (Beall & Restall, 2006, p. 91, italics in the original)

The argument can be reconstructed as follows:

(BR0) A logic is an account of the relation of logical consequence (or validity).

(BR1) An account of logical consequence must be a tool useful for the analysis of the inferential relationships between premises and conclusions expressed in arguments we actually employ.

---

[4]Thus, it is not true that "Beall and Restall's choice what can be dealt with pluralistically is idiosyncratic and not principled." (Bremer, 2014) Also, among other inaccurate claims, it is not true that Beall and Restall incur in an additional "exclusion of logics that fail to meet monotonicity": they only require reflexivity and transitivity, not monotonicity, and indeed monotonicity is nowhere mentioned in the book.

Fifty Shades of Logical Consequence

(BR2) Logical consequence has at least the features of necessity (the truth of the premises in a valid argument necessitates the truth of the conclusions), formality (valid arguments are so in virtue of their logical form), and normativity (rejecting a valid argument is irrational).

(BR3) Logical consequence is forwards truth-preserving.

(BR4) Forwards truth-preservation is a (non-empty)[5] reflexive and transitive relation.

(BR5) Hence, logical consequence is a reflexive and transitive relation.

(BR6) Then, if a relation between the elements of a structure is either non-reflexive or non-transitive, such a relation is not one of logical consequence.

(BR7) Therefore, if no relation on a structure is logical consequence, such a structure is not a logic.

One should not get distracted by the fact that Beall and Restall *defeasibly* draw the line in Reflexivity and Transitivity, as if it were not an important part of their philosophical view for, as the argument shows, they base it on some crucial philosophical motivations and reasons. Of course, they may be willing to drop these requirements, but *pending further argument* they keep them. In this paper I am providing that argument for actually dropping those conditions in terms shared by Beall and Restall and, in general, by those who accept this "core tradition" as a necessary part of logic.

Though nothing is completely uncontentious, I will assume for the sake of the argument that there is hardly any disagreement about what Beall and Restall call "the core tradition" in logic, namely (BR0), (BR1), and (BR2); I do not count disagreements about how to spell out some crucial notions in the above – like formality or normativity, to mention just two – as disagreements with the spirit of these premises.[6] (BR3) has also been widely accepted.

Let me say that a relation R *forwards-preserves* a property ϕ if and only if for any x, y such that Rxy, if x has the property ϕ, y has it too. (BR4)

---

[5] In what follows I will always deal with candidates to be logical consequence that are non-empty relations, so I will omit this qualification. The discussion whether the empty relation can be rightly regarded as logical consequence would require a separate work.

[6] As it has been noted in (Paseau, 2007), there might be disagreement on whether normativity, formality, and necessity are the only settled features of logical consequence – others might be, say, aprioricity or universality –, but not that they are features of logical consequence. In any case, as I have reconstructed the argument, it is not necessary to assume that these are the only ones.

has been proved for certain formal languages, among them the most widely used in logic, for example, in (Hardegree, 2005). (BR5) follows from (BR3) and (BR4) by the transitivity of the predicative 'is'; (BR6) from (BR5) by contraposition and (BR7) by a generalization from (BR0) and (BR6).

Read (2003) has argued that (BR5) and (BR4) can fail, but not exactly by rejecting any of the premises, but by making precisions about (BR3). Recently there has been opposition to (BR3) in the following terms:
- Logical consequence is not forwards truth-preservation because there is nothing like (forwards) truth-preservation in all cases, on pain of triviality. See for example (Beall, 2009; Field, 2009; Murzi & Shapiro, in press).
However, we are interested in the following way of disputing (BR3) and (BR5):
- Logical consequence is not forwards truth-preservation because there are other ways of logically relating premises and conclusions which are not reflexive or not transitive but count nonetheless as accounts of logical consequence as in (BR0), (BR1), and (BR2).

3 Towards "axiomatic emptiness"

My own approach owes a lot to Béziau's ideas on Universal Logic; he rejects (BR5) too, i.e. for him logical consequence is not necessarily reflexive or transitive. However, Béziau does not draw any conclusion about the relation between logical consequence and truth-preservation. Béziau deploys an historical-practical-inductive-analogical argument to show that logical consequence needs not to satisfy any principles (see Béziau, 1994, 2001, 2006, 2010) that can be reconstructed as follows[7]:

---

[7] Although the same ideas are recurrent mentioned here and there through the papers, the argument below has to be extracted by brute force, so to speak, from Béziau's writings, since it is never explicit and its parts appear as disconnected remarks, hidden amongst the discussion of several other topics and often shrouded by strange pieces of rhetoric. Just as an example, here are the views as expressed at some places of (Béziau, 2001):
"Traditionally the principle of contradiction is taken as a fundamental pillar of logic. The idea is that reasoning is not possible without it. Paraconsistency goes against this idea. And if paraconsistent logic is rightly a logic, therefore what are the ground principles of logic, if any?" (p. 5)
"The number of new logics has increased these last years due to the need of computer sciences, artificial intelligence, cognition, and all the stuff of our cybertime." (p. 20)
"My motivation and my terminology were taken from Birkhoff's famous notion of abstract algebra, that I found in *Lattice theory* (cf. Birkhoff 1940), which is just a set with a family of operations. My idea was already that the basic foundations of logic were not more principles for the consequence relation than principles for connectives, like the principle of contradiction.

Fifty Shades of Logical Consequence

(**B1**) Virtually every theorem, principle for connectives, principle for the consequence relation, etc., let us call them collectively 'properties of a logic', has been thrown out or, at least, challenged.

(**B2**) The outcomes of such droppings and challenges have been regarded as logics.

(**B3**) If the properties $P_1, \ldots, P_n$ of a logic can be dropped or challenged, an additional property P_m also can be dropped or challenged and the result will still count as a logic.

(**B4**) The situation is analogous to the case of algebra, where an algebraic structure needs not to satisfy any property in particular.

(**B5**) Hence, a relation of logical consequence can be defined with no reference to a particular property of a logic.

Then Béziau proposes that a logic is a certain kind of mathematical structure and that a logical structure is a structure of the form $LS = \langle S, \vdash_{LS} \rangle$, where S is an arbitrary structure and $\vdash_{LS}$ is also an arbitrary relation on $\wp(S) \times S$ (provided we have means to obtain "powers" on such structure). Equivalently, it could be described as a pair $LS = \langle S, C_{LS} \rangle$, where C_{LS} is an arbitrary mapping $C_{LS} : \wp(S) \longrightarrow \wp(S)$. "Arbitrary" means here that neither the structure nor the relation or mapping need to satisfy any axiom. "Universal logic" would be a discipline whose subject matter is logical structures independently of the features of particular logics, analogous to universal algebra.

(B1) is a premise concerning the history of logic and even if there were properties of a logic that has not been actually challenged, let us grant it. (B2) is a premise based on a kind of observation of practice. But practice might not be enough. It has been challenged, for example, by Quine (1970) and Dummett (1998), to name just two prominent philosophers, on the basis that certain logics are "uninterpreted theories" or "abstract algebras" rather than logics, receiving that name only by partial analogy. However, let us assume that such "partial analogy" suffices.

But even granting this, the problems of this argument are those of any inductive and analogical argument. On one hand, one has to be very careful about how close the analogy between logic and algebra, as in (B4) and

---

I reached the idea that we must throw out all principles altogether, that logic *is not grounded on any principles or laws*." (p. 8; italics in the original)

(B5), is. On the other hand, (B3) is ambiguous between a distributive and a collective sense of dropping properties and still obtaining logics. Thus, (B3) might mean at least the two following claims:

(B3') If the properties $P_1, \ldots, P_n$ of a logic L_A can be dropped or challenged giving rise to another logic L_B, an additional property P_m also can be dropped or challenged from L_A and the result, L_C, will still count as a logic.

(B3") If the properties $P_1, \ldots, P_n$ of a logic L_A can be dropped or challenged giving rise to another logic L_B, an additional property P_m also can be dropped or challenged from L_B and the result, L_C, will still count as a logic.

According to (B3'), for any property P of a logic, there could be another logic such that it lacks it. (B3") says something stronger: There could be a logic such that it lacks any property of another logic, and even any properties of any other logic.

In any case, Beall and Restall would oppose to either version of (B3). Their contention would be that success in dropping or challenging certain properties of a logic does not imply success in dropping or challenging every property of a logic, whether collectively or distributively. Talking specifically about the properties of the relation of logical consequence, from the fact that we can do without the Monotonicity of logical consequence it does not follow that Reflexivity or Transitivity can also be dropped; after analysis, the very notion of logical consequence could require at least the latter two properties. Something about logicality may be lost whether in the analogy with algebra or in the unrestricted dropping of properties. It is not obvious that Béziau's arbitrary structures and arbitrary mappings are still in the traditional of logic and, if they are not, there is no clear warrant for the departure: Generalization from analogy with algebra and induction from the history of logic are not enough. Until it is shown that nothing logical is lost, Béziau's argument is wanting. But although defective, it highlights several interesting points; for my present purposes, it shows that an argument for the conclusion is not completely untenable. I will argue in the following sections that indeed nothing logical in the sense of the core tradition is lost when reflexivity or transitivity are dropped. Although my approach is independent because it relies on none of Béziau's premises, it can be regarded as providing certain missing links in Béziau's argument, at least when "property of a logic" means "property of the parameters in the definition of relation of logical consequence".

Fifty Shades of Logical Consequence

4 Non-Tarskian notions of logical consequence

I will use the label "strongly non-Tarskian" for those notions of consequence that satisfy neither Reflexivity nor Transitivity to distinguish them from "generalized Tarskian logics", or "weakly non-Tarskian" logics, i.e. logics that only drop Monotonicity.

Before going into the actual details of strongly non-Tarskian logics, some notions are needed. Consider a structure $S_L = \langle F, \mathcal{V}, D^+, \{f_c : c \in C\}, \mathfrak{V}, \mathcal{I}\rangle$ which can be used to give a semantics for a logic. F stands for a collection of formulas, C is a collection of connectives and, for each connective c, f_c is the truth-relation it denotes. $\mathcal{V}$ is a non-empty collection of truth values. For simplicity I assume that $\mathcal{V}$ comes with an ordering, $\leq$ (which may be a partial ordering).[8] I also assume that every subcollection of $\mathcal{V}$ has a greatest lower bound (Glb) and least upper bound (Lub) in the ordering. Let us suppose that there is an $x \in \mathcal{V}$ such that for every $y \in \mathcal{V}$, $y \leq x$. Let us call $true$ such x and denote it '⊤'. $\mathcal{I}$ is a non-empty collection of indexes of evaluation (possibly empty sets of conditions at which formulas are evaluated); a family of interpretations, $\mathfrak{V}$, assigns values in $\mathcal{V}$ to propositional parameters at indexes; the values of all formulas can then be computed using the f_cs, and relations of consequence $\models$ determined by the values and the valuations can be defined; values, valuations and indexes can be collectively called *cases*. $D^+ \subseteq \mathcal{V}$ is a collection of *designated values* if it satisfies the following conditions:

(a^+) ⊤ $\in D^+$;
(b^+) for every $x, y \in \mathcal{V}$, if $x \in D^+$ and $y \notin D^+$ then $x \not< y$;
(c^+) q is a logical D^+-consequence from premises Γ, in symbols $\Gamma \models^{D^+} q$, if and only if any case in which each premise in Γ is designated is also a case in which q is designated.

A collection $D^- \subseteq \mathcal{V}$ is going to be called collection of *antidesignated values* if it satisfies the following conditions:

(a^-) There is an $x \in \mathcal{V}$ such that for every $y \in \mathcal{V}$, $x \leq y$. Let us call $false$ such x and denote it '⊥'. ⊥ $\in D^-$;
(b^-) for every $x, y \in \mathcal{V}$, if $x \in D^-$ and $y \notin D^-$ then $y \not< x$;
(c^-) q is a logical D^--consequence from premises Γ, in symbols $\Gamma \models^{D^-} q$, if and only if any case in which q is antidesignated, is also a case in which each of Γ is antidesignated too.

---

[8]I use the idea that every partial order induces a strict order, defined as $x < y = (x \leq y$ and $x \neq y)$.

Luis Estrada-González

The determining factor for obtaining different logics from a structure S_L is the structure of truth values, that is, their exact number and the ordering between them. In the classical case where exactly two truth values are given, we have as consequences that $D^+ = \{\top\}$ and $D^- = \{\bot\}$ (and that $\neg\top = \bot$ and $\neg\bot = \top$). In classical logic D^+ and D^- satisfy thus the following equalities:

(d) $D^+ \cup D^- = \mathcal{V}$
(e) $D^+ \cap D^- = \varnothing$

When more values are given, one has to choose by hand, as it were, what the elements of D^+ are. It has been standard to assume (d)-(e) even for the case where there are more than two elements in $\mathcal{V}$ and, moreover, the coextensionality of D^+-consequence and D^--consequence but, as I will show later, this might not be the case. Thus, an exact, definite number of truth values just varies the collections of logical truths or of valid inferences but the properties of truth values and designated and antidesignated values described in (a^+)-(c^+), (a^-)-(c^-) are not changed and even the properties (d)-(e) may still hold.

Let me consider two of those strongly non-Tarskian notions of consequence, Malinowski's Q-consequence ("Q" for "Quasi"; cf. Malinowski, 1990a, 1990b) and Frankowski's P-consequence ("P" for "Plausible"; cf. Frankowski, 2004a, 2004b).

Q-*consequence*. q is a logical Q-consequence from premises Γ, in symbols $\Gamma \models^Q q$, if and only if any case in which each premise in Γ is not antidesignated is also a case in which q is designated. Or equivalently, there is no case in which each premise in Γ is not antidesignated, but in which q fails to be designated.

P-*consequence*. q is a logical P-consequence from premises Γ, in symbols $\Gamma \models^P q$, if and only if any case in which each premise in Γ is designated is also a case in which φ is not antidesignated. Or equivalently, there is no case in which each premise in Γ is designated, but in which q fails to be not antidesignated.

Suppose that there are three truth values, $\top$, μ, and $\bot$, with the order $\bot < \mu < \top$. Take $\top$ as the only designated value, $\bot$ as the only antidesignated value and μ as neither designated nor antidesignated. Suppose that $\mathfrak{v}(p) = \mu$. Then p is not a logical Q-consequence of p, because Q-consequence requires that if premises are not antidesignated, conclusions must be designated, which is not the case in this example. Q-consequence is not reflexive.

Take the same truth values with the same order and the same partition.

Suppose now that $\mathfrak{v}(p) = \top$, $\mathfrak{v}(q) = \mu$ and $\mathfrak{v}(r) = \bot$. Thus even if q is a logical P-consequence of p and r is a logical P-consequence of q, r is not a P-consequence of p, because P-consequence requires that if premises are designated, conclusions must be not antidesignated, which is not the case in this example. P-consequence is not transitive.

A good signal that we are not very far from what is commonly called 'logic' is that, under minimal classical constraints on the structure of truth values, these notions of consequence are indistinguishable from the Tarskian one. If there are only two truth values, $true$ and $false$ with their usual order, the collections of designated and antidesignated values exhaust all the possible values, hence designated = not antidesignated and not designated = antidesignated. But if this were a feature merely of classical logic, surely Q-consequence and P-consequence would have emerged earlier than they actually did. However, these notions of consequence collapse if the collections of designated and antidesignated values are supposed to be mutually exclusive and collectively exhaustive with respect to the total collection of values given, as is assumed in almost every known logic.

At least technically, strongly non-Tarskian notions of consequence such as Q-consequence and P-consequence are as legitimate as non-classical logics are. More elaborate answers could be given along the lines of non-monotonic logics. For example, Q-consequence would serve to "jump" to conclusions more certain than the premises, and P-consequence would allow to jump to conclusions less certain than the premises. However, more serious arguments than these resemblances can be given.

5 Strongly non-Tarskian notions of logical consequence and the core tradition

There is no prima facie reason as to why it is possible to do without monotonicity but not without reflexivity or transitivity: As properties of a relation of logical consequence they seem to be *pari passu*. Beall and Restall's claim that they are not on equal footing "because preservation of designated values (from premises to conclusions) is a reflexive and transitive relation" begs the question: Non-reflexive and non-transitive logics are precisely asking for ways of logically connecting premises and conclusions others than preservation of designated values.[9]

---

[9] In the discussion at Hejnice, Graham Priest objected that Transitivity is required since one wants to chain (valid) reasonings. Two replies can be given. One, that in the same way one

Now, granting that the core tradition in logic is indeed a tradition, core and about logic (and there is overwhelming evidence for that), the burden is on the proponent of strongly non-Tarskian notions of logical consequence. She has to show either that the core tradition underlying truth-preservation is wrong or that it makes room for non-reflexivity and non-transitivity. I think a strong case can be made for the latter option. Let me discuss first how the strongly non-Tarskian notions of consequence can serve as tools for the analysis and evaluation of arguments we actually employ. It is easy to see that, in spite of the appearances, we are still in the business of logic. If one uses cognitive states like acceptance and rejection (or their linguistic expressions, assertion and denial) to define validity, these different notions of consequence arise almost naturally. For example, Let $\mathcal{V}$ be a collection of "dialogical properties" so that one uses, for example, 'accepted' for D^+ and 'rejected' for D^-:

- An argument is (generalized) Tarskian-valid if and only if, if for every case the premises are *accepted*, then the conclusions are also *accepted*. Equivalently, if the conclusions are *non-accepted*, at least some of the premises are *non-accepted* too. The relevant dialogical property is *accepted*, so there is only one property forwards-preserved. That property determines another property, *non-accepted*, which makes the collections of properties mutually exclusive and collectively exhaustive.

- An argument is Q-valid if and only if, if for every case the premises are *non-rejected*, then the conclusions are *accepted*. Equivalently, if the conclusions are *non-accepted* in some case, at least some of the premises are *rejected*. The relevant dialogical properties are those that count as *non-rejected*, but there is only one of them that has to be forwards-preserved: *accepted*. *Accepted* and *rejected* are mutually exclusive, but probably not collectively exhaustive; *non-accepted* and *non-rejected* are collectively exhaustive, but probably not mutually exclusive.

- An argument is P-valid if and only if, if for every case the premises are *accepted*, then the conclusions are *non-rejected*. Equivalently, if the conclusions are *rejected* in some case, at least some of the premises are *non-accepted*. The relevant dialogical properties are those that count as *non-*

---

would like to keep conclusions reached through valid reasonings (Monotonicity), but in general it is not possible. But that in general it is not possible to chain reasonings does not mean, and this is the second reply, that one never can chain them; weaker forms of transitivity might hold, for example, that one cannot have as last shackle of a chain of reasonings one where the conclusions are antidesignated unless the premises of the first chain shackle are antidesignated too.

rejected, since any of them can be preserved from *accepted*. Again, *accepted* and *rejected* are mutually exclusive, but probably not collectively exhaustive; *non-accepted* and *non-rejected* are collectively exhaustive, but probably not mutually exclusive.

Secondly, *necessity*. The strongly non-Tarskian notions of consequence above demand *preservation of non-antidesignatedness*. Clearly, designatedness is a case of non-antidesignatedness, so Beall and Restall's generalized Tarskian consequence is a case of a more general notion which also encompasses Q-consequence and P-consequence. For example, Q-consequence deals with preservation of non-antidesignated values but in such a strong way that it rather forces passing from non-antidesignated values to designated values. Similarly, P-consequence is preservation of non-antidesignated values but in such a weak way that it allows passing from designated values to some non-designated values (but never from designated to antidesignated values!). Let me call this notion of consequence *TMF-consequence* (for Tarski, Malinowski and Frankowski) and define it as follows:

TMF-*consequence*. q is a logical TMF-consequence from premises Γ, in symbols $\Gamma \models^{TMF} q$, if and only if any case in which each premise in Γ is not antidesignated is also a case in which q is not antidesignated. Or equivalently, there is no case in which each premise in Γ is not antidesignated, but in which q fails to be antidesignated. Again equivalently, if there is a case in which q is antidesignated, then at least one premise in Γ is also antidesignated. In short, in the non-Tarskian notions of consequence the non-antidesignatedness of premises necessitates the non-antidesignatedness of conclusions.

Note that TMF-consequence has a dialogical version too, which illustrates its usefulness for evaluating ordinary arguments:
- An argument is TMF-valid if and only if, if the premises are not rejected, then conclusions are also not rejected. Equivalently, if the conclusions are rejected, premises are rejected too. The relevant dialogical properties are those that count as *not rejected*, so there is more than one property that could be forwards-preserved. The preservation of these properties is required, but the direction is not important since *not rejected* determines another property, whose name depends of what the components of the first property are taken to be, which makes the collections of properties mutually exclusive (but probably not collectively exhaustive).

Now, *normativity*. In a monist framework, with only one logic at play, normativity is not so problematic: There is but one canon saying what is rational and what is irrational concerning argumentation. But even under the (generalized) Tarskian framework, there are many logics, each sanctioning as (in)valid different argument-forms as they might be operating on different specifications of cases. So normativity should not be understood from the outset as "what must be done in all cases *simpliciter*" (or "in absolutely all cases"), but as "what must be done in all cases *of a kind k*". Just as it could be irrational, in spite of what bold Tarskian consequence could say, to reason monotonically in cases in which information does not have the required properties to do so (i.e., it is incomplete, not known but merely probable, etc.), inferring according to Q-consequence (accepting a conclusion on the basis that there is no case of the pertinent kind in which the premises are non-rejected but the premise is non-accepted) in some contexts – whether by time pressure, resource-saving, etc. – might be the rational thing to do, instead of waiting until the information is not merely not-rejected but fully accepted. Ampliative inferences have their contexts of application and establish their own sets of norms for evaluating arguments in all the contexts of the pertinent kind.[10]

About *formality*: Beall and Restall are not very precise on this matter, but still it is fairly innocuous given that their notion of logical consequence is supposed to be a generalization of Tarskian insights and the non-Tarskian notions of consequence are generalizations of this, as I have showed. Additionally, the meanings of no other terms than those accepted by the (generalized) notion of Tarskian consequence are in play here, so there is no threat on formality if the original notion is already formal.[11]

6   Another jump into abstraction

Although my case against a generalized Tarskianism like Beall and Restall's has been completed in the previous section, I would like to discuss further issues which could be used to argue against even weaker forms of Tarskianism.

Note that TMF-consequence has the following features:

---

[10] I want to thank Delia Belleri and Maite Ezcurdia for pressing the point regarding normativity, especially the normativity of Q-consequence, and to Cristian Alejandro Gutiérrez-Ramírez, Pau Luque, and Moisés Macías-Bustos for valuable suggestions on the topic.

[11] Recently Restall has stressed the importance of these cognitive, normative notions for the definition of logical consequence and theories, but his approach is different. Rather than studying separate definitions of consequence stressing either of these cognitive notions, Restall studies interactions between them within a single notion of logical consequence to give a more comprehensive notion of *(bi)theory*; (see Restall, 2013).

Fifty Shades of Logical Consequence

(TMF 1) It preserves a certain (kind of) value.

(TMF 2) This preservation has a *direction* ("forwards"): It is from premises to conclusions.

(TMF 3) The preservation in the other direction ("backwards") does not alter the logical consequences already determined by preservation-forward, so it can be dismissed.

Let me say that a relation R *backwards-preserves* a property ψ if and only if for any x, y such that Rxy, if y has the property ψ, x has it too. If the property ϕ forwards-preserved and the property ψ forwards-preserved are collectively exhaustive and mutually exclusive with respect to the properties that x and y might have, then forwards-preservation and backwards-preservation coincide. But D^+- and D^--consequence distinguish directions, in a way that forward-preservation and backward-preservation might not coincide. When the arrangement of values is such that $D^+ \cup D^- \neq \mathcal{V}$ or $D^+ \cap D^- \neq \varnothing$, D^+-consequence and D^--consequence do not coincide. Suppose, as above, that there are three truth values, $\top$, μ, and $\bot$, with the order $\bot < \mu < \top$ and with $\top$ as the only designated value, $\bot$ as the only antidesignated value and μ as neither designated nor antidesignated. Suppose pretty standard truth-conditions for conjunctions and conditionals, say $(p \wedge q) = \inf(p,q)$ and $(p \Rightarrow q) = \top$ iff $p = \inf(p,q)$, otherwise $(p \Rightarrow q) = q$. Then one has that q is a logical D^+-consequence, but not a logical D^--consequence, of $p \wedge (p \Rightarrow q)$.

This suggests a more abstract notion of logical consequence, where the direction is abstracted and that can be called *WS-consequence*[12] stating the following: Let premises Γ and conclusion q be called *sides* of the logical consequence relation and V and V^* certain (kinds of) values. q is a logical WS-consequence from premises Γ, in symbols $\Gamma \models^{WS} q$, if and only if, if one side has values in V, the other has values in V^*.

In spite of the jump into abstraction, we are still in the business of logic. For its application to the evaluation of arguments, let $\mathcal{V}$ be again a collection of dialogical properties:
- An argument is WS-valid if and only if, if one side of the argument has a dialogical property, then the other also has it. No dialogical property is

---

[12]For (Heinrich) Wansing and (Yaroslav) Shramko, who have studied most of the abstract notions of logical consequence presented here, but whose main contribution to this topic has been the explicit distinction between the directions of D^+- and D^--consequence. (See for example Wansing & Shramko, 1989).

especially relevant nor properties are required to have further structure, like being mutually exclusive; neither preservation of dialogical properties is not required nor a specific direction in which they are connected, just that there is some connection between them.

- An argument is accepted-valid if and only if, if the premises are accepted, then the conclusions are also accepted. The relevant dialogical property to be preserved is *accepted*; this preservation has a preferred direction (forwards).

- An argument is rejected-valid if and only if, if the conclusions are rejected, then the premises are also rejected. The relevant dialogical property that has to be preserved is *rejected*; this preservation has a preferred direction (backwards).

Accepted-validity and rejected-validity need not coincide for acceptance and rejection, even if mutually exclusive, are not collectively exhaustive. Similar variations can be obtained using not only dialogical, but also probabilistic, psychological, epistemic or whatnot specifications of consequence, for example.

An argument is unconceivable-valid if and only if, if it is unconceivable for the conclusions to be true, then it is also unconceivable for the premises to be true. The relevant ("psychological") property that has to be preserved is *unconceivability*; this preservation has a preferred direction (backwards).

- An argument is believed-valid if and only if, if the premises are believed, then the conclusions are also believed. The relevant ("epistemic property") property to be preserved is *believed*; this preservation has a preferred direction (forwards).

Other variations are left to the reader.

As to other elements of the core tradition, formality is kept since it is an abstraction from the previous notions, so WS-consequence is formal if the previous notions are, and necessity is also kept since a (kind of) value on one of the relata necessitates some (kind of) value in the other. The normative aspect is inherited from the other notions: It is not rational not rejecting the premises in all cases of a certain kind but rejecting the conclusions at some of those very cases.

To sum up: Traditionally, logics built upon the (generalized) Tarskian notion of logical consequence allow for variation in the cases in which premises and conclusions are evaluated, thus one obtains different particular logics. That is Beall and Restall's logical pluralism. It keeps the ideas that designated and antidesignated values form collectively exhaustive and mutually exclusive collections of values; that truth is the only designated value and that it is the value that must be preserved; that this preservation has a direction, namely forwards-preservation (i.e. from premises to conclusions) and that backwards-preservation yields the same results as forwards-preservation. An attempt to generalize this is allowing other values than truth to be designated and keeping the rest of elements as before. This is still Tarskian (at least generalized, i.e. reflexive and transitive) consequence nonetheless. Further generalization comes from the previous step yet allowing designated and antidesignated values to be either not collectively exhaustive or not mutually exclusive collections of values and allowing for different types of values to be preserved (only demanding that they are not antidesignated). These notions of consequence are no longer Tarskian, since they may be not reflexive (Q-consequence) or not transitive (P-consequence). WS-consequence abstracts the direction, kind of preservation and also the kind of value preserved: Logical WS-consequence is a relation between values, where the order of the relata is undetermined.

Thus far I could say that WS-logical consequence is a relation with a certain direction between kinds of values – the values of the premises and the values of the conclusions – such that if either of them has a kind of value V, then the other has a kind of value V^*. More formally,

WS-*consequence.* Let premises Γ and conclusion q be called *sides* of the logical consequence relation and V and V^* certain (kinds of) values. q is a logical WS-consequence from premises Γ, in symbols $\Gamma \models^{WS} q$, if and only if, if one side has values in V, the other has values in V^*.

Thus, logical consequence is an octuple $LC = \langle \mathcal{V}; V_1, \ldots, V_n; \Vdash; V^P; V^C; CON; \delta, \mathcal{I} \rangle$ where $\mathcal{V}$ is a collection of values; $V_1, \ldots, V_n$ are kinds of values such that each $V_i \subseteq \mathcal{V}$ and $V_1 \cup \ldots \cup V_n = \mathcal{V}$; V^P and V^C are the values for the sides of the relation, premises and conclusions, respectively; $\Vdash$ is a relation on $\mathcal{V}$; CON is the condition of connection in such relation, stating that if one side of the relation has a value of the kind V_i then the other has a value of the kind V_j; δ is the specification of the direction of such connection; finally, $\mathcal{I}$ is the family of indexes at which values for the sides hold or otherwise. The notions of logical consequence discussed above appear naturally as specifications on the parameters of this octuple.

Béziau's notion of logical consequence results from WS-logical consequence when even the nature of the V_i is left unspecified. "Values" still have many "material" connotations, they have a very specific nature, so it would be better to say that they are structures whatsoever. This does not mean the last bit of logicality, in the sense of the core tradition, is lost. This rather means that the last step into abstraction could be done if a better cartography of logicality is provided in a way that makes sense of both the core tradition and the abstraction suggested by Béziau. Let me sketch very roughly how some familiar distinctions could serve for the last step into abstraction.

Priest has argued explicitly and continuously for a distinction between pure logics, certain mathematical structures, and applied logics, pure logics applied to some end (see for example Priest, 2000, 2003), has stressed the importance of distinguishing between logic as a theory and logic as an object of study (Priest, 2003, p. 207) and finally has introduced the very fruitful notion of 'canonical application' (see again Priest, 2003). A *pure logic* would be a kind of mathematical structure, a logical structure. A *logical structure* would be what Béziau says. *Phenomena with a logical structure* would be phenomena with a specific "enriched" logical structure, for example arguments, the flowing of electricity in a circuit, closures on topological spaces, the functioning of a computer program, etc. *Canonical phenomena with a logical structure* would be specific phenomena with logical structures traditionally regarded as belonging to the proper domain of logic, such as formal or informal arguments made by people. Usually the term 'logic' has been used without distinction to designate a theory as well as for designating what the theory is about. For example, 'logic' has been used to name a certain science, roughly the science of right reasoning, as well as for the presumed subject matter of such science, roughly the structure of norms that rule the right reasoning. Some terminology should mirror this distinction. 'Logic' could be used for the study of pure logics, 'applied logic' for the study of empirical phenomena with a logical structure, and 'canonically applied logic' for the study of canonical empirical phenomena with a logical structure. It has been a common practice to reserve the term without adjectives ('geometry', 'arithmetic') for the pure part, and noting a distinction between object and theory prevents most of the problems in the use of the term 'logic'. If there is some reticence in using the word 'logic' for any of the parts, the study of pure logics might be termed rightly "Universal Logic", as Béziau suggested.

7 Conclusions

I reconstructed an argument due to Beall and Restall against non-reflexive and non-transitive logics, which exhibits very well the kind of philosophical background at stake when judging whether something is a logic or not. After expounding a proposal cousin to mine which I consider inconclusive,

Fifty Shades of Logical Consequence

I presented non-Tarskian notions of logical consequence and showed that, formally, they are not very different from the (generalized) Tarskian notion of logical consequence and, moreover, that they can be considered a part of the "core tradition" of logic because they can be useful for the analysis of the inferential relationships between premises and conclusions expressed in arguments we actually employ and have the features of necessity, formality, and normativity. Then I glanced at more abstract notions of consequence which are arguably still in the business of the core tradition in logic, making stronger the case that logicality might not rest on this or that property of the relation of logical consequence. Finally, I must say that I granted too much to the core tradition, in the sense that I made a question of all or nothing the satisfaction of the traditional criteria, when this surely is a matter of degree (as stressed by Paseau (2007)), and this could block the skeptical stance on the normativity side of the more abstract shades of logical consequence, but I tried to play in the hardest side of the tradition to give my ideas a better test.

References

Aliseda-Llera, A. (2006). *Abductive Reasoning: Logical Investigations into the Processes of Discovery and Evaluation*. Dordrecht: Kluwer Academic Publishers.

Avron, A. (1991). Simple Consequence Relations. *Information and Computation, 92*(1), 105–139.

Beall, J. (2009). *Spandrels of Truth*. Oxford: Oxford University Press.

Beall, J., & Restall, G. (2006). *Logical Pluralism*. Oxford: Oxford University Press.

Béziau, J.-Y. (1994). Universal Logic. In T. Childers & O. Majer (Eds.), *Logica '94 – Proceedings of the 8th International Symposium* (pp. 73–93). Prague: Czech Academy of Sciences.

Béziau, J.-Y. (2001). From Paraconsistent Logic to Universal Logic. *Sorites, 12*, 5–32.

Béziau, J.-Y. (2006). 13 Questions about Universal Logic. *Bulletin of the Section of Logic, 35*(2–3), 133–150.

Béziau, J.-Y. (2010). What is a Logic? – Towards Axiomatic Emptiness. *Logical Investigations, 16*, 272–279.

Bremer, M. (2014). Restall and Beall on Logical Pluralism: A Critique. *Erkenntnis, 79*(2), 293–299.

Dummett, M. (1998). Truth from the Constructive Standpoint. *Theoria*, *64*(2–3), 122–138.
Field, H. (2009). Pluralism in Logic. *Review of Symbolic Logic*, *2*(2), 342–359.
Frankowski, S. (2004a). Formalization of a Plausible Inference. *Bulletin of the Section of Logic*, *33*, 41–52.
Frankowski, S. (2004b). p-consequence Versus q-consequence Operations. *Bulletin of the Section of Logic*, *33*, 197–207.
Gabbay, D. M. (1994). Classical vs Non-classical Logics (The Universality of Classical Logic). In D. M. Gabbay, C. J. Hogger, & J. A. Robinson (Eds.), *Handbook of Logic in Artificial Intelligence and Logic Programming-deduction Methodologies* (Vol. 2, pp. 359–500). Oxford: Clarendon Press.
Hardegree, G. M. (2005). Completeness and Super-valuations. *Journal of Philosophical Logic*, *34*(1), 81–95.
Malinowski, G. (1990a). Q-consequence Operation. *Reports on Mathematical Logic*, *24*, 49–59.
Malinowski, G. (1990b). Towards the Concept of Logical Many-valuedness. *Folia Philosophica*, *7*, 97–103.
Mossakowski, T., Diaconescu, R., & Tarlecki, A. (2009). What Is a Logic Translation? *Logica Universalis*, *3*(1), 95–124.
Murzi, J., & Shapiro, L. (in press). Validity and Truth-preservation. In H. G. T. Achourioti F. Fujimoto & J. Martínez-Fernández (Eds.), *Unifying the Philosophy of Truth*. Berlin: Springer.
Paoli, F. (2002). *Substructural Logics. A Primer*. Dordrecht: Kluwer Academic Publishers.
Paoli, F. (in press). Semantic Minimalism for Logical Constants. *Logique et Analyse*.
Paseau, A. (2007). Review: *Logical Pluralism*. *Mind*, *116*(462), 391–396.
Priest, G. (2000). Review of N.C.A. da Costa's *Logiques Classiques et non Classiques*. *Studia Logica*, *64*, 435–443.
Priest, G. (2003). On Alternative Geometries, Arithmetics, and Logics; A Tribute to Łukasiewicz. *Studia Logica*, *74*, 441–468. (Reprinted with some modifications as Chapter 10 of Priest's *Doubt Truth to be Liar*, Oxford: Oxford Clarendon Press, 2006. Quoted from that edition.)
Quine, W. V. (1970). *Philosophy of Logic*. Englewood Cliffs, New Jersey: Prentice Hall.
Read, S. (2003). Logical Consequence as Truth-preservation. *Logique et Analyse*, *183*, 479–493.

Restall, G. (2013). Assertion, Denial and Non-classical Theories. In K. Tanaka, F. Berto, E. Mares, & F. Paoli (Eds.), *Paraconsistency: Logic and Applications* (pp. 81–99). Berlin: Springer.

Ripley, D. (2012). Conservatively Extending Classical Logic with Transparent Truth. *Review of Symbolic Logic*, *5*(2), 354–378.

Tarski, A. (1928). Remarques sur les Notions Fondamentales de la Méthodologie des Mathématiques. *Annales de la Société Polonaise de Mathématiques*, 270–272.

Tennant, N. (1994). The Transmission of Truth and the Transitivity of Deduction. In D. Gabbay (Ed.), *What Is a Logical System?* (pp. 161–178). Oxford: Oxford University Press.

Wansing, H., & Shramko, Y. (1989). Suszko's Thesis, Inferential Many-valuedness and the Notion of Logical System. *Studia Logica*, *88*(1), 405–429. (See also the erratum in volume 89, p. 147, 2008)

Weir, A. (1998). Naïve Set Theory Is Innocent! *Mind*, *107*(428), 763–798.

Wójcicki, R. (1988). *Theory of Logical Calculi. Basic Theory of Consequence Operations*. Dordrecht: Kluwer Academic Publishers.

Zardini, E. (2008). A Model of Tolerance. *Studia Logica*, *90*(3), 337–368.

Luis Estrada-González
Institute for Philosophical Research, National Autonomous University of Mexico
Mexico
E-mail: loisayaxsegrob@gmail.com

Poincaré and Prawitz on Mathematical Induction

YACIN HAMAMI[1]

Abstract: Poincaré and Prawitz have both developed an account of how one can acquire knowledge through reasoning by mathematical induction. Surprisingly, their two accounts are very close to each other: both consider that what underlies reasoning by mathematical induction is a certain chain of inferences by modus ponens 'moving along', so to speak, the well-ordered structure of the natural numbers. Yet, Poincaré's central point is that such a chain of inferences is not sufficient to account for the knowledge acquisition of the universal propositions that constitute the conclusions of inferences by mathematical induction, as this process would require to draw an infinite number of inferences. In this paper, we propose to examine Poincaré's point – that we will call the *closure issue* – in the context of Prawitz's framework where inferences are represented as operations on grounds. We shall see that the closure issue is a challenge that also faces Prawitz's own account of mathematical induction and which points out to an epistemic gap that the chain of modus ponens cannot bridge. One way to address the challenge is to introduce suitable additional inferential operations that would allow to fill the gap. We will end the paper by sketching such a possible solution.

Keywords: Poincaré, Prawitz, mathematical induction

1 Introduction

Poincaré and Prawitz share a common philosophical interest in the nature of reasoning by mathematical induction. For Poincaré, mathematical induction constitutes mathematical reasoning *par excellence* and thereby takes central stage in his philosophical analysis of mathematical reasoning (Poincaré, 1894) as well as in his critical discussion of the role of logic in the foundations of mathematics (Poincaré, 1905, 1906). For Prawitz, mathematical induction is the archetypal example of a deductive reasoning principle that

---

[1]The author would like to thank the participants at Logica 2014 as well as Jean Paul van Bendegem, John Mumma, and Göran Sundholm for stimulating discussions relative to the content of this paper. The author is a doctoral fellow of the Research Foundation Flanders (FWO).

governs inferences that are valid but not *logically* valid – i.e., whose conclusions are not logical consequences of their premises in the Bolzano-Tarski sense – and thereby appears as the main example of application of his recent account of the validity of deductive inferences (Prawitz, 2009, 2012, 2013).

In their respective analyses, Poincaré and Prawitz are both concerned with the *epistemological* dimension of reasoning by mathematical induction. Poincaré (1894) aims to understand the capacity of mathematical induction to *extend* mathematical knowledge,[2] a feature that would distinguish it from syllogistic reasoning which "can teach us nothing essentially new" (Poincaré, 1894, p. 31).[3] Prawitz (2009, 2012, 2013) aims to provide an account of the validity of deductive inferences that would explain how one can acquire knowledge by drawing valid inferences. When applied to mathematical induction, such an account would explain how one can acquire knowledge by drawing mathematical induction inferences.

Interestingly, the two accounts developed by Poincaré and Prawitz in order to explain how reasoning by mathematical induction can generate (new) knowledge are very close to each other. More specifically, both consider that what underlies an inference by mathematical induction – i.e., an inference of the following form,

$$\frac{H(0) \quad H(p) \to H(p+1)}{H(n)}$$

where n and p are arbitrary natural numbers – is a chain of n inferences by modus ponens which can be described in metaphorical terms as 'starting' from the number 0 and 'moving along' the well-ordered structure of the natural numbers up to the number n: the first inference consists in deducing $H(1)$ from the premises $H(0)$ and $H(0) \to H(1)$, the pth inference ($p > 0$) consists in deducing $H(p)$ from $H(p-1)$ and $H(p-1) \to H(p)$, and the nth inference consists in deducing $H(n)$ from $H(n-1)$ and $H(n-1) \to H(n)$.[4] Thus, such an account of reasoning by mathematical induction explains how one can deduce the proposition $H(n)$ for any $n \in \mathbb{N}$ from the two premises $H(0)$ and $H(p) \to H(p+1)$.

---

[2] Poincaré speaks of the "creative virtue" of mathematical reasoning (Poincaré, 1894, p. 32).

[3] See (Detlefsen, 1992) and (Heinzmann, 1995) for interpretations that take as central the idea that the main difference between mathematical reasoning and logical reasoning for Poincaré is an epistemological one.

[4] In this paper, we will refer to this inferential process as the *chaining operation*.

Yet, at this stage, Poincaré's and Prawitz's analyses diverge in an important way. On the one hand, Prawitz considers that, insofar as the above chaining operation allows to reach any $n \in \mathbb{N}$, it allows to establish that $H(n)$ holds for an arbitrary $n \in \mathbb{N}$, and so to conclude – by universal generalization – that the universal proposition $\forall n H(n)$ holds, resulting thereby in the knowledge acquisition of the universal proposition $\forall n H(n)$. On the other hand, Poincaré considers that, even though the above chaining operation allows to establish $H(n)$ for any $n \in \mathbb{N}$, it does not allow to establish the universal proposition $\forall n H(n)$, as this would require to draw an *infinite* number of inferences. Poincaré concludes that, to obtain knowledge of the universal proposition $\forall n H(n)$, an additional element is required and this is precisely where Poincaré appeals to his own notion of *intuition*.

The aim of this paper is to understand what lies behind this divergence. To this end, we will examine what we will call Poincaré's *closure issue* – i.e., the incapacity of the chaining operation to yield knowledge of the universal proposition $\forall n H(n)$ – within Prawitz's framework. We shall see that Prawitz's account of the validity of inferences offers a very suitable framework for formulating precisely the closure issue. In turn, this will allow us to examine the impact of Poincaré's closure issue on Prawitz's own account of mathematical induction. We will see that the closure issue points out to a certain gap in the inferential operations necessary for the agent to acquire knowledge of the universal proposition $\forall n H(n)$. Thus, our analysis will show a mutually fruitful interaction between Poincaré's and Prawitz's analyses of mathematical induction: on the one hand, Prawitz's account of the validity of inferences provides a framework to formulate and analyze in a precise way the main point of Poincaré's analysis; on the other hand, Poincaré's analysis yields a central insight that can benefit directly to Prawitz's own account of mathematical induction.

The paper is organized as follows. We will begin by presenting Poincaré's and Prawitz's accounts of mathematical induction respectively in sections 2 and 3. In section 4, we will present Poincaré's closure issue, we will show how it can be formulated within Prawitz's framework in which inferences are represented as operations on grounds and we will analyze its impact on Prawitz's account of mathematical induction. We shall see that the closure issue points out to an epistemic gap that the chain of inferences by modus ponens alone cannot bridge. One way to overcome this is to introduce additional inferential operations that would allow to fill the gap. In section 5, we will sketch such a possible solution by introducing two additional inferential operations that would allow to fill the gap.

2 Poincaré on mathematical induction

Poincaré is well known for his opposition against philosophical views that attribute a prominent role for logic in mathematics, views that emerge from the early development of modern logic and set theory in the context of the foundations of mathematics at the beginning of the 20th century.[5] What is maybe less known is that Poincaré already wrote a philosophical study on the nature of mathematical reasoning as early as 1894,[6] i.e., before most of the developments in logic and set theory that he will discuss in (Poincaré, 1905, 1906), and *a fortiori* before he came aware of them. A central theme of (Poincaré, 1894) was already to compare and contrast mathematical reasoning with logical reasoning – where logical reasoning was equated by Poincaré with syllogistic reasoning. Poincaré took as the starting point of his study that logical reasoning is epistemically sterile while mathematical reasoning has a certain "creative virtue" or, in other words, that mathematical reasoning can *extend* mathematical knowledge while logical reasoning cannot. His aim was to provide an explanation of this epistemological difference between the two types of reasoning. To this end, Poincaré suggested to analyze mathematical reasoning in the branch of mathematics where: "mathematical thought [...] has remained pure, that is, in arithmetics" (Poincaré, 1894, p. 34). Poincaré continued by examining some of the most elementary proofs of arithmetic, namely the proofs of the basic properties of addition (associativity, commutativity) and multiplication (distributivity, commutativity). Poincaré noticed a uniform principle of reasoning present in all these proofs, and this principle is precisely mathematical induction. Poincaré concluded that:

> If we look closely, at every step we meet again this mode of reasoning, either in the simple form we have just given it, or under a form more or less modified.

---

[5] See (Goldfarb, 1988) for a discussion of the arguments that Poincaré developed in his debate 'against the logicians', among whom figure Cantor, Peano, Russell, Zermelo, and Hilbert. For an overview of Poincaré's work in the philosophy of mathematics and the philosophy of science, we refer the reader to (Heinzmann & Stump, 2014).

[6] We refer here to the paper (Poincaré, 1894) entitled 'Sur la Nature du Raisonnement Mathématique' ('On the Nature of Mathematical Reasoning'). The paper has been reprinted as the first chapter of the book (Poincaré, 1902) entitled *Science et Hypothèse* (*Science and Hypothesis*), which has been translated into English in (Poincaré, 1929). All the quotations from (Poincaré, 1894) in the present paper are taken from the translation provided in (Poincaré, 1929).

> Here then we have the mathematical reasoning *par excellence*, and we must examine it more closely. (Poincaré, 1894, p. 37)

The conception of reasoning by mathematical induction that Poincaré pushes forwards is then spelled out in the following passage:

> The essential characteristic of reasoning by recurrence is that it contains, condensed, so to speak, in a single formula, an infinity of syllogisms.
>
> That this may the better be seen, I will state one after another these syllogisms which are, if you will allow me the expression, arranged in 'cascade.'
>
> These are of course hypothetical syllogisms.
>
> The theorem is true of the number 1.
>
> Now, if it is true of 1, it is true of 2.
>
> Therefore it is true of 2.
>
> Now, if it is true of 2, it is true of 3.
>
> Therefore it is true of 3, and so on. (Poincaré, 1894, p. 37)

Three central points of Poincaré's conception of mathematical induction are presented here. Firstly, Poincaré considers that reasoning by mathematical induction is composed of 'smaller' inferential steps that take the form of *hypothetical syllogisms*. What Poincaré means by hypothetical syllogisms are simply inferences by modus ponens which, in the context of mathematical induction, take the following form:

$$\frac{H(p) \quad H(p) \to H(p+1)}{H(p+1)}$$

where p denotes an arbitrary natural number. Secondly, these hypothetical syllogisms are organized in a 'cascade'. If we call the above inference or hypothetical syllogism $\mathbf{I}_p$, this means that an inference by mathematical induction consists in *chaining* hypothetical syllogisms of the above form in the sequence $\mathbf{I}_1, \mathbf{I}_2, \ldots, \mathbf{I}_p, \ldots$ etc. Thirdly, and this is maybe the most important point, insofar as such a sequence of hypothetical syllogisms goes *ad infinitum*, an inference by mathematical induction requires to draw an *infinity* of hypothetical syllogisms. As we shall see in section 4, it is precisely this appeal to infinity that lies at the heart of the epistemological difference between mathematical reasoning and logical reasoning in Poincaré's view.

Yacin Hamami

3 Prawitz on mathematical induction

In his most recent work (Prawitz, 2009, 2012, 2013), Prawitz has developed a new account of the validity of deductive inferences that aims to explain how one can acquire knowledge by drawing valid inferences. To this end, Prawitz has proposed to re-conceptualize the notions of inference and the validity thereof in a way that such a desideratum comes out as a conceptual truth. More specifically, Prawitz proposes to think of an *inference* as comprising not only a set of premisses and a conclusion, but also an *operation*[7] that acts on *grounds* for the premisses and which hopefully would yield a ground for the conclusion. An inference is then said to be *valid* in case such an operation does indeed yield a ground for the conclusion when applied to grounds for the premisses. Given such a re-conceptualization, the above desideratum becomes immediately fulfilled: if we consider that one has knowledge of a given proposition when one has a ground for it,[8] then by drawing a valid inference for which one has grounds for its premisses, one automatically obtains a ground for its conclusion and thereby acquires knowledge of it. Of course, to complete such an account requires to make precise the notions of operation and ground, and we shall now see how this is spelled out.

In Prawitz's framework, the notion of ground is intimately connected to the one of *meaning*:

> The line that I shall take is [...] roughly that the meaning of a sentence is determined by what counts as a ground for the judgement expressed by the sentence. Or expressed less linguistically: it is constitutive for a proposition what can serve as a ground for judging the proposition to be true. From this point of view I shall specify for each compound form of proposition expressible in first order languages what constitutes a ground for an affirmation of a proposition of that form. (Prawitz, 2009, pp. 191–192)

Prawitz (2009) provides such specifications for the propositions of first-order logic that are formed using conjunction, implication and universal quantification. The specifications for the different compound forms follow the same scheme: one should specify how a ground for the compound

---

[7]We sometimes use equivalently the term *inferential operation*.

[8]This is the view of knowledge that Prawitz focuses on in this context (see Prawitz, 2012, p. 890).

proposition is obtained from its components, and this is achieved by specifying the *grounding operation* associated to each compound form. For instance, in the case of conjunction, Prawitz proposes the following specification: α is a ground for the conjunction $\varphi \wedge \psi$ if and only if $\alpha = \wedge G(\beta, \gamma)$ for some β and γ such that β is a ground for φ and γ is a ground for ψ. In this specification, the *conjunction grounding operation* $\wedge G$, specifies how grounds for propositions of the form $\varphi \wedge \psi$ are formed given grounds for the components φ and ψ.

In order to be able to deal with inferences drawn from assumptions, Prawitz introduces a distinction between *saturated* and *unsaturated* grounds. An unsaturated ground $\alpha(\xi_1, \ldots, \xi_n)$ for a proposition φ under the assumptions $\varphi_1, \ldots, \varphi_n$ is a function that takes as arguments saturated grounds $\beta_1, \ldots, \beta_n$ for $\varphi_1, \ldots, \varphi_n$ and yields a saturated ground $\alpha(\beta_1, \ldots, \beta_n)$ for φ. Furthermore, in order to deal with inferences involving open propositions, an additional distinction is introduced between *closed* and *open* grounds. An open ground $\alpha(x)$ is simply a ground for an open proposition $\varphi(x)$, where the variable(s) in parentheses denotes the free variable(s) in the considered open proposition. Similarly, a closed ground is a ground for a closed proposition. It is important to notice that the two distinctions saturated-unsaturated and closed-open are independent from each other – one can have closed and open saturated grounds as well as closed and open unsaturated grounds.

The distinction between saturated and unsaturated grounds enables to formulate the *implication grounding operation* $\to G$ which specifies how a saturated ground for a proposition $\varphi \to \psi$ is formed from an unsaturated ground $\beta(\xi)$ for φ under the assumption ψ: α is a ground for $\varphi \to \psi$ if and only if $\alpha = \to G_\xi(\beta(\xi))$ where $\beta(\xi)$ is an unsaturated ground for ψ under the assumption φ. Furthermore, the distinction between closed and open grounds enables to formulate the *universal grounding operation* $\forall G$ which specifies how a closed ground for a proposition of the form $\forall x \varphi(x)$ is formed from an open ground $\beta(x)$ for the open proposition $\varphi(x)$: α is a ground for $\forall x \varphi(x)$ if and only if $\alpha = \forall G_x(\beta(x))$ where $\beta(x)$ is an open ground for $\varphi(x)$.

The other key notion of Prawitz's framework is the one of *operation* on grounds. The three grounding operations $\wedge G$, $\to G$ and $\forall G$ that we have just defined are typical examples of operations on grounds. Indeed, those operations can be seen as corresponding closely to certain inference rules, namely the introduction rules for conjunction, implication and universal quantification in Gentzen's system of natural deduction. Thus, we can now see how inferences that are made according to those inference rules are represented and categorized as valid in Prawitz's account. For instance, drawing an inference with premisses φ and ψ and conclusion $\varphi \wedge \psi$ according to the introduction rule for conjunction is represented as performing the operation $\wedge G$ on grounds β and γ respectively for φ and ψ which results in the obtention of a ground $\gamma = \wedge G(\beta, \gamma)$. This inference is categorized as valid insofar as $\gamma = \wedge G(\beta, \gamma)$ is indeed a ground for $\varphi \wedge \psi$ in virtue of the meaning of $\varphi \wedge \psi$, i.e., in virtue of what constitutes a ground for $\varphi \wedge \psi$.

The interest of Prawitz's account lies in its capacity to define more complex inferential operations, and thereby to account for more complex inferences, than these primitive ones. In particular, we now have all the elements to spell out how inferences by *modus ponens* and by *mathematical induction* – the two types of inferences central to the subject matter of this paper – can be represented within this framework. To this end, we should specify the inferential operations associated to these two types of inferences.

Modus ponens is an inference schema with premisses φ and $\varphi \to \psi$, and conclusion ψ. Prawitz (2009, p. 197) defines the inferential operation MP associated to modus ponens by the following equation

$$\mathsf{MP}(\alpha, \beta(\xi)) = \beta(\alpha)$$

where α is a ground for φ and $\beta(\xi)$ is an unsaturated ground for ψ under the assumption φ.[9] We can immediately see that such an inferential operation, when applied to grounds for φ and $\varphi \to \psi$, yields a ground for ψ since, by definition of unsaturated grounds, the result of saturating the unsaturated ground $\beta(\xi)$ by a saturated ground α for φ is a saturated ground $\beta(\alpha)$ for ψ. Modus ponens is thus an inference schema that generates valid inferences since the inferential operation MP, when applied to grounds for premisses of the forms φ and $\varphi \to \psi$, yields a ground for the conclusion ψ.

The inferential operation associated to mathematical induction has been informally described by Prawitz as follows:

> Let us consider the inference form of mathematical induction, in which it is concluded that a sentence $A(n)$ holds for an arbitrary natural number n, having established the induction base

---

[9] Insofar as the second premiss of an inference by modus ponens is a proposition of the form $\varphi \to \psi$, the correct way to write a ground for such a proposition is $\to G_\xi(\beta(\xi))$ where $\beta(\xi)$ is an unsaturated ground for ψ under the assumption φ. For readability reasons, we abuse notation and just write $\beta(\xi)$ instead of $\to G_\xi(\beta(\xi))$.

Poincaré and Prawitz on Mathematical Induction

> that $A(0)$ holds and the induction step that A holds for the successor n' of any natural number n given that A holds for n. The ground for the induction step may be thought of as a chain of operations that results in a ground for $A(n')$ when applied to a ground for $A(n)$. The operation that is involved in this inference form may roughly be described as the operation which, for any given n, takes the given ground for $A(0)$ and then successively applies the chain of operations given as ground for the induction step n times. (Prawitz, 2013, p. 199)

Prawitz (2012, p. 897) provides a precise formulation of this informal description. In order to state it, it will be useful to introduce some convenient notations. We assume that we are working with the language of first-order Peano Arithmetic. We denote variables of the language by x, y, and constants or numerals by n, p.[10] Prawitz (2012) states mathematical induction as an inference schema with premises $H(0)$ and $H(y) \rightarrow H(y+1)$ and conclusion $H(n)$. We introduce the following notations for grounds of the relevant proposition forms, where '$\alpha : \varphi$' should be read as 'α is a ground for φ':[11]

α_0	:	$H(0)$	$\beta_0(\xi_0)$	:	$H(1)$ under the assumption $H(0)$
α_n	:	$H(n)$	$\beta_p(\xi_p)$	:	$H(p+1)$ under the assumption $H(p)$
$\alpha(x)$	:	$H(x)$	$\beta(y, \xi_y)$	:	$H(y+1)$ under the assumption $H(y)$

We can now state precisely the inferential operation of mathematical induction following (Prawitz, 2012, p. 897):

$$\mathsf{IND}_{\xi y}^n(\alpha_0, \beta(y, \xi_y)) = \begin{cases} \alpha_0 & \text{if } n = 0 \\ \mathsf{MP}\left(\mathsf{IND}_{\xi y}^{n-1}(\alpha_0, \beta(y, \xi_y)), \beta_{n-1}(\xi_{n-1})\right) & \text{if } n > 0 \end{cases}$$

This definition by recursion aims to capture the idea that the inferential operation of mathematical induction consists in a chain of applications of modus ponens as informally described in the previous quote as well as in the introduction of this paper, and corresponding to Poincaré's notion of 'cascade'. Prawitz (2012) shows that the operation $\mathsf{IND}_{\xi y}^n$ does indeed yield a ground

---

[10] We assume that we have numerals in the language available to denote all the natural numbers. We will use the same notation for the numeral and the number it refers to.

[11] For Prawitz, "Grounds are naturally typed by the propositions they are grounds for" (Prawitz, 2009, p. 193). This aspect is rendered here by introducing specific notations for the type of grounds associated to the different proposition forms that will be involved in the present discussion.

for $H(n)$ given grounds α_0 for $H(0)$ and $\beta(y, \xi_y)$ for $H(y+1)$ under the assumption $H(y)$, and this for any $n \in \mathbb{N}$. As we shall now see, Poincaré agrees on this point, but contests that we can infer from this that the agent can obtain a ground for the universal proposition $\forall x H(x)$ in this way.

4 The closure issue

Both Poincaré and Prawitz consider that the chaining operation underlies inferences by mathematical induction. Furthermore, Poincaré agrees with Prawitz that the chaining operation allows one to acquire knowledge of the proposition $H(n)$, and this for any $n \in \mathbb{N}$, as witnessed by the following passage:

> If instead of showing that our theorem is true of all numbers, we only wish to show it true of the number 6, for example, it will suffice for us to establish the first 5 syllogisms of our cascade; 9 would be necessary if we wished to prove the theorem for the number 10; more would be needed for a larger number; but, however great this number might be, we should always end by reaching it, and the analytic verification would be possible. (Poincaré, 1894, p. 38)

This observation of Poincaré – that the chaining operation allows one to reach knowledge of the proposition $H(n)$ however great n might be – can be shown formally within Prawitz's framework:

Proposition 1 *If an agent has at her disposal (i) a ground α_0 for $H(0)$ and a ground $\beta(y, \xi_y)$ for $H(y+1)$ under the assumption $H(y)$ and (ii) the inferential operation* MP, *then the agent can obtain a ground for $H(n)$ by drawing n inferences, and this for any $n \in \mathbb{N}$.*

Proof. This can be shown by a simple induction on n.
Base case: The agent can obtain a ground for $H(1)$ by applying MP to the grounds α_0 for $H(0)$ and $\beta_0(\xi_0)$ for $H(1)$ under the assumption $H(0)$, as this would yield the ground $\beta_0(\alpha_0)$ which is a ground for $H(1)$.
Induction case: Assume that the agent can obtain a ground for $H(n)$ by drawing n inferences. Let α_n be such a ground for $H(n)$. The agent possesses then a ground α_n for $H(n)$ and a ground $\beta_n(\xi_n)$ for $H(n+1)$ under the assumption $H(n)$ (obtained from $\beta(y, \xi_y)$ by universal instantiation). Thus, the agent can apply MP to α_n and $\beta_n(\xi_n)$, obtaining thereby a ground $\beta_n(\alpha_n)$ for $H(n+1)$. □

However, Poincaré considers that the chaining operation is insufficient to reach knowledge of the universal proposition $\forall x H(x)$, as Poincaré puts it:

> And yet, however far we thus might go, we could never rise to the general theorem, applicable to all numbers, which alone can be the object of science. To reach this, an infinity of syllogisms would be necessary; it would be necessary to overleap an abyss that the patience of the analyst, restricted to the resources of formal logic alone, never could fill up. (Poincaré, 1894, p. 38)

Thus, what prevents the chaining operation to reach knowledge of the universal proposition $\forall x H(x)$, according to Poincaré, is precisely that this would require to draw an *infinite* number of inferences, and this is what we refer to as the *closure issue*. As expressed in the last sentence of the above passage, Poincaré attributes this limitation to the resources of logical reasoning.

The closure issue can be established formally within Prawitz's framework. To this end, we first need to prove the following lemma:

Lemma 1 *If an agent only has at her disposal (i) a ground α_0 for $H(0)$ and a ground $\beta(y, \xi_y)$ for $H(y+1)$ under the assumption $H(y)$ and (ii) the inferential operation* MP, *then the only grounds the agent can obtain by drawing n inferences are grounds for the propositions $H(1), \ldots, H(n)$.*

Proof. The proof goes by induction on the number of inferences n.
Base case: We have to show that the only ground the agent can obtain by drawing a single inference is a ground for $H(1)$. We first notice that, in order to apply the inferential operation MP, the agent needs to be in possession of a ground for a proposition φ and a ground for a proposition of the form $\varphi \to \psi$. By assumption, the two grounds that the agent has initially are grounds α_0 for $H(0)$ and $\beta(y, \xi_y)$ for $H(y+1)$ under the assumption $H(y)$. Thus, the only grounds to which the agent can apply MP are α_0 for $H(0)$ and $\beta_0(\xi_0)$ for $H(1)$ under the assumption $H(0)$, and this would yield $\beta_0(\alpha_0)$ which is a ground for $H(1)$. Hence, the only ground the agent can obtain by drawing a single inference is a ground for $H(1)$.
Induction case: Assume that the only grounds the agent can obtain by drawing n inferences are grounds for the propositions $H(1), \ldots, H(n)$. Again, in order to apply the inferential operation MP, the agent needs to be in possession of a ground for a proposition φ and a ground for a proposition of the form $\varphi \to \psi$. By induction hypothesis, the only grounds the agent can

possess after n inferences are grounds for the propositions $H(1), \ldots, H(n)$, in addition to the grounds α_0 for $H(0)$ and $\beta(y, \xi_y)$ for $H(y+1)$ under the assumption $H(y)$ that the agent possessed initially. Thus, the only pairs of grounds to which the agent can apply the inferential operation MP are $(\alpha_0, \beta_0(\xi_0)), \ldots, (\alpha_n, \beta_n(\xi_n))$. It follows that the ground the agent can obtain by applying MP in this case is necessarily a ground for one of the propositions $H(1), \ldots, H(n+1)$. □

Poincaré's closure issue is a direct consequence of Lemma 1 and can be formalized as follows:

Proposition 2 (**Closure Issue**) *If an agent only has at her disposal (i) a ground α_0 for $H(0)$ and a ground $\beta(y, \xi_y)$ for $H(y+1)$ under the assumption $H(y)$ and (ii) the inferential operation* MP, *then the agent cannot obtain a ground $\alpha(x)$ for $H(x)$ by drawing a finite number of inferences.*[12]

Proof. This follows directly from Lemma 1. □

This proposition is the direct formalization of Poincaré's closure issue within Prawitz's framework. Given that Prawitz's informal description of the inferential operation of mathematical induction in the quote of section 3 is similar to the 'cascade' operation provided by Poincaré, the closure issue does apply as well to Prawitz's account of mathematical induction. Indeed, we can even show that the inferential operations $\text{IND}_{\xi_y}^n$ defined in section 3 provide insufficient means to obtain a ground for the universal proposition $\forall x H(x)$:

Proposition 3 *If an agent only has at her disposal (i) a ground α_0 for $H(0)$ and a ground $\beta(y, \xi_y)$ for $H(y+1)$ under the assumption $H(y)$ and (ii) the inferential operations $\text{IND}_{\xi_y}^n$ for all $n \in \mathbb{N}$, then the agent cannot obtain a ground $\alpha(x)$ for $H(x)$ by drawing a finite number of inferences.*

Proof. This can be proved by induction in a similar fashion as in Lemma 1 and Proposition 2. □

---

[12] Since one can easily pass from a ground $\alpha(x)$ for $H(x)$ to a ground $\beta = \forall G_x(\alpha(x))$ for $\forall x H(x)$ and the other way around by applying the inferential operations corresponding respectively to the introduction and elimination rules for universal quantification in natural deduction, we do not necessarily distinguish between the two cases in the present discussion.

5 A possible solution

One could expect that Poincaré's conclusion from the closure issue is a skeptical one – namely that it is simply impossible to acquire knowledge of the universal proposition $\forall x H(x)$. This is not the case. Poincaré does think that it is possible to acquire knowledge of the universal proposition $\forall x H(x)$. His conclusion from the closure issue is rather that *logical* or *syllogistic* reasoning provides *insufficient* means to acquire knowledge of $\forall x H(x)$. Poincaré explains then our capacity to acquire knowledge of $\forall x H(x)$ by appealing to another mean of knowledge acquisition, namely *intuition*:

> Why then does this judgment force itself upon us with an irresistible evidence? It is because it is only the affirmation of the power of the mind which knows itself capable of conceiving the indefinite repetition of the same act when once this act is possible. The mind has a direct intuition of this power, and experience can only give occasion for using it and thereby becoming conscious of it. (Poincaré, 1894, p. 39)

Unfortunately, this passage contains all what Poincaré has to say on the role of intuition in mathematical induction in (Poincaré, 1894). In the remainder of this paper, our goal is not to propose a detailed analysis of what Poincaré could mean here by intuition. Rather, we propose to sketch a solution to the closure issue based on a straightforward interpretation of the above passage.

Poincaré says that the mind has the capacity to conceive "the indefinite repetition of the same act when once this act is possible". Taken at face value, this statement states that the mind is in possession of an inferential operation that allows to take the limit or closure, so to speak, of the chaining operation. This could be formalized in Prawitz's framework by introducing an inferential operation $\mathsf{IND}_{\xi y}^{\infty}$ which would correspond to what would be achieved by carrying out the chaining operation *ad infinitum*:[13]

$$\mathsf{IND}_{\xi y}^{\infty}(\alpha, \beta(y, \xi_y)) = \lim_{n \to \infty} \mathsf{IND}_{\xi y}^{n}(\alpha, \beta(y, \xi_y)) = \bigcup_{n \in \mathbb{N}} \mathsf{IND}_{\xi y}^{n}(\alpha, \beta(y, \xi_y))$$

Yet, this inferential operation would still be insufficient to acquire a ground for the universal proposition $\forall x H(x)$, as shown by the following proposition:

---

[13] We use here the set-theoretic notion of limit. For readability reasons, we identify the ground $\mathsf{IND}_{\xi y}^{n}(\alpha, \beta(y, \xi_y))$ with the singleton $\{\mathsf{IND}_{\xi y}^{n}(\alpha, \beta(y, \xi_y))\}$.

Yacin Hamami

Proposition 4 *If an agent only has at her disposal (i) a ground α_0 for $H(0)$ and a ground $\beta(y, \xi_y)$ for $H(y+1)$ under the assumption $H(y)$ and (ii) the inferential operation $\mathsf{IND}_{\xi y}^\infty$, then the agent cannot obtain a ground $\alpha(x)$ for $H(x)$ by drawing a finite number of inferences.*

Proof. The operation $\mathsf{IND}_{\xi y}^\infty$ only produces grounds for propositions of the form $H(n)$, and so will never be able to output a ground for a proposition of the form $H(x)$. This can be shown by a straightforward induction on the number of inferences in a similar fashion as in the proof of Lemma 1. □

Nevertheless, by carrying out the operation $\mathsf{IND}_{\xi y}^\infty$ on grounds for $H(0)$ and $H(y) \to H(y+1)$, one would already be in possession of grounds for *all* the propositions of the form $H(n)$ with $n \in \mathbb{N}$. To reach a ground for $\forall x H(x)$, it suffices to introduce a further inferential operation which would take as inputs the grounds $(\alpha_n)_{n \in \mathbb{N}}$ for $(H(n))_{n \in \mathbb{N}}$ and which would output a ground for $H(x)$. Surprisingly, such an operation corresponds exactly to what is known as the ω-Rule in the field of proof theory. This inferential operation would take grounds $(\alpha_n)_{n \in \mathbb{N}}$ respectively for the propositions $(H(n))_{n \in \mathbb{N}}$ and be defined by the following equation:

$$\omega\text{-R}\left((\alpha_n)_{n \in \mathbb{N}}\right) = \alpha(x)$$

We can now show that, when the agent disposes of the inferential operations $\mathsf{IND}_{\xi y}^\infty$ and ω-R, she has the capacity to acquire knowledge of $\forall x H(x)$:

Proposition 5 *If an agent has at her disposal (i) a ground α_0 for $H(0)$ and a ground $\beta(y, \xi_y)$ for $H(y+1)$ under the assumption $H(y)$ and (ii) the inferential operations $\mathsf{IND}_{\xi y}^\infty$ and ω-R, then the agent can obtain a ground $\alpha(x)$ for $H(x)$ by drawing a finite number of inferences.*

Proof. By applying the inferential operation $\mathsf{IND}_{\xi y}^\infty$ to α_0 and $\beta(y, \xi_y)$, the agent will obtain the set of grounds $\bigcup_{n \in \mathbb{N}} \mathsf{IND}_{\xi y}^n (\alpha, \beta(y, \xi_y))$. Given that $\mathsf{IND}_{\xi y}^n (\alpha, \beta(y, \xi_y))$ is a ground for $H(n)$, the agent is then in possession of a ground α_n for $H(n)$ and this for all $n \in \mathbb{N}$. The agent can then apply the operation ω-R to $(\alpha_n)_{n \in \mathbb{N}}$, obtaining thereby a ground $\alpha(x)$ for $H(x)$. □

6 Conclusion

Poincaré's and Prawitz's accounts of mathematical induction are facing the same challenge raised by the closure issue. This challenge must be addressed for the two accounts to reach their objective, namely to account for knowledge acquisition through reasoning by mathematical induction. We have provided a possible solution to this challenge by introducing two inferential operations $\text{IND}_{\xi y}^{\infty}$ and ω-R. However, this would qualify as a potential solution only insofar as it is *epistemologically acceptable* for the agent to possess and carry out such inferential operations. To determine whether this is so requires an epistemological analysis that is left to further investigations.

References

Detlefsen, M. (1992). Poincaré against the Logicians. *Synthese*, *90*(3), 349–378.

Goldfarb, W. (1988). Poincaré against the Logicists. In W. Aspray & P. Kitcher (Eds.), *History and Philosophy of Modern Mathematics* (pp. 61–81). Minneapolis: University of Minnesota Press.

Heinzmann, G. (1995). *Zwischen Objektkonstruktion und Strukturanalyse. Zur Philosophie der Mathematik bei Jules Henri Poincaré*. Göttingen: Vandenhoek & Ruprecht.

Heinzmann, G., & Stump, D. (2014). Henri Poincaré. In E. N. Zalta (Ed.), *The Stanford Encyclopedia of Philosophy* (Spring 2014 ed.). http://plato.stanford.edu/archives/spr2014/entries/poincare/.

Poincaré, H. (1894). Sur la Nature du Raisonnement Mathématique. *Revue de Métaphysique et de Morale*, 2, 371–384. Reprinted in (Poincaré, 1902). Cited as it appears in the English translation (Poincaré, 1929).

Poincaré, H. (1902). *La Science et l'Hypothèse*. Paris: Ernest Flammarion.

Poincaré, H. (1905). Les Mathématiques et la Logique. *Revue de Métaphysique et de Morale*, *13*(6), 815–835.

Poincaré, H. (1906). Les Mathématiques et la Logique (Suite et Fin). *Revue de Métaphysique et de Morale*, *14*(6), 17–34.

Poincaré, H. (1929). *The Foundations of Science: Science and Hypothesis, The Value of Science, Science and Method*. New York and Garrison, N.Y.: The Science Press.

Prawitz, D. (2009). Inference and Knowledge. In M. Peliš (Ed.), *Logica Yearbook 2008* (pp. 175–192). London: College Publications.

Prawitz, D. (2012). The Epistemic Significance of Valid Inference. *Synthese*, *187*(3), 887–898.

Prawitz, D. (2013). Validity of Inferences. In M. Frauchiger (Ed.), *Reference, Rationality, and Phenomenology: Themes from Føllesdal* (pp. 179–204). Heusenstamm: Ontos Verlag.

Yacin Hamami
Centre for Logic and Philosophy of Science, Vrije Universiteit Brussel
Belgium
E-mail: yacin.hamami@vub.ac.be

Points, Gunk, Boundaries, and Contact

GEOFFREY HELLMAN AND STEWART SHAPIRO

Abstract: We look at the metaphysical difficulties encountered by thinking of space or space-time as composed of points.

Keywords: Boundaries, contact, points, Hudson, metaphysics, objects, receptacles, Varzi, Suárez, Brentano, Bolzano, Whitehead

Some analytic metaphysicians have occupied themselves with the nature (or the possible nature) of space and time (or space-time), and with the relationship between physical objects and the regions of space or space-time they occupy. Some of the issues concern the boundaries of objects and the notion of contact. Much, but not all, of the metaphysical literature just assumes, without comment, that the underlying space or space-time is punctiform, following the now standard Dedekind-Cantor picture.[1] That is, many writers assume that space or space-time is composed of points – it is structured like $\mathbb{R}^3$ or $\mathbb{R}^4$. Physical objects thus occupy regions that are nothing but sets of points. Many of the authors then get tangled up over what to say about point-sized regions, or even point-sized objects, and what to say about boundaries and contact, given that the boundaries are themselves nothing but sets of points. The purpose of this note is to give a somewhat biased overview of some of this literature, arguing that many of the issues are much easier to negotiate if we assume a "gunky" regions-based space or space-time (see, for example Hellman & Shapiro, 2012, 2013).

1 The questions; the perspective

One primary question concerns the *nature* of space or space-time. We will not address issues concerning whether it is Euclidean. We take that to be a matter for the relevant sciences. Our questions concern points.

---

[1] There are some more or less dismissive references to point-free spaces, often with reference to Tarski (1983) or, more likely, Whitehead (e.g. 1917, 1919, Part III, or 1929, Part IV, Chapter 2). Some authors do take the possibility of gunky space (or space-time) seriously.

Geoffrey Hellman and Stewart Shapiro

One other preliminary concerns just what sort of "objects" we are talking about. Most metaphysicians are aware of the fact that physical objects in the actual world do not have the sort of boundaries envisioned in their metaphysical theorizing. They know, perhaps on authority from their science educations, that tables, baseballs, planets, and the like, are mostly empty or nearly empty space. Ordinary, garden-variety objects contain molecules that are themselves composed of atoms; and each atom has a nucleus around which electrons orbit (at least on one crude model). With quantum mechanics, perhaps understood second-hand, our metaphysicians know that subatomic particles like electrons simply do not have sharp boundaries. They are more like probability densities that fade off over space (at least on some understandings). Contemporary metaphysicians also know that actual physical objects do not come into contact with each other in anything like the manner of their theorizing. There are forces that repel the objects long before they can actually touch each other (assuming that touching even makes sense for probability densities).

So what are these contemporary metaphysicians talking about, concerning boundaries, contact, and the like? There are two different orientations. Some of the literature explicitly takes itself to be exploring logical space (so to speak). Our theorists ask what boundaries and contact are like (or would be like) in those possible worlds where physical objects do have precise locations, and where objects do come into direct and full contact with each other (and not just "near contact", or overlapping probabilities, or something like that). As noted, members of this group are fully aware that the actual world is not one of those worlds; they profess interest in the other worlds.

One underlying theme seems to be that assumptions concerning mereology – the part-whole relationship – entail constraints on what location and contact are. So, perhaps the best way to put this orientation is that the metaphysicians explore what is possible and what is necessary given that such-and-such mereological theory holds, and given that such intuitive principles are correct. They take themselves to be articulating entailments that illuminate connections between various mereological assumptions and the structure of space or space-time.[2]

A second orientation is that our questions are addressed in what P. F. Strawson (1959, p. 9) calls "descriptive metaphysics", a project of capturing "the actual structure of our thought about the world". Descriptive metaphysics aims "to lay bare the most general features of our conceptual

---

[2]Thanks to Gabriel Uzquiano here.

structure", and not to revise this conceptual structure – and presumably not to figure out how the world really is.

Who is the "our" here? *Whose* thought is the target of this literature? *Whose* conceptual structure is the descriptive metaphysician after? Like above, it is not that of the scientist, or the scientifically informed layperson, certainly not Penelope Maddy's (2007) Second Philosopher. *That* theorist uses the findings of every relevant science help her figure out what the world is like, including the structure of space-time, the nature of physical objects, the extent to which such objects have determinate boundaries, and the ways they manage to interact with each other. Our descriptive metaphysicians do not take themselves to be doing *that*. Presumably, they are exploring "our" ordinary, everyday concepts of boundaries and contact. They want to figure out what is behind ordinary talk of whether a given stone has a rough or smooth surface, or an umpire's decision concerning whether a given baseball came into contact with a given bat, and thus whether it is a wild pitch or a foul ball.

One might take descriptive metaphysics to just be a study of the semantics of the English words "boundary", "surface", "touch", and "contact", or perhaps a preliminary to such study. In that case, we might get different results if we focused instead on how similar words behave in different languages. Strawson (1959, pp. 9–10) writes that up "to a point, the reliance upon a close examination of the actual use of words is the best, and indeed the only sure, way in philosophy. But the discriminations we can make, and the connections we can establish in this way, are not general enough and not far-reaching enough to meet the full metaphysical demand for understanding". The relation between Strawson's descriptive metaphysics and what Emmon Bach (1986) calls "natural language metaphysics" is interesting, but would take us far afield to explore.

The two orientations – exploring logical space and articulating descriptive metaphysics – are sometimes combined. John Hawthorne (2000, p. 623) presents a thought experiment involving an infinite series of walls placed parallel to each other in a finite amount of space (and so all but finitely many of the walls are much thinner than the supposed diameter of any molecule). He anticipates an objection:

> Some may complain that the world is too distant to be worth being interested in. Actual walls do have extra repulsive forces, don't get to be rigid and impenetrable at any thickness, and so on. Such a reaction is far too hasty. Distant worlds can

often be either revealing or therapeutic with regard to our actual conceptual scheme.

Hawthorne thus seems to hold that by examining what seems to happen in, or, better, what he and others are inclined to say concerning, bizarre, physically impossible situations – the playground of our first orientation above – we learn something about "our actual conceptual scheme", the main goal of descriptive metaphysics.

The enterprise thus seems to presuppose that there is such a thing as "our actual conceptual scheme", and that is monolithic, consistent, and applicable in any and all situations. This presupposition is of-a-piece with what Mark Wilson (2006) derides as the "classical picture", the view that the concepts we deploy are precisely delimited in all possible situations – that in any conceivable situation, there is one and only one correct way to apply our concepts.

An opposing view begins with the observation that our own "concepts" evolved to help us negotiate the world we find ourselves in, and they are applicable, at most, to the sorts of situations that come up here. It would be something of a miracle if *any* of the everyday concepts we deploy could somehow apply smoothly, directly, and univocally to any conceivable situation whatsoever. Moreover, as new situations arise, and as new deep theories of the world emerge, sometimes our concepts evolve. We sometimes have to improvise. That may take us into what Strawson calls "revisionary metaphysics".

Of course, we do not intend to resolve these matters of philosophical methodology here. With this background in place, let us turn to the issues at hand.

2 Boundaries and contact

Although we wish to avoid the question of whether to understand physical objects as three dimensional or four dimensional (or something else), it is tedious to keep repeating phrases like "space or space-time". For the sake of simplicity, we will phrase our questions and accounts in terms of three (or fewer) dimensions. Not much turns on this (or, to be precise, we will avoid matters that do turn on this).

It is common to define a *receptacle* to be a chunk of space that a physical object can exactly occupy. If a given receptacle is occupied by a given object, then no part of the object lies outside the receptacle, and every part

Points, Gunk, Boundaries, and Contact

of receptacle is occupied by a part of the object. Of course, articulating this more precisely depends on what counts as a "part" of a physical object and, more central here, what counts as a "part" of a region of space. These are matters of mereology. It seems to be assumed that each object has a unique receptacle.

Our first and, indeed, primary question concerns just which chunks of space are (or can be) receptacles. On a gunky, regions-based account of space, a natural thesis would be that *any* region is a receptacle. If we wish to avoid what Richard Cartwright (1975) calls "scattered objects", like a chess configuration, a university campus, or The State of Hawaii, then we can restrict attention to connected regions. If we wish to avoid objects that are infinitely large, relative to, say, a given stone, then we can restrict attention to bounded regions.

On the more standard Dedekind-Cantor accounts of space, regions are sets of points. So the task at hand is determine which sets of points are receptacles. Is any non-empty set of points a receptacle? Or any Lebesgue measurable set (to avoid Tarski-Banach worries)? Or perhaps only open sets, or only regular open sets, or only closed sets, or only regular closed sets ... are receptacles. Opinions vary.

Once we decide or figure out what the receptacles are, we then have to say something about what the boundaries of the objects are (if they have boundaries), and what it is for objects to be in contact with each other. Presumably, these questions are to be answered in terms of the receptacles the objects occupy. Consider, for example, an ice-cube that exactly occupies a closed cube. Then, it seems, the *object* – the ice cube itself – somehow includes its boundary. The ice-cube itself has an interior and also a boundary. If, instead, an ice-cube exactly occupies an open cube, then it – the ice-cube – does not include its boundary. Every point occupied by the ice-cube is contained in a neighborhood, say a sphere, that lies properly *inside* the ice-cube, not on one of its sides.

Can we really make physical sense of, say, two ice cubes, one that contains its boundary and one that does not? Well, perhaps there are different kinds of objects, some of which contain their boundaries, and some of which do not. Questions like these become pressing if one thinks of space as composed of points in the now standard Dedekind-Cantor manner.[3]

---

[3] This is not to say that there are no issues concerning boundaries and contact on a regions-based account of space. We briefly address such issues below.

Geoffrey Hellman and Stewart Shapiro

When it comes to receptacles in a Dedekind-Cantor space, Hud Hudson (2005) is the most liberal. He claims that *any* non-empty set of points can be a receptacle. Thus he envisions point-sized objects, scattered objects whose receptacles are finite or countably infinite sets of points, objects in 3-space that are planar squares, and the like. So, a fortiori, Hudson envisions open objects, those whose receptacles are open sets of points, and he envisions closed objects, those whose receptacles are closed sets of points.

Among other things, Chapter 3 of (Hudson, 2005) is devoted to an analysis of the notion of *contact*, of what it is for two objects to touch each other (and, it seems, for what it is for an object to touch itself). One preliminary proposal is that two objects are in contact just in case (a part of) the receptacle for one of the objects shares a boundary point with (a part of) the receptacle of the other. So two open spheres that are next to each other, separated by a single point, would qualify as in being contact, since the point between them is a common boundary – even though that point is not a member of either receptacle. Against this, however, Hudson takes it as a criterion of "perfect contact" that "whenever two objects stand in that relation, it is not even possible that any two objects be more deserving of the description 'in perfect contact' than they". Our two open spheres fail that test. The spheres are separated by a point. Indeed, one can get an entire plane between them (and so, on Hudson's view, one can get an entire *object* between them). Contrast this with an open sphere that is next to a closed sphere. These two objects are supposedly "more deserving of the description 'in perfect contact'" than the open spheres (not in terms of distance, of course), since we can't get *anything* to come between our second pair of spheres, not even a point or a point-sized object. So the early thought about contact is rejected.

Hudson then gives an extended series of thought experiments, involving things like "objects" that consist of a countable infinity of point-sized parts that converge to a point along a straight line (but does not include the limit point), along with zero, one, two, and three dimensional objects that occupy the missing point. Hudson restricts attention to those objects which do not, and, indeed, cannot interpenetrate, in the sense that they do not, or cannot, share any parts, including point-sized parts. His considered definition of contact, the fifth candidate, is this (p. 65):

Necessarily, x touches y if and only if $\exists r_1, \exists r_2, \exists w, \exists v, \exists p$ (i) w is a part of x, whereas v is a part of y; (ii) w exactly occupies r_1, whereas v exactly occupies r_2; (iii) p is a boundary point of both r_1 and r_2; (iv) p is a member of exactly one of r_1 and r_2; and (v) $w \neq v$.

We would think that our everyday concepts of "contact" and "touching" – the relevant bits of "our conceptual scheme" – are just not equipped to deal with things like "physical objects" consisting of a countable infinity of point-sized parts. The everyday notions evolved to deal with things like rocks, baseballs, and bats, not such bizarre entities. Of course, one can always speculate as to how our everyday concepts could be, or should be, extended to cover such bizarre things, but one can also wonder whether there is a unique way to extend the notions, consistent with the meaning of the terms. Hudson seems to presuppose that there is such a unique way to apply our notions to such "objects", or at least he takes that as a goal of the enterprise. One can also question the purpose of the extension, or of this piece of metaphysics. Why does it matter what we say concerning whether Hudson's "objects" are in contact or not. Our notions seem to work fine for the objects we encounter. Perhaps Hudson accepts Hawthorne's (2000, p. 623) claim that "[d]istant worlds can often be either revealing or therapeutic with regard to our actual conceptual scheme". Presumably, he takes the "distant worlds" here – those with point-sized objects – to be revealing.

Be that as it may, Hudson (Chapter 3, §6) articulates a "knot" for which he has no satisfying resolution, by his own lights. Notice, first, that on Hudson's view, certain kinds of objects simply *cannot* touch each other (in his articulated sense of "perfect contact"). Two open cubical objects, for example, cannot share a boundary point occupied by one of them, for the simple reason that neither of them can contain any of its boundary point-sized parts. However, we can know how close two such objects can get to each other – they can get in position where only boundary points separate them. In contrast, consider two closed cubical objects. They also cannot touch each other, but for a different reason. If they did touch, they would have to have a common boundary point, but, since the cubes are closed, this point would have to be occupied by *both* objects, and so they would interpenetrate. On Hudson's metaphysics, this is not possible. He insists that in at least some cases, the receptacles of distinct objects cannot overlap, even on a single point.

So here is the "knot": Suppose that, say, two closed cubical objects are moving toward each other at a constant velocity. Suppose they are oriented so that a side of one of them is always parallel to a side of the other. Let the two objects start out exactly ten meters from each other, with one of them moving at one meter per second relative to the other. What happens as we approach the ten second limit? The objects cannot end up touching, since they cannot touch (as a matter of metaphysical necessity). But no matter how close they get, the objects can always get closer without bumping into each other. So what does happen? Of course, in the actual world, no object can exactly occupy a closed region. Moreover, the particles that make up each object exhibit repulsive forces that prevent them from touching. But here we are (supposedly) exploring metaphysical possibilities here, not doing actual-world physics. Hudson explains:

> ... with the exception of pairing a closed object ... with an open object ..., any other pairing can be so oriented that both interpenetration and touching are not options; the items paired must ... either slow down, jump, stop, turn, or otherwise deviate from their courses ... Given the discussion above, we can clearly explain *that* such deviant behavior will occur, but just what explains *why*? ... *Why* must approaching objects change speed or direction? I don't know – that's why the title of this section advertises a knot rather than a knot untied. (p. 82)

Clearly, Hudson encounters this "knot" only because of his super-liberal views on receptacles and, thus, on what kinds of objects are possible. Other metaphysicians are not as liberal as Hudson in allowing just any set of points to constitute a receptacle – a region that can be exactly occupied by a physical object. But at least some of the difficulties associated with points and such extend beyond this liberal orientation. Indeed, many such views face versions of Hudson's "knot".

Achille Varzi (1997) provides a nice, mathematically informed instance of descriptive metaphysics, focused on matters of contact and boundaries. The article opens thus:

> The world of everyday experience is mostly a world of physical things separated in various ways from their environment: things with surfaces, skins, crusts, boundaries of some sort. These may not always be sharply defined, or so some would argue. Clouds, for instance, seem to have hazy boundaries, ... But, generally speaking, boundaries seem to belong to the palette of basic ontological tools with which we commonly describe the middle-size reality of ordinary experience. ... This general picture is so natural and pervasive that it is hard to deny boundaries a

central place in our conceptual scheme. Boundaries are intrinsically connected with whatever entities they bound – in a way boundaries are even more important, since they are the first and primary things with which we seem to be directly acquainted. At the same time, the folk theory of boundaries is far from complete, and indeed several puzzles threaten any attempt to provide a comprehensive and systematic account of this notion. For instance, common sense is unaccustomed to the point-set topological distinction between open and closed entities. But then, how are the spatiotemporal relations between an extended object and its complement to be explained? And how is the relation of contact between contiguous substances to be accounted for? (p. 26)

Much of Varzi's article deals with what he calls "Peirce's puzzle": "Which color is the line of demarcation between a black spot and a white background? Does the line belong to the spot, or to the background?" (p. 29). If one wants to take ordinary talk of boundaries seriously, and literally, questions like these demand answers (supposedly). One can give the boundary to only the spot, or to only the background, or else one can think of the boundary as somehow shared, as belonging to both of them. As Varzi puts it, take "any entity x. Does the boundary (or any piece thereof) belong to x or to [the complement of x]? Does [the boundary] inherit the properties of the object or those of the complement?"

Varzi argues that at least some instances of this question are straightforward:

> ... we may find good reasons to maintain that material bodies such as stones or soap bars are the owners of their boundaries – their surfaces, in fact. Thus, where the complement meets an object of this sort, it will be open. (The object, in turn, will be the closed complement of the complement.) We may also ... argue that immaterial bodies such as holes are not the owners of their boundaries: these belong to the material bodies that host the holes. Thus, where the two meet, the complement (host) is closed and the entity (hole) open."

So let's start with a body of water, with respect to the air that surrounds it (an instance of our problem that Varzi attributes to Leonardo da Vinci). Is the boundary of the body of water part of the water or is it part of the air?

Suppose the boundary goes with the water. Now consider a block of wood floating on the water, partially submerged. According to Varzi's intuition, the block of wood contains *its* boundary. Since the block is thus closed, the parts of the water that touch the block are open (there), while, as above, the parts of the water that are in contact with the air are closed. So, it seems, throwing the block on top of the water causes some of the boundary points of the water to be moved away (to make room for the boundary points on the block of wood). Of course, a similar issue confronts the option of holding that the water is open and the air closed. There we would wonder about the parts of air that are in contact with the wood.

There are also analogues of Hudson's "knot" for this view. Imagine two objects, both of which include their boundaries, say a baseball and a bat. Suppose the ball is thrown toward the bat; the bat is swung and makes contact with the ball, which then goes off toward the outfield. What happens to the boundaries of the bat and the ball at the moment, or during the period, when they come into contact? It follows from a suggestion that Varzi makes that a bit of at least one of those boundaries is temporarily pushed aside so that the boundary of the other can fit in that place. Well, which boundary gets pushed aside, the boundary of the bat or the boundary of the ball?

Varzi concedes that some of these issues are difficult, perhaps irresolvable, but he suggests that this is not due to the notion of contact:

> I think we may respond that such dilemmas are real, and yet insist on keeping the theory as it is. For, we may argue, the actual ownership of a boundary is not an issue that the theory must be able to settle. The theory only needs to explain what it means for two things to be connected. It doesn't need to give a full explanation of the underlying metaphysical grounds ... By the same token, we can say that every instance of Peirce's puzzle ... is truly problematic and yet extrinsic to our present concerns. Give me a theory of black spots, and make sure to tell me who gets the boundary – the spot or the background. Give me a theory of events, and make sure to tell me which gets the boundary – the movement or the rest. If we accept this response, we have a way of disposing of the puzzle in its general form ...

We respectfully submit that it might be better to adopt a view of space, time, and thus of receptacles, on which problems like this do not arise. Of course, Varzi is correct that any complete descriptive metaphysics should have something to say about surfaces and boundaries, or at least about our everyday *thought* and *talk* of surfaces and boundaries, but perhaps this can be done without having to worry about which objects are open and which closed (see below). *That* distinction simply makes no sense on the regions-based accounts of space.

Part of the problem here may be a thesis that the boundary of a closed object is made up of the same kind of stuff as the rest of the object. Suppose, for example, that both the boundary and the interior of a (closed) block of wood consists of uncountably many point-sized bits of wood (pretending that this makes sense, at least in some possible worlds). Dean Zimmerman (1996a, p. 20) provides a thought experiment – an instance of Hudson's "knot" – designed to show that two closed objects cannot share their boundaries. It seems to be based on the foregoing thought about the constitution of boundaries:[4]

> Take two closed cubes, for example, which move continuously toward one another until they touch by having faces which occupy the same plane. What prevents their motion from continuing, so that the cubes pass right through one another? Since each cube is the mereological sum of a set of two-dimensional objects lying in planes parallel to the coincident surfaces, and since these internal two-dimensional objects differ from the coincident surfaces only in being inside the cubes, why can *they* not coincide with two-dimensional parts of other objects as well? If extended objects are made up out of simples, and distinct simples can be in the same place at the same time, then does it not follow that any two distinct extended objects could be in the same place at the same time?

So one way out of at least some of the problems here would be to postulate that the interior of an object is not composed of the same kind of stuff as its boundaries.

Some metaphysicians, notably Francisco Suárez (1597, disp. 40), Franz Brentano (e.g. 2010), and Roderick Chisholm (e.g. 1983; 1989, Chapter 8), hold that every extended, three-dimensional object, has a boundary consisting of point-sized parts, and that two such objects are in contact with each

---
[4] A "simple" is something that has no proper parts (i.e. no parts that are not identical to itself). In Dedekind-Cantor space, the only non-empty simples are singleton points. A gunky space has no simples.

other when (and only when) at least one boundary point of one *coincides* with a boundary of the other.[5] Suárez and Brentano, at least, also held that, as Zimmerman (1996b, p. 158) put it, "the three-dimensionally extended parts of a thing are not made up out of indivisibles alone but also contain some 'atomless gunk', a substance all of whose parts have proper parts." The *boundary points* of different objects can coincide – they can be in the same place at the same time – but the *interior, gunky parts* of (at least some) different objects cannot coincide. So, in the above thought experiment, from Zimmerman (1996a, p. 20), the two cubical objects would continue moving until "they touch by having faces that occupy the same plane". At that instant, some of the points on one of the touching faces coincide with points on a face of the other one. The motion would then stop (or perhaps the objects would bounce and move apart), since the interior, gunky parts of the two objects cannot coincide.

So far, so good, we suppose. This resolves these instances of Hudson's "knot". However, there are further issues concerning what we may call "interior boundaries". Consider any cubical object, say an ice cube. There is a plane surface within the object that separates its left half from its right half (in a given orientation). According to Suárez, Brentano, and Chisholm, that plane is an *interior boundary*, and it also consists of point-sized parts. Indeed, there are point-sized parts of the object located *everywhere* inside it – along every plane that can divide the object. Presumably, there are continuum-many such interior potential boundary points. And for Suárez and Brentano, and, possibly also Chisholm, all of that is in addition to the gunky parts of the cube. So it is kind of crowded in there.

Suárez held there is only one point-sized part of the cube at each interior spatial point. Zimmerman (1996a, 1996b) argues that this thesis has awkward, and, indeed, preposterous consequences. Suppose our cubical object is cut along a plane, to produce two separate half-cubes. The separated parts each have a face that was interior to the original object. Where did the point-sized parts of those two faces come from? One possibility is that the points on the original, internal plane constitute the surface of one of the pieces, and a new boundary – a new batch of point-sized parts – is *created* for the other piece. That seems weird, and arbitrary. It leaves the question of determining which half gets the point-sized pieces from the original, erstwhile internal plane and which gets the new boundary. Suárez, apparently, thought it best

---

[5] Zimmerman (1996a, especially §III) is an illuminating account of the relevant views of Suárez and Brentano. Smith (1996) contains a formal rendering of Chisholm's views on boundaries.

to maintain that the original internal plane is *destroyed* by the break, and *two* new surfaces are created (see Zimmerman, 1996b, p. 159).[6]

Brentano and Chisholm held, instead, that each internal spatial point of each extended object is occupied by *infinitely many* (perhaps even continuum many) co-located point-sized parts, one for each way the object can be divided into parts through that point. So suppose the object is cut along a given plane, and let p be a point on this plane. In the original object, there are infinitely many point-sized parts of the object all located at p. When the object is cut, some of those point-sized objects go to one of the pieces, to be part of that new surface, and the rest of the point-sized objects at p go to the other piece.[7] Infinitely many of these boundary points go to each piece. Suppose that one of the pieces is cut again along another plane. Then the point-sized parts (along the original plane) are split again, some going to one of the new pieces and some to the other.

Brentano complained that the mathematician and philosopher Bernard Bolzano "was led ... to his monstrous doctrine that there would exist bodies with and without surfaces, the one class containing just so many as the other, because contact would be possible only between a body with a surface and another without." Bolzano's "monstrous doctrine" is, of course, related to Hudson's "knot". Apparently, monstrosity is in the eye of the beholder.

To be sure (and, as conceded above), advocates of a gunky, regions-based account of space should also have something to say about boundaries and contact, or at least about ordinary talk of boundaries and contact. One can show how to define "points" as certain equivalence classes of "convergent" sequences (or sets or pluralities) of regions (in a few different ways). Then "lines" can be defined as equivalence classes of pairs of distinct points. It is just as straightforward to define "planes" in a three-dimensional regions-based space (and so on and on for higher dimensions), as equivalence classes of non-colinear triples of "points". So one option for an advocate of a regions-based space is to define the surface of a region as a certain set of "points". For example, on one definition, a convergent sequence s of regions is on the boundary of a region r just in case every member of s overlaps r, and no member of s is part of r. This resembles

---

[6] Perhaps one can think that each of the original point-sized interior boundary parts somehow splits into two, like an amoeba. Thanks to Graham Priest here.

[7] One might soften the blow a bit by insisting that the internal boundaries are only "potential" (see, for example, Smith, 1996). Varzi (1997) does not countenance internal boundaries at all, even potentially. He suggests instead that when an object is cut, its boundary somehow stretches along the cut, to form the boundary surfaces of the separated parts.

(or, better, is an instance of) Alfred North Whitehead's (1917) method of "extensive abstraction".

This, of course, makes boundaries into highly abstract objects. Officially, a boundary is a set of sets of sequences of regions. Varzi (1997) argues that this does not make sense of ordinary talk of surfaces and such. For example, one might say that one *sees* the surface of a block of wood or that the surface of a given car has been *painted red*. Surely, we do not *see* sets of sets of sequences, and we cannot paint such things either (Maddy, 1990, notwithstanding).[8]

So, for some purposes having to do with descriptive metaphysics, it may be better to avoid reliance on this Whiteheadian "extensive abstraction". For what it is worth, we see little harm in speaking, informally, of the edges, surfaces, and boundaries of regions. For example, in presenting a two dimensional theories, one can speak of the top, bottom, left, or right side of a given quadrilateral (even before we define points and lines), although one might put those locutions in scare quotes. Of course, we insist that edges and the like are not *parts* of regions, in any sense of that term. The mantra of the enterprise is that the only parts of regions are other regions. We'd go further and insist that for purposes of descriptive metaphysics, there is no reason to reify edges and boundaries at all; no reason to think of them as *things*. The right side of a quadrilateral can no more be separated from the quadrilateral than the smile of the Cheshire cat can be separated from the cat. In speaking, informally, of the sides or edges or boundaries of a region, we are just calling attention to certain aspects of the region, for example, where it is located with respect to other regions. Talk of "boundaries" can be made as rigorous as one likes, without reifying such things as existing independently of the host region.

From a regions-based perspective, contact between objects is particularly easy to understand. Two objects touch if the regions they occupy are contiguous, in Aristotle's sense: regions are contiguous just in case there is a direction on which no region lies between them. Consider, for example, an object A that exactly occupies a cubic space, and another object B that exactly occupies the space of a cube that is congruent to the first and adjacent to it. Clearly, A is in contact with B: one side of A is snug against one side of B. If we define "sides" via "extensive abstraction", then we can say that

---

[8] We presume that contemporary advocates of the foregoing point-based metaphysical views agree that one does not literally *see* zero or one-dimensional things either. Can light reflect off a point or a line? Can paint stick to a single point? How? Does a paint-point somehow get stuck on a wood-point?

these two "sides" are located in the same "place": the convergent sequences along the indicated side of A are, in fact, exactly the same as the convergent sequences along the indicated side of B. But there should be no temptation to worry about co-located anythings (at least not here). The indicated side of A is the side *of A*, and is not to be thought of as an object somehow floating free of the "rest" of A.[9] If we don't reify "edges", there is at least less temptation to think otherwise.

3 Metaphysical possibilities

Hudson (2001, pp. 84–89) launches an argument against the very (metaphysical) possibility of "material atomless gunk". That is, he argues that there *must be* "material simples", physical objects that have no proper parts. Hudson allows that there are (or might be) two different kinds of material simples, "pointy objects", those that exactly occupy a single point of space or space-time, and "maximally continuous objects".[10] Present concern, of course, is with the former. Hudson notes that the following five theses are inconsistent with each other:

> (1) The Doctrine of Arbitrary Undetached Parts: if a material object x exactly occupies a region r, and if r^* is any "exactly occupiable subregion" of r, then x has a material object part that exactly occupies r^*;
>
> (2) Necessarily, no hunk of material atomless gunk exactly occupies a point-sized region;
>
> (3) Necessarily, any hunk of material atomless gunk exactly occupies some region or other;
>
> (4) Necessarily, any region has at least one point-sized subregion; and
>
> (5) Necessarily, any point-sized region is exactly occupiable.

---

[9]Thanks to Sarah Broadie for suggesting something along present lines in the context of Aristotle's conception of "contact" and "contiguity".

[10]Following Markosian (1998), Hudson defines an object x to be "maximally continuous" if "x is a continuous object (neither spatially nor temporally gappy), and there is no region of spacetime, s, such that (i) the region occupied by x is a proper subset of s, and (ii) every point in s falls within some object or other". This seems to presuppose that spacetime is itself punctiform.

Hudson supports (5) on the basis of a premise that material simples are point-sized objects, together with "the claim that material simples are possible". It seems to us that this just begs the question. Someone who doubts that there are, or that there can be, point-sized objects will also doubt that "material simples are possible" (if all material simples are point-sized).[11] Hudson suggests that "the most promising strategy for the gunk theorist is to take aim at premise (4) on the grounds that 'gunky space' is possible" (p. 89). Hudson notes that

> [r]ecent and intriguing defenses of a Whiteheadian theory of space – a view that might deserve the description of "gunky space" – are available in the literature ... [S]trictly speaking, questions about the occupiability of point-sized regions could not even arise, for there would be no point-sized regions to have questions about.

Apparently, Hudson thinks that all such studies (including Hellman & Shapiro, 2012, 2013) violate a metaphysical law. He concludes:

> The mathematical project of constructing points out of sets of infinitely many, converging, nested, extended regions does not guarantee the metaphysical possibility of gunky space any more than the formal consistency of geometries of arbitrary many dimensions establishes the metaphysical possibility of four-dimensional space ... More pressing, however, it seems to me that the claim that "space is gunky" must have its truth value as a matter of necessity. But then, if we ground our belief in the possibility of material atomless gunk with an appeal to gunky space, we will effectively rule out the possibility of material simples ... And that consequence, I submit, is too high a price to pay. (pp. 89–90)

One might retort that just because one can describe space as punctiform, along standard Dedekind-Cantor lines, does not guarantee that space *is* punctiform, as a matter of metaphysical necessity (or even metaphysical possibility). Who has the burden of proof here? We are not metaphysicians,

---

[11] As Hudson notes, denying premise (2) would saddle a metaphysician with point-sized objects that have point-sized proper parts, a dire conclusion (supposedly). A Brentano-Chisholm view of borders would entail (2) (it seems), but Hudson rejects it because he insists that distinct objects cannot be co-located.

and we don't put forward a thesis that space or space-time is really gunky. Presumably, the actual structure of space or space-time is a scientific matter or, perhaps better, is up to scientists to decide which description of space or space-time best suits their needs. In other work, we take ourselves to have demonstrated that the various Euclidean and non-Euclidean spaces can be described, mathematically, either way: the regions-based and punctiform theories are mutually translatable. Hudson informs us that the "cost" of holding that space is really gunky is "too high", but we are not given much of an economic analysis. Above, when we mentioned Brentano's comment on Bolzano, we suggested that what counts as "monstrous" is in the eye of the beholder. Apparently, the same goes for what is "too high a price to pay". For what it is worth, our own "intuition" is that eschewing the very possibility of point-sized objects is not really a cost at all, at least not on the metaphysical front, let alone a high cost – but perhaps that just begs the question.

References

Bach, E. (1986). Natural Language Metaphysics. In R. B. Marcus, G. Dorn, & P. Weingartner (Eds.), *Logic, Methodology, and Philosophy of Science VII* (pp. 573–595). Amsterdam: North Holland.

Brentano, F. (2010). *Philosophical Investigations on Space, Time and the Continuum*. New York: Routledge.

Cartwright, R. (1975). Scattered Objects in Analysis and Metaphysics. In K. Lehrer (Ed.), (pp. 153–171). Dordrecht: Reidel.

Chisholm, R. M. (1983). Boundaries as Dependent Particulars. *Grazer Philosophische Studien, 20*, 87–95.

Chisholm, R. M. (1989). Boundaries. In *Metaphysics by R. M. Chisholm* (pp. 83–89). Minneapolis: University of Minnesota Press.

Hawthorne, J. (2000). Before-effect and Zeno Causality. *Noûs, 34*, 622–633.

Hellman, G., & Shapiro, S. (2012). Towards a Point-free Account of the Continuous. *Iyyun, 61*, 263–287.

Hellman, G., & Shapiro, S. (2013). The Classical Continuum without Points. *Review of Symbolic Logic, 6*, 488–512.

Hudson, H. (2001). Touching. *Philosophical Perspectives, 15*, 119–128.

Hudson, H. (2005). *The Metaphysics of Hyperspace*. Oxford: Oxford University Press.

Maddy, P. (1990). *Realism in Mathematics*. Oxford: Oxford University Press.

Maddy, P. (2007). *Second Philosophy: A Naturalistic Method*. Oxford: Oxford University Press.

Markosian, N. (1998). Simples. *Australasian Journal of Philosophy*, 76, 213–228.

Smith, B. (1996). Boundaries. In L. Hahn & L. Salle (Eds.), *The Philosophy of Roderick M. Chisholm* (pp. 533–561). Open Court.

Strawson, P. F. (1959). *Individuals: An Essay in Descriptive Metaphysics*. London and New York: Routledge.

Suárez, F. (1597). *Disputationes Metaphysicae, in Quibus et Universa Naturalis Theologia Ordinate Traditur, et Quaestiones Omnes ad Duodecim Aristotelis Libros Pertinentes Accurate Disputantur*. Salamanca, Spain. (Reprinted by Georg Olms, Hildesheim, 2009.)

Tarski, A. (1983). *Logic, Semantics, and Metamathematics: Papers from 1923 to 1938* (J. Corcoran, Ed.). Indianapolis: Hackett Publishing Company. (First published in 1956.)

Varzi, A. (1997). Boundaries, Continuity, and Contact. *Noûs*, 31, 26–58.

Whitehead, A. N. (1917). *The Organisation of Thought, Educational and Scientific*. London: Williams and Norgate.

Whitehead, A. N. (1919). *An Enquiry Concerning the Principles of Natural Knowledge*. Cambridge: Cambridge University Press.

Whitehead, A. N. (1929). *The Aims of Education and Other Essays*. New York: MacMillan.

Wilson, M. (2006). *Wandering Significance*. Oxford: Oxford University Press.

Zimmerman, D. (1996a). Could Extended Objects Be Made Out of Simple Parts? An Argument for 'Atomless Gunk'. *Philosophy and Phenomenological Research*, 56, 1–29.

Zimmerman, D. (1996b). Indivisible Parts and Extended Objects: Some Philosophical Episodes from Topology's Prehistory. *The Monist*, 79, 148–180.

Points, Gunk, Boundaries, and Contact

Geoffrey Hellman
Department of Philosophy, University of Minnesota
United States of America
E-mail: `hellm001@umn.edu`

Stewart Shapiro
Department of Philosophy, The Ohio State University
United States of America
E-mail: `shapiro.4@osu.edu`

The Principle of Reflection via Nested Sequents

HIDENORI KUROKAWA

Abstract: Sambin, Battilotti, and Faggian (2000) propose a means of introducing logical constants called "the principle of reflection," which can broadly be understood as belonging to proof-theoretic semantics. The idea works well for conjunction and disjunction. But handling implication in this manner is more complicated. We will propose a means of treating implication by using a proof theoretic framework called "nested sequents," following Sambin et al.'s conceptual motivations. Nested sequents have been used in modal logic as a representation of Kripke semantics. We will argue, however, that there are better means of motivating nested sequents, both technically and philosophically. On the philosophical side, conceptual problems arising in intuitionistic logic point towards the adoption of weaker logics. To formulate these weak logics, it is often convenient to use nested sequents. Once this step is undertaken, we can have a technical advantage that nested sequents uniformly formulate an extensive variety of non-classical logics.

Keywords: the principle of reflection, nested sequents, proof theoretic semantics, strict implication, basic propositional logic, substructural logic

1 Introduction

The problem of what logical constants are has been widely discussed in philosophy of logic. In proof-theoretic semantics broadly considered, in particular, the problem of how to characterize logical constants from a proof-theoretic point of view is one of its main concerns.

The tradition of proof-theoretical characterization of logical constants goes back to (Gentzen, 1935), in which Gentzen considered the introduction rule of a logical constant in his natural deduction system to give "as it were", a "definition" of a logical constant. Since then the main tool in Dummett-Prawitz-style proof-theoretic semantics has been natural deduction. However, since a large variety of non-classical logics, including substructural logics, strict implication logics, etc. has been proposed, philosophical considerations about the logical constants should not only discuss the issue of whether there is only one correct logical system but also make sense of common features and differences of the logical constants occurring in these non-classical logics.[1] For this purpose, sequent calculi can be better tools, since structural and operational rules are separated.

In this paper, we take Sambin et al.'s principle of reflection (Sambin et al., 2000) to be a basis for introducing logical constants in "minimal" logic (in the sense that rules or conditions other than basic operational rules are minimal, not in the sense of Johansson's minimal logic). This is originally based on sequents, but handling implication in Sambin's principle of reflection is more complicated than conjunctions and disjunction. To resolve this complication on implication, we propose a motivated way of modifying Sambin's principle of reflection by using a proof-theoretic framework more general than traditional sequents, which is called "nested sequents." (We take this as a basic framework for solving the forgoing problem.) We then give a few motivating discussions for adopting nested sequents and present nested sequent calculi for various logics and a few results.

2 Sambin's principle of reflection

The fundamental idea of the principle of reflection due to (Sambin et al., 2000) can be described as follows. First, we assume a few metatheoretical notions called "link," "$\vdash$" ("yields"), "," ("and"). For the latter, we assume that this is a connecting operation satisfying the conditions of commutative monoid operation.[2] For the former, we assume reflexivity ($\varphi \vdash \varphi$) and generalized transitivity ($\Gamma \vdash \Delta, \varphi$ and $\Pi, \varphi \vdash \Theta$ yields $\Gamma, \Pi \vdash \Delta, \Theta$.[3]), where Γ, Δ, etc. are multi-set of formulas. This is supposed to be a semi-formal framework in which a logical constant is introduced. Against this background, Sambin et al. (2000) consider semi-formal (meta-) equivalence

---

[1] In the traditional approach, the *normative* aspect in its treatment of meaning of logical constants is crucial. In the current approach, although it is related to a partial specification of the meaning of logical constants (cf. the notion of "operational meaning" in (Paoli, 2003)), the focus is primarily put on (at least partial) *syntactic characterization* of logical constants.

[2] We do not initially assume that the reading of "," as contextual, depending on whether it is on the left or the right of $\vdash$ (the former is conjunctive and the latter is disjunctive.) These are given after the official nested sequent calculi are defined. For the moment, we assume that "," just combines expressions. Indeed, that is all we need in "solving the definitional equation."

[3] This can be taken to be an abstract consequence relation in the sense of (Avron, 1991), which makes the metatheoretical setting rigorous, but its intended meaning would be changed. Incidentally, note that we will not necessarily be precise about the use/mention distinction in the following prose.

schemata, called "definitional equations." Definitional equations have the form $\Gamma, A \circ B \vdash \Delta$ iff $\Gamma, A, B \vdash \Delta$, where $\circ$ is a logical constant to be introduced. (This equivalence schema is inspired by Tarski's inductive clause in truth theory, e.g., $A \wedge B$ is true iff A is true and B is true.) Third, they introduce the notion of "solving a definitional equation," which means (semi-formally) deriving the traditional operational rules in sequent calculi. Fourth, since solving a definitional equation is carried out by deductive resources available only in a "minimal" logic[4] and a variety of non-classical logics are differentiated by structural rules, they can be "structural variants."[5]

Before proceeding, let us make a methodological remark that we make a distinction between two different presentations of logics. We call one of them (the one discussed above) a "structural presentation" of logics, in which we formulate logics as structural variants. We call the other one "G3-style presentation" (Troelstra & Schwichtenberg, 2000), in which structural rules are absorbed in invertible "operational rules" (cf. §5). The former may be conceptually prior from a substructural point of view, but apparently we need both presentations of a logic, not only because of practical reasons such as proof search, but also because both are related to the nature of logic.[6]

Here we give examples. For conjunctions and disjunctions, we have the following definitional equations.[7] (We use the notation of (Paoli, 2002).)

1. Multiplicative conjunction:
    $\Gamma, A, B, \vdash \Delta$ iff $\Gamma, A \otimes B \vdash \Delta$.
2. Additive conjunction:
    $\Gamma \vdash \Delta, A$ and $\Gamma \vdash \Delta, B$ iff $\Gamma \vdash \Delta, A \wedge B$.

3. Multiplicative disjunction:
    $\Gamma \vdash A, B, \Delta$ iff $\Gamma \vdash A \oplus B, \Delta$.
4. Additive disjunction:
    $\Gamma, A \vdash \Delta$ and $\Gamma, B \vdash \Delta$ iff $\Gamma, A \vee B \vdash \Delta$.

---

[4]Our philosophical position is to take this solvability of definitional equation to be at least *a positive criterion* for a logical constant, but directly arguing for this is not the purpose of this paper. (We are skeptical about the idea that a negative criterion can be provided at all.) Also, the program of "basic logic" has two guiding concepts other than reflection: symmetry; visibility. But here we focus only on reflection.

[5]In this sense, the program of basic logic is congenial to so-called Došen's principle often summarized by the phrase "logical constants as punctuation marks. (Došen, 1989)

[6]We leave the details on this topic to another occasion.

[7]Note that introducing conjunction/disjunction in this way has an advantage of making intuitive the multiplicative/additive distinction.

We show the case of multiplicative disjunction $\oplus$ in order to show that in its derivation itself, we do not use the disjunctive feature of ",".

L$\oplus$ is immediate. $\dfrac{\Gamma \vdash A, B, \Delta}{\Gamma \vdash A \oplus B, \Delta}$

R$\oplus$ is derived via reflexivity and Cut.

$$\dfrac{\dfrac{\dfrac{A \oplus B \vdash A \oplus B}{A \oplus B \vdash A, B} \quad \Gamma, A \vdash \Delta}{\Gamma, A \oplus B \vdash \Delta, B} \quad B, \Sigma \vdash \Pi}{\Gamma, \Sigma, A \oplus B \vdash \Delta, \Pi}$$

This gives the binary rule R$\oplus$. $\dfrac{\Gamma, A \vdash \Delta \quad \Sigma, B \vdash \Pi}{\Gamma, \Sigma, A \oplus B \vdash \Delta, \Pi}$

After introducing these logical constants in this way, we should formulate official sequent calculi formally and prove the cut-elimination theorem. Proving cut-elimination is an essential part of the program, partly because cut-elimination guarantees soundness of this procedure and "separation" of logical constants.[8]

One could apply this method to implication as follows.[9]

Multiplicative implication (MI1): $\Gamma, A \vdash \Delta, B$ iff $\Gamma \vdash \Delta, A \to B$

$R \to$ is immediate. $L \to$ is derived as follows.

$$\dfrac{\Gamma \vdash A \quad \dfrac{\dfrac{A \to B \vdash A \to B}{A \to B, A \vdash B}}{\Gamma, A \to B \vdash B} \quad \Gamma', B \vdash C}{\Gamma, \Gamma', A \to B \vdash C}$$

This works to some extent,[10] but it is another matter whether or not this is completely satisfactory from the original conceptual view of "reflection."

---

[8]Compare analytic cut and cut-elimination. To do proof-search, these two may not be much different. However, if one considers the issue of (at least partial) specification of (operational) meaning of logical constants via the principle of reflection, then cut may be more useful to keep the separation property, since having only analytic cut may not guarantee that some subformulas of the proven formula may be once introduced and disappear to derive another subformula.

[9]An additive implication can be handled by $\Gamma \vdash \Delta, A$ and $\Gamma, B \vdash \Delta$ iff $\Gamma, A \to B \vdash \Delta$.

[10]Avron (1991) and Došen (1988) adopt this derivation, since it is enough for their purpose.

The Principle of Reflection

3 Potential problems about implication and a proposal of their solution via nested sequents

We point out both an internal problem about implication to the approach itself and an external problem. The former is concerned with the differences between disjunction (conjunction) and implication. Reflection of "," always occurs on the same side as "," occurs, i.e., the logical constant is introduced on the same side as ",", whereas reflection of "⊢" by "→" occurs by mixing up levels in the sense that "→" being regarded as reflecting "⊢", the lower sequent of $R \to$ appears to have two items related to "⊢" ("⊢" and "→"). However, the upper sequent has only one "⊢". One may wonder whether this is conceptually acceptable or not. The external one is that some foundationally important (cf. §4) logics cannot be formulated by traditional cut-free sequent calculi.

The former problem is actually addressed in (Sambin et al., 2000).

> The peculiarity of implication is that it reflects a link *yields*, that is the turnstile sign ⊢ itself, and this is what makes it different from other connectives. ... So to see that implication follows the same conceptual pattern as all other connectives, a richer metalanguage is needed, in which the link *yields* to be reflected appears inside the scope of another link *yield*. In terms of the shorthand notation, also nested occurrences of ⊢ must be considered. Then the definitional equation for → is simply
>
> for all $\Gamma, \Gamma \vdash A \to B$ if and only if $\Gamma \vdash (A \vdash B)$.
>
> It can be solved following the same pattern as other connectives, but it needs two new forms of composition, that is composition of formulae inside two occurrences of ⊢. Then one reaches the rules
>
> $$\frac{\Gamma \vdash (A \vdash B)}{\Gamma \vdash A \to B} \qquad \frac{\Gamma \vdash A \quad B \vdash \Delta}{A \to B \vdash (\Gamma \vdash \Delta)}$$
>
> which however cannot be expressed in the traditional shape of sequent calculus, where nested occurrences of ⊢ are not considered. (p. 990 Sambin et al., 2000)

Our main contention in this paper is to take this possible option rather than dismissing it. This is not only because such a nested construction, called "nested sequents" (Brünnler, 2010)[11] but also because there are motivations for adopting nested sequents in formulating non-classical logics.

Let us now give certain details about the construction we would like to adopt in nested sequents. 1. Nested sequents (in a two-way formulation (Fitting, 2014)) have the following form $\Gamma \vdash \Delta, ()$, where "$\vdash$" and "," mean the same as before, but here "()" means a place holder (or a hole) in which we can plug another sequent (on a succedent "()" can occur more than once). And this can be iterated finitely many times. 2. In addition to "$\vdash$" and ",", we use additional expressions "()". The behavior of this expression is governed by rules for $\rightarrow$ (and in some cases by certain structural rules). Also, we use schematic expression $\mathcal{N}[]$, which is used to formulate a general rule which can be applied to an arbitrary finite nesting of sequents. For a binary rule, we handle only cases where contexts $\mathcal{N}[]$ are identical.[12] 3. Accordingly, reflexivity and transitivity are reformulated with nested contexts: $\mathcal{N}[\varphi \vdash \varphi]$; $\mathcal{N}[\Gamma_1 \vdash \Delta_1, \varphi]$ and $\mathcal{N}[\Gamma_2, \varphi \vdash \Delta_2]$ yields $\mathcal{N}[\Gamma_1, \Gamma_2 \vdash \Delta_1, \Delta_2]$.

We can now formulate the definitional equation for "$\rightarrow$" as follows.[13]

Multiplicative implication: $\mathcal{N}[\Gamma \vdash \Delta, (A \vdash B)]$ iff $\mathcal{N}[\Gamma \vdash \Delta, A \rightarrow B]$.

The solution for $R \rightarrow$ is immediate. $R \rightarrow \dfrac{\mathcal{N}[\Gamma \vdash \Delta, (A \vdash B)]}{\mathcal{N}[\Gamma \vdash \Delta, A \rightarrow B]}$

Then the result of the solution of the definitional equation is as follows.

$L \rightarrow \dfrac{\mathcal{N}[\Gamma_1 \vdash \Delta_1, (\Pi_1 \vdash \Sigma_1, A)] \quad \mathcal{N}[\Gamma_2 \vdash \Delta_2, (B, \Pi_2 \vdash \Sigma_2)]}{\mathcal{N}[\Gamma_1, \Gamma_2, A \rightarrow B \vdash \Delta_1, \Delta_2, (\Pi_1, \Pi_2 \vdash \Sigma_1, \Sigma_2)]}$

The latter can be derived by the following derivation.

$\dfrac{\dfrac{\mathcal{N}[A \rightarrow B \vdash A \rightarrow B]}{\mathcal{N}[A \rightarrow B \vdash (A \vdash B)]} \quad \mathcal{N}[\Gamma_1 \vdash \Delta_1, (\Pi_1 \vdash \Sigma_1, A)]}{\dfrac{\mathcal{N}[\Gamma_1, A \rightarrow B \vdash \Delta_1, (\Pi_1 \vdash \Sigma_1, B)] \quad \mathcal{N}[\Gamma_2 \vdash \Delta_2, (B, \Pi_2 \vdash \Sigma_2)]}{\mathcal{N}[\Gamma_1, \Gamma_2, A \rightarrow B \vdash \Delta_1, \Delta_2, (\Pi_1, \Pi_2 \vdash \Sigma_1, \Sigma_2)]}}$

In this way, Sambin et al.'s fundamental idea can be expressed by nested sequents without essentially distorting the original idea.[14]

---

[11] Other recent sources are (Poggiolesi, 2010) and (Kashima, 1994). Similar constructions have been rediscovered several times with different motivations.

[12] This can be generalized for some logics, but the situation depends on which logic is formulated. For the logics we handle in this paper, we use the shared nested context $\mathcal{N}[]$.

[13] It is only "$\rightarrow$" that matters here. For the others, we can straightforwardly modify the foregoing definitional equations and the method of solving them to nested sequent counterparts.

[14] One may be worried about the trivial solvability, i.e., for any candidate of logical constants,

The Principle of Reflection

4 Philosophical motivations for adopting nested sequents

We have just set up a proof-theoretic framework in which implication can be handled in a way faithful to the original idea of reflection. However, the construction is quite complicated. Hence, one may ask whether we really need this way of introducing → or not. Reflection via traditional sequents are fine for introducing classical or intuitionistic implication. But we claim that, by using the nested version, we can keep both i) the original motivation for the reflection better and ii) opportunities to handle logics that can be formulated only via a generalized proof-theoretic framework.

Taking these possibilities seriously, we propose to adopt nested sequents as a uniform proof-theoretic framework for formulating quite an extensive variety of logics. Apart from the solution of the definitional equation for → presented above, our arguments for adopting nested sequents are indirect in the following sense. We first discuss a few conceptual problems of intuitionistic implication. Second, we point out that some (relatively weak) non-classical logics may be helpful to analyze these conceptual problems. Third, we point out that one of these weak logics has some connection to strict implication and that one precursor of the principle of reflection actually yields strict implication. However, most of these logics do not have cut-free formulations in traditional sequent calculi. Hence, some generalizations of the traditional sequent calculi must be desired.[15]

4.1 Foundations of intuitionistic logic

Let us first take up an issue related to the foundations of intuitionstic logic. The intended interpretation of intuitionistic logical constants is known as the Brouwer-Heyting-Kolmogorov (BHK) interpretation.[16] A conceptual problem on this interpretation of implication has been pointed out since 1930's. The issue can be described as either a circularity or an "impredicativity" of the notion of construction used in the interpretation. The proof condition for intuitionistic implication in BHK interpretation goes as follows.

---

we can solve the definitional equation. Thus, let us give an example for which it is hard even to formulate the definitional equation. The example is Girard's exponential !. We conjecture that this might suggest that the exponential is more a "structural constant" than a logical constant.

[15]Disclaimer: we are not committed to any *one* of the logics presented here as *the only correct logic*, nor we have (yet) committed to logical pluralism. The current work is a technical preliminary for discussing such philosophically substantial discussions on these topics.

[16]This is not a traditional formal semantics, since it uses the notion of "mathematical (mental) constructions."

Hidenori Kurokawa

A proof of $A \to B$ consists in a construction which transforms *any* proof of A into a proof of B.

The problem occurs here due to the universal quantifier over all proofs. Gödel complained against this as follows.

> So the Heyting's axioms concerning absurdity and similar notions differ from the system A [a finitistic system (HK)] only by the fact that the substrate on which the consequences are carried out are proofs instead of numbers or other enumerable sets of mathematical objects. But by this very fact they do violate the principle, which I stated before, that the word "any" can be applied only to those totalities for which we have a finite procedure for generating all their elements. For the totality of all possible proofs certainly does not possess this character, and nevertheless the word "any" is applied to this totality in Heyting's axioms, (...) Totalities whose elements cannot be generated by a well-defined procedure are in some sense vague and indefinite as to their borders. And this objection applied particularly to the totality of intuitionistic proofs because of the vagueness of the notion of constructivity. (p. 53 Gödel, 1995)

In a somewhat different context, Gentzen also comments on a circularity of intuitionistic implication.

> What does $\mathfrak{A} \supset \mathfrak{B}$ mean? Suppose, for example, that there exists a *proof* in which the proposition $\mathfrak{B}$ is proved on the basis of the assumption $\mathfrak{A}$ by means of inferences that have already been recognized as permissible. From this we infer, by $\supset$-introduction: $\mathfrak{A} \supset \mathfrak{B}$. This proposition is merely intended to express the fact that *a proof is available* which permits a proof of the proposition $\mathfrak{B}$ from the proposition $\mathfrak{A}$, once the proposition $\mathfrak{A}$ is proved. ... In interpreting $\mathfrak{A} \supset \mathfrak{B}$ in this way, I have presupposed that the available proof of $\mathfrak{B}$ from the assumption $\mathfrak{A}$ contains merely inferences *already recognized as permissible*. On the other hand, such a proof may itself contain other $\supset$-*inferences* and then our interpretation *breaks down*. For, it is *circular* to justify the $\supset$-inferences on the basis of a $\supset$-interpretation which itself already involves the presupposition of the admissibility of *the same* form of inference. (p. 167 Gentzen, 1936)

The Principle of Reflection

Comments on these quotes must be in order. First, note that Gödel and Gentzen talk about slightly different issues, i.e., the BHK interpretation and a formalized treatment of intuitionistic logic (HQC). Still, they both talk about a conceptual problem underlying intuitionistic implication.[17]

Second, although we do not intend to go into the thoroughgoing discussion on the issues of impredicativity or circularity, let us just point out that it seems difficult to avoid the issues only by appealing to normalization. In their passages they only talk about logic, so it is natural to think that this circularity or impredicativity can be simply avoidable provided that we consider normal proofs in HQC, since by the subformula property we do not have to worry about a proof that has a detour of proofs of arbitrarily more complicated formulas than the subformulas of the proven formula. But if we consider how intuitionistic implication works in a stronger theory (arithmetic or set theory) in which normalization is not provable in an unquestionably finite manner, then appealing to normalization to avoid circularity or impredicativity may not be a plausible option. Or at least we can say that some conceptual problem may remain. (Note that Gentzen's discussion is given in the context of proving consistency of arithmetic.)

Against this background, it may not be absurd to start adopting those logics weaker than intuitionistic logic which are free from the problem of impredicativity or circularity. Let us mention two attempts in the literature to solve this conceptual problem by taking a sublogic of intuitionistic logic. One is weak implication logic (WI) in (Okada, 1987), in which the author claims that the introduction rules of implication in HQC is too general and proposes a logic with a restricted $R\to$ rule in a modified sequent as follows.

$$\frac{\Gamma_\to, A \vdash B}{\Gamma_\to \vdash A \to B} \quad (\Gamma_\to \text{ means that formulas in } \Gamma \text{ has the form } A \to B.)$$

Hence, this logic can be expressed in a traditional sequent calculus only with a substantial modification on the idea of operational rules. $R\to$ looks like an operational rule with restriction on the context (and hence, not entirely schematic). Intuitively, this is intended to restrict the context to the formulas that are assumed to be "proven" in the weaker system in the sense that "$\to$" is considered to be a result of reflecting "$\vdash$". Using the system, Okada gives a kind of provability semantics for the logic. Since $R\to$ is a sequent counterpart of the deduction theorem, this system is formulated by

---

[17] Figuring out a precise difference of these remarks is beyond the scope of this paper. Hence, in the following, we do not take into account this difference.

restricting the rule corresponding to the deduction theorem. Note that this restriction is common in (traditional) sequel calculi for modal logics. This is no wonder since WI turns out to be equivalent to S4 strict implication.

The other approach worth mentioning here is Ruitenberg's logic Basic Propositional Logic (BPL) (Ruitenburg, 1991). BPL is related to K4 modal logic in the same as HPC is related to S4 modal logic via Gödel translation.[18] Ruitenburg claims that BPL is free from the impredicativity of intuitionistic implication. Ruitenburg's argument for the claim is as follows. The impredicativity at issue consists in its unrestricted modus ponens. Ruitenburg formulate the logic via natural deduction, where $\to$-elimination is

$$\frac{A \vdash B \to C}{A \wedge B \vdash C}$$

This rule yields $A \wedge (A \to B) \vdash B$ (from reflexivity $A \to B \vdash A \to B$). However, Ruitenburg points out that this is a problematic consequence of the general modus ponens since the derivation of B seems to be done only by assuming $A \to B$ on the antecedent whose intuitive meaning should be given as a "reflection" of $A \vdash B$ but then the explanation of the meaning of $\vdash$ is circular. As a remedy, Ruitenburg proposes to drop the context A, i.e., a restriction of elimination rule for $\to$.

Interestingly, each of these two logics suggests how to restrict rules corresponding to the introduction rule and elimination rule for $\to$ (or the deduction theorem and modus ponens) to avoid a problem in intuitionistic implication. It is desirable to compare these logics (and the logic with both restrictions) in a *uniform* proof-theoretic framework with the restricted rules, even before we go into more substantial philosophical discussions on whether or not these logics are really useful to avoid the problem of impredicativity or circularity. This is difficult to do in the traditional sequent calculi,[19] but it is possible to formulate all of these logics by nested sequents (cf. §5).

Not necessarily independent of the point discussed above (since it is related to the restriction of modus ponens), we will make another point concerning why nested sequents are particularly useful for expressing different kinds of implication. BPL has the structural rule of contraction in its sequent

---

[18] BPL had been considered by Visser (1981) based on a different motivation. Visser formulated propositional logics called FPL and BPL, which are related to modal logic GL and K4 in the same way as HPC is related to S4.

[19] This logic is formulated in a modified sequent calculus (Ishii, Kashima, & Kikuchi, 2001), but the system was somewhat contrived and is not helpful in comparing BPL with other logics.

formulation (Ishii et al., 2001), but its object language formulation of contraction $(A \to (A \to B)) \to (A \to B)$ fails to hold. Due to this feature, the traditional method of formulating substructural logics does not work well to formulate BPL. We need to take more care in order to formulate this logic in a cut-free manner. It turns out that BPL can be formulated in a nested sequent, but this does not seem to be merely accidental but has a conceptual reason. Ruitenburg points out in his paper that there is a systematic translation between Myhill's system for levels of implication (Myhill, 1975) and BPL. Myhill's idea of level distinctions among implication can be expressed as $\to^n$ (implication is indexed by a level or a "type"). The notation strongly suggests that there is a natural analogy between levels of implication and nested sequents, namely the level of an implication corresponds to the number of nestings. Thus, the failure of contraction in BPL can be most easily explained as distinctions of levels (or "types") in implication. From this point of view, intuitionistic logic can be seen as a logic in which all level (type) distinctions of implication collapse. It is then natural to think that implication with type distinctions collapsed allow contraction.[20]

4.2 "Engendering an illusion of understanding"

Recall that Okada's proposal for modifying intuitionistic logic contains an adoption of "strict implication" logic. Strict implication logic is one of the oldest non-classical logics, but it has not been often discussed these days. Here we discuss a precursor of Sambin et al.'s idea of reflection, which suggests that one way of "reflecting" $\vdash$ result in introducing strict implications.

This topic goes back to Dana Scott's papers written in 1970's (Scott, 1971, 1974).[21] Scott first discusses a problem of how to properly express what "modus ponens" are. In particular, Scott asks what are differences of the following formulations of modus ponens.[22]

---

[20] A similar view is (independently) argued for in (Weber, 2014), where the author cites "Once we have any contraction-free logic, we have *got* levels of implication. It is just that Myhill is explicit about it and other authors are not" (Akama, 1996). This comment applies to BPL best.

[21] Scott's main purpose is to defend C. I. Lewis' strict implication against Quine's critique.

[22] MP2' is from (Peregrin, 2008). How to interpret these inferences ultimately depends what kind of interpretations we give to "$\vdash$". However, some differences among these must be self-explanatory. Note that MP2 is occasionally used to formulate a rule equivalent to MP1. Here we assume that our interpretation of $\vdash$ is a consequence relation, whose conditions we give below.

(MP1) $\vdash A \to B$ (MP2) $A, A \to B \vdash B$ (MP2') $\Gamma \vdash A \to B$
$\quad\quad\; \vdash A$ $\quad\quad\quad\quad\quad\quad\quad\quad\quad\quad\quad\quad\quad\quad\; \Gamma \vdash A$
$\quad\quad\; \vdash B$ $\quad\quad\quad\quad\quad\quad\quad\quad\quad\quad\quad\quad\quad\quad\; \Gamma \vdash B$

In order to make clear the differences, Scott introduces the following rule "horizontalization." Note that although Scott does not discuss the circularity issue discussed in the previous section, the same kind of difference among modus ponens as discussed in the last section is at issue here. If

(V)
$$\Gamma_0 \vdash \Delta_0$$
$$\Gamma_1 \vdash \Delta_1$$
$$\vdots$$
$$\frac{\Gamma_{n-1} \vdash \Delta_{n-1}}{\Gamma_n \vdash \Delta_n}$$

is derivable, then the horizontal form:

(H) $\bigwedge \Gamma_0 \to \bigvee \Delta_0, \bigwedge \Gamma_1 \to \bigvee \Delta_1, \ldots, \bigwedge \Gamma_{n-1} \to \bigvee \Delta_{n-1} \vdash \bigwedge \Gamma_n \to \bigvee \Delta_n$ is derivable.

Scott then showed a meta-metatheorem: horizontalization is derivable in a system if and only if it contains the rule (C) and the other rules below.

$$\frac{A \vdash B}{\vdash A \to B} \; (C)$$

1. Rules for $\vdash$: reflexivity (R); monotonicity(M); transitivity (T).[23]
2. Rules for implication $\to$.
   (MP$_\to$) $A \to B, (A \to B) \to (C \to D) \vdash C \to D$
   (T$_\to$) $A \to B, C \to D \vdash (B \to C) \to (A \to D)$
   (T$_{\land,\lor}$) $A \to (B \lor C), (A \land B) \to C \vdash A \to C$

3. Rules for $\top, \bot, \land, \lor$.
   ($\land$) $A, B \vdash A \land B; A \land B \vdash A, A \land B \vdash B;$
   ($\top$) $\vdash \top;$
   ($\lor$) $A \lor B \vdash A, B; A \vdash A \lor B, B \vdash A \lor B;$
   ($\bot$) $\bot \vdash$

Scott's system S turns out to formalize a strict implication logic exactly corresponding to its modal counterpart K4 + density and a system obtained

---
[23] Here we also assume that Γ, Δ are sets of formulas.

by adding MP2 to S is equivalent to S4 strict implication logic (Došen, 1980). MP1 is actually derivable in S, but MP2 is not derivable in S, since MP2 is equivalent to T axiom in strict implication logic. (Note that MP2′ derives MP2. Contexts matter here.) Thus, Scott successfully made a distinction between two different versions of modus ponens by formulating two systems formalizing strict implications via "reflecting" "⊢".

Scott's result gives some morals on our discussion of the principle of reflection and nested sequents. First, Scott's (C) seems to anticipate the definitional equation of Sambin et al., although the context is confined to the empty one and the extra conditions for "⊢" are given by the other rules. Second, Scott gives a prototypical argument for obtaining strict implications by reflecting a consequence relation, in particular one strictly weaker than S4 strict implication. The question of how Scott's observation on modus ponens is related to Ruitenburg's observation on the circularity is worth further pursuing. (Note that in strict implication T axiom is sufficient for deriving contraction. Scott's logic seems to provide an interesting case in which only a special form of contraction $(\top \to (\top \to A)) \to (\top \to A)$ is derived but not for an arbitrary formula.[24]) Third, although one may doubt that this is directly related to nested sequents, look at the rules from (Došen, 1980).

$$\frac{A \vdash^1 B}{\emptyset \vdash^1 A \to B} \quad \frac{\{A_1 \vdash^1 B_1, \ldots, A_{n-1} \vdash^1 B_{n-1}\} \vdash^2 \{A_n \vdash^1 B_n\}}{\emptyset \vdash^2 \{A_1 \to B_2, \ldots, A_{n-1} \to B_{n-1} \vdash^1 A_n \to B_n\}}$$

These two rules in (Došen, 1980) axiomatize the implication of S. This looks so close to our construction of nested sequents as to make it significant to study a systematic relationship between these, although strictly speaking the rules are based on a guiding principle ("horizontalization") slightly different from the principle of reflection as we have used it in this paper.[25]

5 Cut-free nested sequent calculi for a variety of logics

Now we present nested sequent calculi for logics including the aforementioned ones. Ideally, we should formulate all the logics following the aforementioned "structural presentation." However, due to the limitation of space, we only state the main results directly related to strict implication logics,

---

[24] Exactly what kind of contraction is the source of impredicativity of intuitionistic logic is an interesting question.

[25] As far as we know, no cut-free nested sequent calculus for Scott's S has been formulated. For proof systems for modal logics based on horizontalization, see (Došen, 1985).

which can easily have a formulation absorbing structural rules.[26] Equivalence between this formulation and the formulation using structural rules + operational rules can be easily shown. (For these logics, due to the implicit use of structural rules, we use only $\wedge$, $\vee$.)

Nested sequent system for StK (StK + persistence)

I. Axioms $\qquad \mathcal{N}[\Gamma, p \vdash p, \Delta] \qquad\qquad \mathcal{N}[\Gamma, \bot \vdash \Delta]$.

II. Operational rules for conjunction and disjunction

$$L\wedge \quad \frac{\mathcal{N}[\Gamma, \varphi, \psi \vdash \Delta]}{\mathcal{N}[\Gamma, \varphi \wedge \psi \vdash \Delta]} \qquad R\wedge \quad \frac{\mathcal{N}[\Gamma \vdash \Delta, \varphi] \quad \mathcal{N}[\Gamma \vdash \Delta, \psi]}{\mathcal{N}[\Gamma \vdash \Delta, \varphi \wedge \psi]}$$

$$L\vee \quad \frac{\mathcal{N}[\Gamma, \varphi \vdash \Delta] \quad \mathcal{N}[\Gamma, \psi \vdash \Delta]}{\mathcal{N}[\Gamma, \varphi \vee \psi \vdash \Delta]} \qquad R\vee \quad \frac{\mathcal{N}[\Gamma \vdash \Delta, \varphi, \psi]}{\mathcal{N}[\Gamma \vdash \Delta, \varphi \vee \psi]}$$

III. Nested rules for basic implication (for StK)

$$L\to \quad \frac{\mathcal{N}[\Gamma, A \to B \vdash \Delta, (\Pi \vdash \Sigma, A)] \quad \mathcal{N}[\Gamma, A \to B \vdash \Delta, (\Pi, B \vdash \Sigma)]}{\mathcal{N}[\Gamma, A \to B \vdash \Delta, (\Pi \vdash \Sigma)]}$$

$$R\to \quad \frac{\mathcal{N}[\Gamma \vdash \Delta, (A \vdash B)]}{\mathcal{N}[\Gamma \vdash \Delta, A \to B]}$$

IV. Persistence $\quad \dfrac{\mathcal{N}[\Gamma, A \vdash \Delta, (\Pi, A \vdash \Sigma)]}{\mathcal{N}[\Gamma, A \vdash \Delta, (\Pi \vdash \Sigma)]}$

V. Various rules for implication

1) The implication rule for St4

$$\frac{\mathcal{N}[\Gamma, A \to B \vdash \Delta, (\Pi, A \to B \vdash \Sigma)]}{\mathcal{N}[\Gamma, A \to B \vdash \Delta, (\Pi \vdash \Sigma)]}$$

2) The implication rule for StT

$$\frac{\mathcal{N}[\Gamma, A \to B \vdash \Delta, A] \quad \mathcal{N}[\Gamma, A \to B, B \vdash \Delta]}{\mathcal{N}[\Gamma, A \to B \vdash \Delta]}$$

---

[26] We leave to another occasion a more comprehensive treatment of logics (including substructural logics) by nested sequents. Also, we could use a multi-set formulation in G3-style, but for a technical reason we have chosen this formulation.

3) The implication rule for StB

$$\frac{\mathcal{N}[\Gamma \vdash \Delta, A, (\Pi, A \to B \vdash \Sigma)] \quad \mathcal{N}[\Gamma, B \vdash \Delta, (\Pi, A \to B \vdash \Sigma)]}{\mathcal{N}[\Gamma \vdash \Delta, (\Pi, A \to B \vdash \Sigma)]}$$

4) The implication rule for GL

$$\frac{\mathcal{N}[\Gamma \vdash \Delta, A \to B, (A \to B, A \vdash B)]}{\mathcal{N}[\Gamma \vdash \Delta, A \to B]}$$

Combining some of these rules, we can formulate various logics, such as K, KT, KB, K4, S4 (=KT4), S5 (KTB4), GL = (K + GL) strict implication logics (cf. Corsi, 1987; Ishigaki & Kashima, 2008), BPL (K4 + persistence), FPL (GL + persistence), intuitionistic logic (e.g., intuitionistic logic = StK + 4 + T + persistence). Let us state the main theorem.

Theorem 1 *All the nested sequent calculi for logics mentioned above are complete (without cut) w.r.t. appropriate Kripke semantics.*

Corollary 1 *Cut is admissible for all nested sequent calculi for the aforementioned logics.*

An outline of the current version of a proof goes as follows. We first formulate prefixed tableau systems for these logics and prove completeness without cut. Then, by following (Fitting, 2012), we can translate these prefixed tableau systems into nested sequents. The translation does not require any use of cut. Hence, we can prove the admissibility of cut. [27]

6 Conclusion

In this paper, we gave a reformulation of Sambin's principle of reflection by using nested sequents. This is initially motivated by the fact that the traditional sequent calculi may not be suitable for handling implication from the original viewpoint of the principle of reflection. We provided further indirect reasons why adopting nested sequents in formulating various non-classical logics is potentially important by pointing out that relatively weaker logics that have foundational significance can be uniformly formulated only via nested sequents. We then presented nested sequent calculi for a variety of logics and state the result of cut-admissibility for these calculi.

---

[27]Let us now give philosophical comments on the two notions of conservativeness and uniqueness in nested sequents. Both conditions are satisfied in nested sequents that we have presented above. However, we do not want our claim of appropriateness of adopting nested sequents for characterizing logical constants to rest on their satisfying the two conditions, for we are not sure whether these conditions can be characteristic conditions for logical constants.

References

Akama, S. (1996). Curry's Paradox in Contractionless Constructive Logic. *Journal of Philosophical Logic*, 25(2), 135–150.

Avron, A. (1991). Simple Consequence Relations. *Information and Computation*, 92, 105–139.

Brünnler, K. (2010). *Nested Sequents (Habilitationsschrift)*. Retrieved from http://arxiv.org/abs/1004.1845v1

Corsi, G. (1987). Weak Logics with Strict Implication. *Zeitschr. f. math. Logik und Grundlagen d. Mathematik*, 33, 389–406.

Došen, K. (1980). *Logical Constants: An Essay in Proof Theory*. Unpublished doctoral dissertation, University of Oxford.

Došen, K. (1985). Sequent-systems for Modal Logic. *Journal of Symbolic Logic*, 50(1), 149–168.

Došen, K. (1988). Sequent-systems and Groupoid Models. I. *Studia Logica: An International Journal for Symbolic Logic*, 47(4), 353–385.

Došen, K. (1989). Logical Constants as Punctuation Marks. *Notre Dame Journal of Formal Logic*, 30(3), 362–381.

Fitting, M. (2012). Prefixed Tableaus and Nested Sequents. *Annals of Pure and Applied Logic*, 163, 291–313.

Fitting, M. (2014). Nested Sequents for Intuitionistic Logics. *Notre Dame Journal of Formal Logic*, 55(1), 41–61.

Gentzen, G. (1935). Untersuchungen Über das Logisches Schließen. *Mathematische Zeitschrift*, 1, 176–210.

Gentzen, G. (1936). Die Widerspruchfreiheit der reinen Zahlentheorie. *Mathematische Annalen*, 112, 493–565. (English translation in (1970), The Collected Papers of Gerhard Gentzen, Amsterdam: North-Holland Publishing Company)

Gödel, K. (1995). The Present Situation in the Foundations of Mathematics. In S. Feferman, J. W. Dawson, Jr., S. C. Kleene, G. Moore, R. Solovay, & J. van Heijenoort (Eds.), *Kurt Gödel, Collected Works, vol. III*. Oxford: Oxford University Press.

Ishigaki, R., & Kashima, R. (2008). Sequent Calculi for Some Strict Implication Logics. *Logic Journal of the IGPL*, 16(2), 155–174.

Ishii, K., Kashima, R., & Kikuchi, K. (2001). Sequent Calculi for Visser's Propositional Logics. *Notre Dame J. of Formal Logic*, 42(1), 1–22.

Kashima, R. (1994). Cut-free Sequent Calculi for Some Tense Logics. *Studia Logica*, 53(1), 119–136.

Myhill, J. (1975). Levels of Implication. In A. R. Anderson, R. C. Barcan-

Marcus, & R. M. Martin (Eds.), *The Logical Enterprise*. New Haven: Yale University Press.

Okada, M. (1987). A Weak Intuitionistic Propositional Logic with Purely Constructive Implication. *Studia Logica*, *46*(4), 371–382.

Paoli, F. (2002). *Substructural Logics: A Primer*. Dordrecht: Kluwer Academic Publishers.

Paoli, F. (2003). Quine and Slater on Paraconsistency and Deviance. *Journal of Philosophical Logic*, *32*(2), 531–548.

Peregrin, J. (2008). What Is the Logic of Inference? *Studia Logica*, *88*(2), 263–294.

Poggiolesi, F. (2010). *Gentzen Calculi for Modal Propositional Logic*. Berlin: Springer.

Ruitenburg, W. (1991). Constructive Logic and the Paradoxes. *Modern Logic 1*, *4*, 271–301.

Sambin, G., Battilotti, G., & Faggian, C. (2000). Basic Logic: Reflection, Symmetry, Visibility. *Journal of Symbolic Logic*, *65*(3), 979–1013.

Scott, D. (1971). On Engendering an Illusion of Understanding. *Journal of Philosophy*, *68*, 787–807.

Scott, D. (1974). Rules and Derived Rules. In S. Stenland (Ed.), *Logical Theory and Semantic Analysis* (pp. 147–161). Dordrecht: D. Reidel.

Troelstra, A., & Schwichtenberg, H. (2000). *Basic Proof Theory* (Second ed.). Cambridge: Cambridge University Press.

Visser, A. (1981). A Propositional Logic with Explicit Fixed Points. *Studia Logica*, *40*, 155–175.

Weber, Z. (2014). Naive Validity. *Philosophical Quarterly*, *64*(254), 99–114.

Hidenori Kurokawa
University of Helsinki
Finland
E-mail: `hidenori.kurokawa@gmail.com`

Two Approaches to Philosophically Analysing Language

Pavel Materna

Abstract: According to some philosophers we can distinguish two trends in dealing with (especially natural) language. One of them is older and uses explications that simplify the richness of the language, so that the result of its efforts is an artificial image of language not corresponding to its real shape. The more recent trend tries to capture all the richness of the language together with all its irregularities and is represented mainly by Quine's and later Wittgenstein's philosophy. The older trend (I call it *analytic group*, **AG**, here) is sometimes criticized as being somehow obsolete while the more recent trend (called *Q-W group*, **Q-W**, here) is then evaluated as more promising (more 'progressive'). I try to show that **AG** is incomparable with **Q-W** because both try to answer distinct questions, solve distinct problems. (A comparison could be realized on the higher level of evaluating the *choice of problems itself*, which is another topic.)

Keywords: sense, meaning, reference, denotation, explication, abstraction

1 Motivation

A characteristic (together with an implicit evaluation) of the two approaches that will be dealt with in this article can be found in (Beran, 2013, p. 240). (My translation from Czech:)

> "*Tractatus* is ... a document of far-reaching simplifications of talking about "language", so common among the pioneer generation of analysts. A happy juggling with the words "language", "sentence", "thinking" or "meaning", not very much taking into account their current (actual) sense and motivated largely by the endeavor to let the results of these "analyses" fit in (by means of analogously stultifyingly treating the words "possible", "necessary" or "only") is to certain extent characteristic of a part of analytic philosophy up to now." (Wittgenstein's transient years

are) "a story of waking up from this blissful insanity." Instead of a perfectly fitting system of "language" Wittgenstein "turned to an incoherent and chaotic landscape of language full of irregularities..."

True, both trends I want to analyze are characterized in this quotation, but this characteristic is surely not 'objective', emotionally neutral: it tries to evaluate both approaches and to show that one of them is passé while the other is 'waking up' from 'blissful insanity'.

One of the key notions enabling the author to write such an evaluation is the notion of *simplification*. I will show that the author's standpoint is based on a very essential *misunderstanding*.

2 Frege's legacy

In (Frege, 1884) and articles about nineties, especially in (Frege, 1892), Frege recognized that understanding expressions of a given language has to be *explained*. He proposed his 'semantic triangle', i.e. the scheme where between the *expression* and its *reference* (*denotation*) there is inserted an abstract object called *Sinn* (*sense*, we will also use *meaning*). The idea was sound: without something like Frege's *sense* one could not explain why some series of sounds or letters can denote an object, a property, a relation etc. and why distinct expressions can denote one and the same object.

Remember: to *explain* a fact we have to introduce some *abstract notion(s)* which *explicate* (in Carnap's sense) some intuitively understood idea.

Frege was underestimated and ignored by his contemporaries but the development of (not only mathematical) logic, analytic philosophy and logical analysis of (natural) language in the 20th century is unthinkable without him. Even the original Warsaw-Lemberg school has been influenced (Ajdukiewicz) and the nominalists from the Vienna Circle had to admit that the idea of semantics was attractive – Carnap in 1947.

3 The Golden Age of semantics

Frege did not *define* his *sense*, he just characterized it as *the mode of presentation* (see Black & Geach, 1952, p. 57), which suggests an important role of the sense but which cannot replace a definition / explication. Therefore Frege himself met a problem with semantics of some subordinate sentences (and tried to solve it, unfortunately accepting contextualism), and Carnap (1947) had to solve the same problem. This circumstance has led however to a kind of blossoming semantics. Kaplan (1970) writes on page 13:

> "During the Golden Age of Pure Semantics we were developing a nice homogenous theory, with language, meanings, and entities of the world each properly segregated and related one to another in rather smooth and comfortable ways. This development probably came to its peak in Carnap's *Meaning and Necessity* (from 1947). ... There is great beauty and power in this theory."

According to Kaplan there were some gaps in this beautiful theory. Kaplan speaks about *proper names, demonstratives, quantification into intensional contexts*, but we can add some important points:

Carnap himself recognized that his "method of intensions and extensions" cannot solve some semantic problems ("indirect contexts" – like Frege (1892), who reacted by 'reference shift' and became a contextualist).

Thus Carnap can be said to have discovered the phenomenon of hyperintensionality (described by Cresswell, 1975). Here we can see the inspiration for Alonzo Church, who began to look for a finer criterion of equivalence of expressions than Carnap's *intensional isomorphism*: see (Church, 1993), where Church defines *synonymous isomorphism*, and Anderson's (1998) appraisal of Church's role.

(Church's results can be used when *procedural semantics* is discovered as a tool of solving problems of contemporary analyses.)

4 Analytic theories of language

The development of the analytic theories of language continued in various directions:

Montague's influential school of logical analysis of natural language, based on lambda calculus and presenting particular analyses of a fragment of English. Montague's abstraction works with translations from English to intensional logic and using the standard process of interpreting. See (Gamut, 1991; Montague, 1970, 1974).

Systems using the idea of possible worlds helped modal logic understand modalities and add to C.I. Lewis-like axioms semantic making it possible to also define empirical modalities due to Kripke's accessibility relations (see Kripke, 1963, 1980).

20th century is full of outstanding philosophers and logicians offering their contributions to solving problems first formulated by Frege. We find here so many great names of philosophers and logicians who continued working at least partially on the Frege's legacy (Hintikka, Marcus, D. Lewis, Kaplan, not to forget B. Russell essentially inspired by Frege, Woleński, Partee, Schiffer etc. etc.).

This fact can be explained: The problems that were gradually discovered when the Frege's work was finally appraised gave birth to further problems. The endeavor to closer characterize Frege's idea of *Sinn* always led to new sub-problems.

An important attempt in this respect had it that *intensions* (as PW-intensions) could play this role. This step evoked new problems: would the mathematical expressions not possess meaning? Or: what is denoted by empirical expressions? In what sense can we speak with Cresswell about 'structured meaning'? Is meaning dependent on context?

Among the names of the 'wise (wo)men' working in the 20th century on solving the problems of *meaning* of the NL expressions one name must be certainly adduced: Pavel Tichý, who founded a semantic system as a hyperintensional system generated by a top-down approach.

Here we adduce Tichý's characteristic of *explication* because we should be aware of the fact that we do not claim "Meaning is ..." but rather "The optimum explication of *meaning* is ...". The quotation will also contribute to our main task: showing that the evaluation of the situation concerning the notions of analyzing language, as quoted at the outset of this paper, is based on an essential misunderstanding.

The following quotation can be found in (Tichý, 1988, pp. 194–195).

> "The purpose of theoretical explication is to represent intuitions in terms of rigorously defined entities. It is to Frege that we owe the insight that the mathematical notion of function is a universal medium of explication not just in mathematics but in general. To explicate a system of intuitive, pretheoretical notions is to assign to them, as surrogates, members of the functional hierarchy over a definite objectual base. Relations between the intuitive notions are then represented by the mathematically rigorous relationships between the functional surrogates."

Tichý's characteristic of explication is a concretization of what Carnap

(1950) said about explication. It is clear that every explication has to use abstractions and is therefore a kind of *simplification*. We will return to this point below.

5 Two kinds of criticism

Whenever some shortcomings or simply unsolved problems were detected in the work of analytic philosophers / logicians two kinds of reaction could have been expected:

a) *Collegial*: A way of removing the shortcomings is proposed so that the explanatory character of the text were preserved.

Examples:
Being dissatisfied by Frege's definition of concept Church (1956) redefines concept. His proposal preserves the realistic character of concept and makes it possible to continue Frege's logical treatment of concepts and of the relation between sense and denotation. See Church's reaction(s) to Carnap's intensional isomorphism.

When Tichý detects that Frege's 'reference shift' is untenable he defines suppositions *de re* and *de dicto* and modifies Frege's semantic triangle.

b) *Hostile*: The purpose itself of the criticized text is made dubious.

Examples:
Two examples are sufficient here because they essentially influenced the way in which an illusion arose according to which the analytic endeavor to *explain* some logical puzzles connected with superficial use of language is obsolete and should be replaced by a faithful description of the real use of language.

The first example is the well-known Quine's criticism (1951) of Carnap's attempt at defining semantics of natural language in (1947). According to Quine Carnap's proposal is a simplification. His attempt to define the triple *meaning, analyticity, synonymy* necessarily breaks down because *meaning* is an obscure notion and language behaves holistically.

Holistic behavior means: "the distinction between the linguistic and the empirical factor (that makes it possible to distinguish between analytic and synthetic sentences) is not significantly traceable into the statements of science taken one to one" (Quine, 1951, p. 42).

Quine's critique was not oriented as a rectification of Carnap's attempt, unlike Church's critique: it resulted in turning down Carnap's project of explicating semantic notions. Quine replaces this project by a pragmatic description of the way language *actually* behaves (see Quine, 1960).

Instead of the 'obscure' notion of meaning in the spirit of Frege's *Sinn* a behaviorist generalization *stimulus meaning* is leaving semantics and becomes a naturalistic causal notion.

It seems that while Carnap simplifies reality Quine describes it using empirical methods.

The second example is also famous. Wittgenstein's *Tractatus* was a great attempt at explicating some notions used to explain the way in which language corresponds to the world. This time it was Wittgenstein himself who began to speak about heavy errors in this remarkable book and to abandon any attempt to use abstractions that should *explicate* used notions to find deep regularities helping to *explain* linguistic behavior from the viewpoint of its underlying logic. Instead we find now collections of witty observations of particular "linguistic games", full of interesting ideas and stimuli.

Summarizing, in the light of Quine's refusing analytic methods and later Wittgenstein's shift towards describing "language games" an impression arose that

the *analytic approach* to studying and analyzing language, which tried to *explain* some facts connected with using language and relevant from the viewpoint of *logic*,

explicated therefore some more or less unclear notions and used frequently *abstraction*,

is *too artificial*, and *simplifies* the complicated world of language use, whereas

the philosophers inspired by Quine and the later Wittgenstein, who do not explain the above mentioned facts but *describe* the way a language is used and for whom *generalization* is a sufficient kind of abstraction, represent a more 'progressive' form of doing philosophical analyses because they do not need to simplify the complicated reality of language use.

6   Comments and summarization

A. Abstraction

Reality – including the reality of language use is of course complex and chaotic. We can become aware of some features of this reality when using

Two Approaches to Philosophically Analysing Language

a somehow systematic description and taking into account some generalizations. A more systematic way of detecting some features that are interesting from some relevant viewpoint consists however in *abstracting something from something* and *explicating* the notions that arose due to abstracting. When we want to *explain* some phenomenon then multiple abstractions and so multiple simplifications are necessary.

Imagine how such a successful science as physics would develop if it did not systematically use abstraction. It would probably be a boring discipline describing physical phenomena known to everybody.

Nobody will probably criticize physics although it demonstrably simplifies reality. Everybody knows or at least guesses that there would be in practice no physics if simplifications were forbidden. Logical analysis of language is not as 'popular' as physics, so people are tolerated when they comply that it contains so many abstract objects and so many simplifications – see the quotation at the outset of this article.

As an illustration of the role of (simplifying) abstraction we adduce a quotation from (Tichý, 1995, pp. 183–184) (reprinted in Tichý 2004, pp. 883–884), where Tichý answers the question, when Gallileo did mathematics discovering the law of gravity:

> "while plotting the values of the function against its arguments Gallileo was not doing mathematics. He was just taking down what was dictated to him by nature. ...But Gallileo not only *identified* the free-fall function. He also noted that there is quite a straightforward way of *calculating* the values of the function from its arguments. Given an argument, all one needs to do is multiply it by two, divide the result by 9.7, and then take a square root. It was when he made this second discovery, a discovery concerning a complex involving functions and numbers, that he was doing mathematics."

By the way, the simplification given by this calculation is clear: no object will obey the result because the resistance of the given environment has been abstracted from...Thus abstracting we always simplify but this does not mean that we negate complexity. We can show that as "simplifying" analysts we know that, necessarily, we simplify but that we know very well that this necessary simplification leads after all to acquisition of new knowledge. In (Duží, Jespersen, & Materna, 2010, p. 1) we write:

> "The way we understand the enterprise of logical analysis of

(natural) language, it is neither eliminative nor reductive, but selective. The analysis selects particular features of language, leaving all the remaining untouched and unscathed. We obviously acknowledge the pragmatic categories of (act of) assertion, language acquisition, communication, speaker's intention, etc. And we acknowledge no less the full range of pragmatic paraphernalia that keep natural language lubricated and running, including non-verbal winks and nods, hints and clues. But while they exist in their own right, they are immaterial to the project of, ideally, isolating all and only logically salient features of (natural) language. So we blot out what is in effect the vast bulk of natural language in order to zoom in on the remaining fragment and blow it, as it were, with a view to studying it in more detail."

B. Philosophy

We have suggested the goal of our analytic studies concerning the logical character of some abstract complexes underlying natural language and enabling us to explain the fact that we can clarify the meaning of NL expressions and the deep character of logical laws explained in terms of those complexes. This is in my opinion the task to be gradually fulfilled by our studies and we are convinced that we do something what can be classified with *philosophical* studies, since it transcends the immediate level of observed facts. Let us speak about the *analytic group (AG)*.

Now a following question can emerge:

Can we ascribe philosophical character to studies that are worked out in the spirit of Quine and, e.g., the later Wittgenstein (Q-W)?

There are two possibilities:

I. The work of the Q-W group is essentially not a philosophical work. In this case it is impossible to compare the *Q-W group* with *AG*, let alone to claim that the former is somehow more progressive than the latter.

II. The Q-W group is also a philosophically oriented group. Then it is clear that the kind of philosophy applied by it is essentially distinct from that applied by AG. Then, of course, the comparison of both groups is impossible as well.

Two Approaches to Philosophically Analysing Language

Ad I.

Quine applied philosophy when he criticized Carnap, viz. by using holistic objection. Building up his positive theory of language he created an interesting theory in (Quine, 1960). I think however that his theory is based on experience and is restricted to describing phenomena. As a smart description using more generalization than other kinds of abstraction it could be hardly called *explication*. An empirical description – even be it as good as Quine's – is in my opinion not yet a philosophical analysis.

As for Wittgenstein, let him speak:

> "Philosophy simply puts everything before us, and neither explains nor deduces anything. Since everything lies open to view there is nothing to explain. For what is hidden, for example, is of no interest to us." (Wittgenstein, 1968, §126)

This can be interpreted as either a denial of philosophy or a formulation of philosophy incompatible with the philosophical background of AG.

Ad II.

Well, it would be rather laborious to characterize the philosophy that underlies the views shared by the members of the Q-W group. It will be probably better to mention some points where AG will always dissent with the Q-W people.

The main such point consists in willingness to assume that abstract complexes definable within an explication should be taken seriously: while the members of AG are happy with such complexes the Q-W people warn us against abstract 'commitments'.

By defending the use of abstract complexes (like abstract procedures) members of AG do not necessarily speak about 'existence' because if to exist means to be principally spatio-temporally localizable then abstract objects *do not* exist, as our great Platonist Bernard Bolzano explicitly stated in his (1837, I., p. 224).

Therefore AG people are convinced that *explaining* some phenomena whose necessary character is admitted (albeit reluctantly) even by the Q-W people can be realized only taking in abstract surrogates for not very clear intuitions. They oppose the inclination to sell this necessity for the necessity of norms, which we can observe in some formulations typical for the members of Q-W. Therewith is connected the well-known reluctance to admit necessities even in logic, as we can read in Quine's (1951).

The notion of necessity (to be explained by means of using abstract complexes in explications) makes it possible to distinguish empirical (contingent) knowledge from non-empirical knowledge. The borderline is definite (definable) for AG while the Quinean relativization thereof lives in the followers of Q-W. Thus the AG people warn us against the inclination to accept something like 'absolute empiri(ci)sm', according to which all what can be said about, e.g., language can be said within empirical study pursued, e.g., by linguistic semantics.

Suggesting these differences between AG and Q-W we can state that the *problems which have to be solved by AG are distinct from the problems to be solved by Q-W*. In this sense we can say that the quotation from Beran's book claims nonsense: it tries to compare (and even evaluate) incomparable disciplines.

There is a level on which we perhaps *could* compare, and maybe that the author of that quotation had in mind this higher level, namely the level where we want to compare and evaluate the *choice itself*: would you prefer to choose AG, or Q-W? To consider this problem means to compare two *comparable* approaches if we assume that we want to solve the same problems, and claim that one of these choices is better than the other. Yet I tried to show that AG and Q-W actually solve distinct problems. Thus we can say nothing rational to this second interpretation of Beran's 'Conclusion'.

References

Anderson, C. A. (1998). Alonzo Church's Contributions to Philosophy and Intensional Logic. *The Bulletin of Symbolic Logic*, 4(2), 129–171.
Beran, O. (2013). *"Střední" Wittgenstein: cesta k fenomenologii a zase zpátky*. Červený Kostelec: Pavel Mervart.
Black, M., & Geach, P. (Eds.). (1952). *Translations from the Philosophical Writings of Gottlob Frege*. Oxford: Basil Blackwell.
Bolzano, B. (1837). *Wissenschaftslehre*. Sulzbach: Seidel.
Carnap, R. (1947). *Meaning and Necessity*. Chicago: Chicago University Press.
Carnap, R. (1950). *Logical Foundations of Probability*. Illinois: University of Chicago Press.
Church, A. (1956). *Introduction to Mathematical Logic*. Princeton: Princeton University Press.

Church, A. (1993). A Revised Formulation of the Logic of Sense and Denotation. Alternative (1). *Noûs, 27*(2), 141–157.
Cresswell, M. J. (1975). Hyperintensional Logic. *Studia Logica, 34*, 25–38.
Duží, M., Jespersen, B., & Materna, P. (2010). *Procedural Semantics for Hyperintensional Logic*. Berlin: Springer.
Frege, G. (1884). *Die Grundlagen der Arithmetik*. Breslau: W. Koebner.
Frege, G. (1892). Über Sinn und Bedeutung. *Zeitschrift für Philosophie und Philosophische Kritik, 100*, 25–50.
Gamut, L. T. F. (1991). *Logic, Language and Meaning, vol. II*. Chicago, London: The University of Chicago Press.
Kaplan, D. (1970). Dthat. *Syntax and Semantics, 9*, 221–243.
Kripke, S. (1963). Semantical Considerations on Modal Logic. *Acta Philosophica Fennica, 16*, 83–94.
Kripke, S. (1980). *Naming and Necessity*. Oxford: Basil Blackwell.
Montague, R. (1970). English as a Formal Language. *Linguaggi nella Societa et nella Technica*, 188–221.
Montague, R. (1974). *Formal Philosophy: Selected Papers of Richard Montague* (R. H. Thomason, Ed.). New Haven: Yale University Press.
Quine, W. V. O. (1951). Two Dogmas of Empiricism. *The Philosophical Review, 60*, 20–43. (Reprinted in Quine, W. V. O. (1980). *From a Logical Point of View*. Cambridge, USA: Harvard University Press.)
Quine, W. V. O. (1960). *Word and Object*. Cambridge: MIT Press.
Tichý, P. (1988). *The Foundations of Frege's Logic*. Berlin, New York: De Gruyter.
Tichý, P. (1995). Constructions as the subject-matter of mathematics. In W. DePauli-Schimanovich and E. Köhler and F. Stadler (Ed.), *The Foundational Debate: Complexity and Constructivity in Mathematics and Physics* (pp. 873–885). Dordrecht: Kluwer Academic Publishers.
Wittgenstein, L. (1968). *Philosophical Investigations*. Oxford: Basil Blackwell.

Pavel Materna
Institute of Philosophy of the Czech Academy of Sciences
The Czech Republic
E-mail: maternapavel@seznam.cz

Structures, Homomorphisms, and the Needs of Model Theory

PETER MILNE

Abstract: When we look closely at textbooks on model theory, we find that there are three different accounts of what a model or structure is. One of these is highly language dependent, so that the same structure cannot be the interpretation of two different languages or signatures. The other two definitions do not fall foul of that dependence but *all* textbooks tie the notion of homomorphism so closely to language (signature) that only structures interpreting the same language (signature) are isomorphic. Although this follows the practice in universal algebra, it is highly unnatural. The aim here is to present a notion of homomorphism better consonant with intuition and with what the less cautious authors of textbooks say when they speak informally.

Keywords: Model theory, structures, homomorphisms

1 Introduction

There is an informal notion of a *structure* which one finds in many texts on model theory, a notion borrowed from algebra. There is also, in many, a formal notion supposedly motivated by the informal one. Depending on which text one reads, these two notions may not match up: it depends on the way in which the formal notion involves a language.[1] To see what I'm getting at, begin by considering Wilfrid Hodges' characterization of contemporary practice in model theory:

> Model theorists are forever talking about symbols, names, and labels. A group theorist will happily write the same abelian group multiplicatively or additively, whichever is more convenient for the matter in hand. Not so the model theorist: for him

---

[1] In some texts, such as the classic *Model Theory* (Chang & Keisler, 1990), 'structure' is used only informally and does not appear in the index; on the other hand, 'model' is used both formally and informally.

or her the group with '·' is one structure and the group with '+' is a different structure. Change the name and you change the structure. (Hodges, 1993, p. 1; Hodges, 1997, p. 1)[2]

If we read this literally, it has an odd consequence, for many of the texts in model theory published in English over the past fifteen years or so started life in other languages: (Manzano Arjona, 1999), Spanish; (Poizat, 2000), French; (Rothmaler, 2000), (Prestel & Delzell, 2011), German; (Tent & Ziegler, 2012), has its origin in lecture notes in German; (Marcja & Toffalori, 2003) is not a literal translation but has some overlap with (Marcja & Toffalori, 1998). So many languages – so many different structures? However odd that may seem, on closer inspection of what these and other authors say, we find that there are in practice *three different accounts* of what a model-theoretic structure (or model) is.

Hodges continues,

> This must look like pedantry. [...] Nevertheless there are several good reasons why model theorists take the view that they do. [...]
>
> In the first place, we often want to compare two structures and to study the homomorphisms from one to the other. What is a homomorphism? [...A] homomorphism from structure A to structure B is a map which carries each operation of A to *the operation with the same name in B*. (Hodges, 1993, p. 1; Hodges, 1997, p. 1; emphasis in the originals)

It is an immediate and surely unwanted – and, I think, largely unrecognised – consequence of what Hodges tells us here that where the algebraist sees only one group Hodges' model theorist must see two *non-isomorphic* groups. But worse is to come. Even those authors who can readily agree with the algebraist that there is only one structure give the language (or the signature) used in the specification of a structure such a role in the definition of *homomorphism* that structures specified using different languages/signatures simply *cannot* be isomorphic! We have, then, an unhappy situation: either we have (i) a notion of structure compatible with the notion of homomorphism but which renders structure so parochial that the same model theorist speaking in different languages cannot, by her own lights, be speaking of the same structures, or we have (ii) a notion of structure consonant with common sense and a notion of homomorphism so parochial that what should

---

[2](Hodges, 1997) takes a lot of material unchanged from (Hodges, 1993).

count as the *same* model specified in different languages has to be seen as a family of necessarily non-isomorphic models.

The definition of homomorphism that we find in (Hodges, 1993, 1997; Marcja & Toffalori, 2003; Marker, 2002; Poizat, 2000; Prestel & Delzell, 2011; Rothmaler, 2000; Tent & Ziegler, 2012) takes homomorphisms to be defined *only* between structures with a common signature. Despite this, Hodges says that it 'is meant to take in, with one grand sweep of the arm, virtually all the things that are called "homomorphism" in any branch of algebra'! And truth to tell, textbooks in universal algebra conform. But intuitively the notions of homomorphism and isomorphism that these authors arrive at fall *far* short of the mark. Indeed, going by the letter of the definitions of structure and homomorphism they offer, we can have languages interpreted in the same domain with the same distinguished individuals, functions and relations but the resulting structures are *not* isomorphic exactly because no allowance for difference of language is made in the definition of homomorphism. Authors who do not tie structures to languages – *e.g.*, Bell and Slomson (1969), Sacks (1972), Manzano Arjona (1999) – do no better, for they too, as we'll see, tie homomorphisms to signatures (similarity types).

My aims are twofold: to tease out the various notions of structure in the textbooks and, rather more importantly, to arrive at a notion of homomorphism compatible with the less parochial reading of structure. For help in this, I turn to Joseph Goguen and Rod Burstall's theory of institutions (Goguen & Burstall, 1992); here we find a stepping-stone towards a definition of homomorphism between structures that better matches what the less cautious of our authors say when they speak informally. (Though the morals drawn have wider application, I shall, as is common in textbook presentations of model theory, confine attention to the model theory of first-order languages.)

2 What is model theory?

Let us begin by taking a look at what some texts on model theory say model theory is. In J. L. Bell and A. B. Slomson's *Models and Ultraproducts* (1969) we find

> Model theory ...can be described briefly as the study of the relationship between formal languages and abstract structures.
> (Bell & Slomson, 1969, p. 1)

Peter Milne

David Marker, *Model Theory: An Introduction* (2002):

> Model theory is a branch of mathematical logic where we study mathematical structures by considering the first-order sentences true in those structures and the sets definable by first-order formulas. (Marker, 2002, p. 1)

Annalisa Marcja and Carlo Toffalori, *A Guide to Classical and Modern Model Theory* (2003):

> Model Theory is – or, more precisely, was at its beginning – the study of the relationship between mathematical formulas and structures satisfying or rejecting them. (Marcja & Toffalori, 2003, p. 1)

I fully expect that you find nothing controversial in these claims. What I want to point to here is that they are naturally read as agreeing with what Bell and Slomson say explicitly:

> In ch. 3 we described the language L of predicate calculus with equality and then we looked at interpretations, i.e. realizations, of this language. In this chapter and for most of the rest of this book we are going to look at things the other way round. We regard relational structures as the objects of primary interest and we introduce a first order predicate language in order to be able to say things about them. This is the most natural viewpoint in mathematics; *it is the structures that present themselves first.* (Bell & Slomson, 1969, p. 72, my emphasis)

As Gerald Sacks says pithily, 'The "objects" of model theory are the structures' (Sacks, 1972, p. 10). On the face of it, this is all very much at odds with what we saw Hodges saying.

Now, I certainly do not claim that there is conceptual confusion, far less error, in the actual day-to-day practice of model theory. But there are issues concerning basic notions, issues which become especially pressing if one thinks of model theory as providing a paradigm for semantics. That the identity of what we talk about be in part determined by the language in which we talk, for example, would render translation impossible, as it might seem to be, on Hodges' characterization, in the case of our model theory textbooks, for we cannot refer in one language to what we speak of in another.

Structures, Homomorphisms, and the Needs of Model Theory

3   Structures in the intuitive sense

Let's take a quick look at what some of the textbooks say about structures. Chang and Keisler begin *Model Theory* by saying

> Let us now take a short introductory tour of model theory. We begin with the models which are structures of the kind that arise in mathematics. For example, the cyclic group of order 5, the field of rational numbers, and the partially-ordered structure consisting of all sets of integers ordered by inclusion, are models of the kind we consider. (Chang & Keisler, 1990, p. 1)

In similar fashion, in his chapter on first-order logic in the model theory section of the *Handbook of Mathematical Logic*, Jon Barwise gives us groups, a group being a triple $\langle G, +, 0 \rangle$ satisfying certain conditions, namely associativity of the operation $+$, the existence of an identity, 0, for $+$, and the existence of inverses with respect to $+$ and the identity, and the ordered field $\mathfrak{R} = \langle \mathbb{R}, +, \cdot, <, 0, 1 \rangle$ of real numbers among other examples (Barwise, 1977, pp. 7-8, 11).

Philipp Rothmaler (*Introduction to Model Theory*, 2000) offers

> By specifying certain neutral elements and operations, we may view the set $\mathbb{Z}$ of integers as an additive group, as a multiplicative semigroup, or as a ring. [...] We could also add the inverse operation $-$ or the ordering relation $<$. (Rothmaler, 2000, p. 3)

Marker says

> Intuitively a structure is a set that we wish to study equipped with a collection of distinguished functions, relations, and elements. We then choose a language where we talk about the distinguished functions, relations, and elements and nothing more. For example, when we study the ordered field of real numbers with the exponential function, we study the structure $(\mathbb{R}, +, \cdot, \exp, <, 0, 1)$ where the underlying set is the set of real numbers, and we distinguish the binary functions addition and multiplication, the unary function $x \mapsto e^x$, the binary order relation, and the numbers 0 and 1. To describe this structure we could use a language where we have symbols for $+$, $\cdot$, $\exp$, $<$, $0, 1 \ldots$.

For another example, we might consider the structure $(\mathbb{N}, +, 0, 1)$ of the natural numbers with addition and distinguished elements 0 and 1. The natural language for studying this structure is the language where we have a binary function symbol for addition and constant symbols for 0 and 1. (Marker, 2002, p. 7)

Marcja and Toffalori say,

Structures are an algebraic notion. [...] What is a structure? Basically, it is a non-empty set A, with a collection of distinguished elements, operations, and relations. For instance, the set **Z** of integers with the usual operations of addition + and multiplication · is a structure, as well as the same set **Z** with the order relation $\leq$. Note that in these cases the underlying set is the same (the integers), but, of course, the structure changes: in the former case we have the ring of integers, in the latter the integers as an ordered set. (Marcja & Toffalori, 2003, p. 1)

But some authors do follow Hodges' line. Like Hodges, Bruno Poizat (2000), Alexander Prestel and Charles Delzell (2011), and Katrin Tent and Martin Ziegler (2012) are careful to spell out the language first when presenting a structure – in flat out contradiction, it must be said, of what Bell and Slomson called the most natural viewpoint in mathematics.

Hodges, however, rather gives the game away with one of his examples – either that or he credits naming with magical power! He says:

Linear orderings Suppose $\leq$ linearly orders a set X. Then we can make $\langle X, \leq \rangle$ into a structure A as follows. The domain of A is the set X. There is one binary relation symbol R, and its interpretation R^A is the ordering $\leq$. (Hodges, 1993, p. 3; Hodges, 1997, p. 3)

For Hodges, going by what he says informally, the ordered pair $\langle X, \leq \rangle$ is *not* a structure, not, at least, for the model theorist – it might be for the algebraist. *We* make it into one by taking a two-place relation symbol and assigning $\leq$ as its interpretation. This gives rise to a bizarre profusion of entities. Suppose you pick one relation symbol and I pick another – we now have two structures where previously we had none. How many structures built over $\langle X, \leq \rangle$ are there? Hodges is, quite explicitly, not at all fussed about what counts as a language.

Structures, Homomorphisms, and the Needs of Model Theory

> Aha – says the group theorist – I see you aren't really talking about *written* symbols at all. For the purposes you have described, you only need to have formal labels for some parts of your structures. It should be quite irrelevant what kinds of thing your labels are; you might even want to have uncountably many of them.
>
> Quite right. In fact we shall follow the lead of A. I. Mal'tsev (1936) and put no restrictions at all on what can serve as a name. For example, any ordinal can be a name, and any mathematical object can serve as a name of itself. (Hodges, 1993, p. 2; Hodges, 1997, p. 2)

Now, if that is right, $\langle X, \leq \rangle$ *is* a structure with $\leq$ autonymously naming itself and we are not needed to make it into one (which, to be fair, I don't really suppose Hodges thinks we are) and, assuming we aren't really needed to associate names and what they name, there are at least as many structures built over $\langle X, \leq \rangle$ as there are items in your favourite set-theoretic hierarchy. And – the main point, to be developed below – they are all isomorphic in a way that model-theorists do *not* capture with their definitions of isomorphism between structures!!!

4 Structures in the formal sense

Minor variations aside, there are *three* styles of definition for models or structures in textbooks on model theory.

Definitions of the first kind: labelling (1) Let L comprise the non-logical vocabulary of a first-order language. A model/interpretation/relational structure $\mathfrak{A}$ for L is an ordered pair $\langle A, \mathscr{I} \rangle$, where

- A is a non-empty set;

- $\mathscr{I}$ assigns elements of A to the constants in L, n-ary operations on A to the n-place function-symbols of L, and n-place relations over A to the n-place relation-symbols of L.

We call $\mathscr{I}$ an *interpretation function*. We call L the *signature* of $\mathfrak{A}$.

Definitions of the second kind: labelling (2) Let L and $\mathscr{I}$ be as above. A model/interpretation/relational structure $\mathfrak{A}$ for L is an ordered pair $\langle A, \mathscr{I}[L]\rangle$, where $\mathscr{I}[L]$ is the image of L under $\mathscr{I}$.

As $\mathscr{I}$ is a function from L into $A \cup \bigcup_{n \in \mathbb{N}+} A^{A^n} \cup \bigcup_{n \in \mathbb{N}+} \mathscr{P}(A^n)$, $\mathscr{I}[L]$ is some non-empty subset of $A \cup \bigcup_{n \in \mathbb{N}+} A^{A^n} \cup \bigcup_{n \in \mathbb{N}+} \mathscr{P}(A^n)$.

The difference – a difference that has, of course, no real significance for the practice of model theory but is nonetheless present right in the very characterization of model theory's objects of study! – is this: for some authors a model comprises a domain A and a *function* from a language/signature into the set $A \cup \bigcup_{n \in \mathbb{N}+} A^{A^n} \cup \bigcup_{n \in \mathbb{N}+} \mathscr{P}(A^n)$; for others it is the domain together with the *image* of the language/signature under such a function. Often enough the distinction is collapsed notationally: while a model is said to comprise a domain and function, it is written out as a domain together with the image of the language/signature under that function. For example, Annalisa Marcja and Carlo Toffalori tell us

> A structure $\mathcal{A}$ for L is a pair consisting of a non empty set A, called the **universe** of $\mathcal{A}$, and a function mapping
>
> (i) every constant c of L into an element $c^{\mathcal{A}}$ of A,
>
> and, for any positive integer n,
>
> (ii) every n-ary function symbol f of L into an n-ary operation $f^{\mathcal{A}}$ of A (hence a function from A^n into A),
>
> (iii) every n-ary relation symbol R of L into an n-ary relation $R^{\mathcal{A}}$ of A (hence a subset of A^n).
>
> The structure $\mathcal{A}$ is usually denoted as follows
>
> $$\mathcal{A} = (A, (c^{\mathcal{A}})_{c \in L}, (f^{\mathcal{A}})_{f \in L}, (R^{\mathcal{A}})_{R \in L}).$$

(Marcja & Toffalori, 2003, p. 2)[3]

Chang and Keisler, more careful than most, distinguish between a model and its "displayed form" (Chang & Keisler, 1990, p. 20).

---

[3] $(c^{\mathcal{A}})_{c \in L}$, $(f^{\mathcal{A}})_{f \in L}$, and $(R^{\mathcal{A}})_{R \in L}$ are indexed families (or indexed sets). On some accounts, indexed families are functions (with the indexing set as domain and the set of indexed elements as the range). But Sacks is clear that, *e.g.*, $\{c_k^{\mathcal{A}} | k \in K\}$ in his notation is a *subset* of the domain A (Sacks, 1972, p. 8) and Hodges is likewise so (Hodges, 1993, p. 2, Hodges, 1997, p. 2). Moreover, what would be the point of replacing one function with three?

We should note that there are two ways to think of an indexed family such as $(c^{\mathcal{A}})_{c \in L}$. One

Structures, Homomorphisms, and the Needs of Model Theory

Definitions of the third kind: indexing A (similarity) type or signature is a quintuple $\langle I, J, K, \mu, \nu \rangle$ such that $\mu : J \to \mathbb{N}$ and $\nu : K \to \mathbb{N}$.

A model/relational structure $\mathfrak{A}$ of signature/type $\langle I, J, K, \mu, \nu \rangle$ consists of

- a non-empty set A;

- for each $i \in I$, an element $c_i^{\mathfrak{A}}$ of A;

- for each $j \in J$, a $\mu(j)$-ary operation $f_j^{\mathfrak{A}}$ defined on A;

- for each $k \in K$, a $\nu(k)$-place relation $R_k^{\mathfrak{A}}$ defined over A.

The sets I, J and K index the distinguished individuals, functions and relations in the model, respectively. We can again think of there being an interpretation function, or, better, three interpretation functions, from the indexing sets into A, $\bigcup_{n \in \mathbb{N}^+} A^{A^n}$, and $\bigcup_{n \in \mathbb{N}^+} \mathscr{P}(A^n)$, respectively.

Gerald Sacks (*Saturated Model Theory*) takes the signature/type to be the quintuple; María Manzano (*Model Theory*) takes it to be just $\langle \mu, \nu \rangle$ – she takes individuals to be functions of zero arguments but notes that we could distinguish them from functions.

We have a 2×2 classification of what structures/models are said to be: they divide on whether the signature σ is the non-logical vocabulary L of a language $\mathscr{L}$ or comprises arbitrary indexing sets and on whether the second component of a model is an interpretation function $\mathscr{I}$ or the image of the signature under that function $\mathscr{I}[\sigma]$. In the case of indexing, the model is given as the images of the two (Manzano) or three (Sacks) indexing functions; what's important is that it's the images, not the functions themselves, that are constituents of the model. Table 1 encapsulates where various authors stand.

Here σ is the type or signature. When $\mathfrak{A} = \langle A, \mathscr{I} \rangle$, we have Hodges' "different language, different structure". When $\mathfrak{A} = \langle A, \mathscr{I}[\sigma] \rangle$, the same structure can be specified using different signatures.[4]

---

way respects the multiplicity of multiple occurrences, with different indices, of the same item. This way, we would obtain a *multiset*. The other is to ignore repetitions, thus taking $(c^A)_{c \in L}$ and its ilk to be sets (as I have in speaking of the image of the signature under the interpretation function). None of our authors explicitly endorses the multiset reading; some are explicit in insisting on sets. On multisets, see, *e.g.*, (Blizard, 1989, 1991; Hickman, 1980; Singh, 1994; Singh, Ibrahim, Yohanna, & Singh, 2007).

[4]Textbooks on universal algebra mostly fall into the lower right quadrant (*e.g.* Bergman, 2012, pp. 3-4; Cohn, 1981, pp. 48 & 188-9; Denecke & Wismath, 2002, p. 4; Grätzer, 1979,

	$\sigma = L$ (labelling)	$\sigma \neq L$ (indexing)
$\mathfrak{A} = \langle A, \mathscr{I} \rangle$	Chang and Keisler 1990 Barwise 1977 Poizat 2000 Marcja and Toffalori 2003	
$\mathfrak{A} = \langle A, \mathscr{I}[\sigma] \rangle$	Hodges 1993, 1997 Rothmaler 2000 Marker 2002 Prestel and Delzell 2011 Tent and Ziegler 2012	Bell and Slomson 1969 Sacks 1972 Manzano Arjona 1999

Table 1: What is a structure?

The placing of Hodges here may strike the reader as odd given our starting point. What Hodges says is

> A **structure** A is an object with the following four ingredients.
>
> (1.1) A set called the **domain** of A. ...
>
> (1.2) A set of elements of A called **constant elements**, each of which is named by one or more **constants**. If c is a constant, we write c^A for the constant element named by c.
>
> (1.3) For each positive integer n, a set of of n-ary relations ... each of which is named by one or more n-ary **relation symbols**. If R is a relation symbol, we write R^A for the relation named by R.
>
> (1.4) For each positive integer n, a set of of n-ary operations ... each of which is named by one or more n-ary **function symbols**. If F is a function symbol, we write F^A for the function named by F.

(Hodges, 1993, p. 2, Hodges, 1997, p. 2)

I leave it to the reader to judge the accuracy of my placing.[5]

---

pp. 8 & 223-4). A couple speak of symbols and look more like they belong in the bottom left (*e.g.* Plotkin, 1994, pp. 36-7; Wechler, 1992, pp. 5-6). I owe to Göran Sundholm the suggestion to investigate what textbooks in universal algebra say.

[5] Hodges defines signatures thus:

Structures, Homomorphisms, and the Needs of Model Theory

5 Homomorphisms

Every one of our authors ties homomorphisms to signatures (types).

Let $\mathfrak{A}$ and $\mathfrak{B}$ be two models of the same signature L with domains A and B respectively, determined by two interpretation functions, $\mathscr{I}$ and $\mathscr{J}$, respectively (both with domain L).

A function $h : A \to B$ is a homomorphism from $\mathfrak{A}$ to $\mathfrak{B}$ iff

- $h(\mathscr{I}(c)) = \mathscr{J}(c)$, for each constant c in L;

- for each $\langle a_1, a_2, \ldots, a_n \rangle$ in A,
  $h(\mathscr{I}(f)(a_1, a_2, \ldots, a_n)) = \mathscr{J}(f)(h(a_1), h(a_2), \ldots, h(a_n))$, for each n-ary function-symbol f in L;

- for each $\langle a_1, a_2, \ldots, a_n \rangle$ in A, if $\mathscr{I}(R)(a_1, a_2, \ldots, a_n)$ then $\mathscr{J}(R)(h(a_1), h(a_2), \ldots, h(a_n))$, for each n-place relation-symbol R in L.

Let $\mathfrak{A}$ and $\mathfrak{B}$ be two models of the same signature/type $\langle I, J, K, \mu, \nu \rangle$ with domains A and B respectively.

A function $h : A \to B$ is a homomorphism from $\mathfrak{A}$ to $\mathfrak{B}$ iff

- $h(c_i^{\mathfrak{A}}) = c_i^{\mathfrak{B}}$, for each i in I;

- for each $\langle a_1, a_2, \ldots, a_{\mu(j)} \rangle$ in A,
  $h(f_j^{\mathfrak{A}}(a_1, a_2, \ldots, a_{\mu(j)})) = f_j^{\mathfrak{B}}(h(a_1), h(a_2), \ldots, h(a_{\mu(j)}))$, for each j in J;

The **signature** of a structure A is specified by giving

> the set of constants of A, and, for each $n > 0$, the set of n-ary relation symbols and the set of n-ary function symbols of A.

We shall assume that the signature of a structure can be read off uniquely from the structure.

The symbol L will be used to stand for signatures. Later it will also stand for languages – think of the signature of A as a rudimentary language for talking about A. If A has signature L, we say A is an L-structure. (Hodges, 1993, pp. 4-5; Hodges, 1997, pp. 4-5)

Given that Hodges allows each item – object, n-ary relation, n-ary operation – in a structure to be named by more than one element of the signature, one might wonder how we read the signature off the structure. I guess the thought is that in *presenting* a structure we cannot but associate the items comprising the structure with elements of a signature/language.

Peter Milne

- for each $\langle a_1, a_2, \ldots, a_{\nu(k)} \rangle$ in A, if $R_k^{\mathfrak{A}}(a_1, a_2, \ldots, a_{\nu(k)})$ then $R_k^{\mathfrak{B}}(h(a_1), h(a_2), \ldots, h(a_{\nu(k)}))$, for each k in K.[6]

So even when $\langle A, \mathscr{I}[\sigma] \rangle = \langle A, \mathscr{J}[\sigma'] \rangle$, the identity function on A cannot count as a homomorphism unless $\sigma = \sigma'$.

In the intuitive sense, a structure is a set A, its domain, together with some subset of $A \cup \bigcup_{n \in \mathbb{N}^+} A^{A^n} \cup \bigcup_{n \in \mathbb{N}^+} \mathscr{P}(A^n)$ comprising the distinguished individuals, functions, and relations. With that in mind, there are two ways we might go from here. One is to broaden the model-theorist's notion of homomorphism to allow homomorphisms between structures *in the formal sense* with different signatures. The other is to define a notion of homomorphism appropriate to structures *in the intuitive sense*. Although textbooks in model theory ignore both, it's fairly clear how to go about getting both.

Firstly, we borrow from Joseph Goguen and Rod Burstall's theory of *institutions* the idea of a *signature morphism* (Goguen & Burstall, 1992). A signature morphism is a function between signatures mapping constants to constants, n-ary function-symbols to n-ary function-symbols, and n-place relation-symbols to n-place relation-symbols.[7] What we're after is a con-

---

[6]In their definition of *homomorphism*, Bell and Slomson (1969, p. 73) strengthen this to an 'if, and only if', thus defining what Manzano Arjona (1999, p. 24) calls a 'strong homomorphism'. *Cf.* the definitions of *monomorphism* in (Sacks, 1972, p. 9), where additionally injectivity of h is required.

[7]I owe thanks to Roy Dyckhoff for drawing the literature on institutions to my attention. Although signature morphisms are the essential stepping-stone to where we want to get to here, Goguen and Burstall's aims are different; they do not define a notion of homomorphism between structures with different signatures.

Goguen and Burstall are interested in the categorial structure you get when you put signature morphisms together with homomorphisms of structures (in the formal sense) for fixed signatures. In particular, if i is a signature morphism from L to L' then there's an induced mapping from the category of models with signature L' into the category of models with signature L – notice the reversal of the order. (The morphisms in the categories are the homomorphisms in the formal sense between structures with the same signature – L' and L, respectively.)

Given a signature morphism $i : L \to L'$ and a model with domain A and signature L', determined by interpretation function $\mathscr{J}$, we define the associated model with domain A and signature L by defining the interpretation function $\mathscr{I}$ as follows:

- $\mathscr{I}(c) = \mathscr{J}(i(c))$, for all constants c in L;
- $\mathscr{I}(f) = \mathscr{J}(i(f))$, for all n-ary function-symbols f in L;
- $\mathscr{I}(R) = \mathscr{J}(i(R))$, for all n-place relation-symbols R in L.

It now turns out that if there's a homomorphism (in the already defined sense) from the model with domain A determined by interpretation-function $\mathscr{J}$ to the model with domain B

Structures, Homomorphisms, and the Needs of Model Theory

ception of a homomorphism from a model with signature L to a model with signature L'. We can get that by combining our previous definition of a homomorphism for models of the same signature with a signature morphism. What we get is the following:

$h : A \to B$ is a homomorphism from the model $\mathfrak{A}$, with domain A, determined by signature L and interpretation function $\mathscr{I}$, to the model $\mathfrak{B}$, with domain B, determined by signature L' and interpretation function $\mathscr{J}$ if there's a signature morphism i from L to L' and

- $h(\mathscr{I}(c)) = \mathscr{J}(i(c))$, for each constant c in L;

- for each $\langle a_1, a_2, \ldots, a_n \rangle$ in A,
  $h(\mathscr{I}(f)(a_1, a_2, \ldots, a_n)) = \mathscr{J}(i(f))(h(a_1), h(a_2), \ldots, h(a_n))$, for each n-ary function-symbol f in L;

- for each $\langle a_1, a_2, \ldots, a_n \rangle$ in A, if $\mathscr{I}(R)(a_1, a_2, \ldots, a_n)$ then $\mathscr{J}(i(R))(h(a_1), h(a_2), \ldots, h(a_n))$, for each n-place relation-symbol R in L.[8]

The guiding thought is that we can have a homomorphism not only between structures that *are* labelled by the members of the same signature but between structures that *can* be labelled by the members of the same signature. We can limit attention to the case when the signature morphism is surjective; indeed, we had better if we are not to lose sight of reducts and expansions. (Obviously, we can draw up an analogous definition for the case of structures defined by types. So too in what follows immediately below.)

A structure in the intuitive sense *serves as an interpretation* of the signature L if there is an interpretation function $\mathscr{I}$ from L onto the set of distinguished items of the structure. The structure *serves as a duplication-free interpretation* if $\mathscr{I}$ is a bijection. We can then say: there is a homomorphism from the structure with domain A to the structure with domain B if (*i*) they serve as duplication-free interpretations of the same signature L (under the interpretation functions $\mathscr{I}$ and $\mathscr{J}$, respectively) and (*ii*) there

---

determined by interpretation-function $\mathscr{J}'$, both with signature L', then the same function from A to B is a homomorphism from the associated model with domain A and signature L induced by i to the associated model with domain B and signature L induced by i – and, if you think about it, that's obvious.

[8] For a strong homomorphism from $\mathfrak{A}$ into $\mathfrak{B}$ we require that the last clause be tightened to an 'if and only if' (Manzano Arjona, 1999, p. 24; Rothmaler, 2000, p. 5); for an embedding (monomorphism) of $\mathfrak{A}$ into $\mathfrak{B}$ we require further that h be injective. For an isomorphism we require, additionally, that it be surjective.

is a homomorphism from the structure in the formal sense with domain A, determined by L and $\mathscr{I}$, into the structure in the formal sense with domain B, determined by L and $\mathscr{J}$.

(Given a domain A, let S be some subset of $A \cup \bigcup_{n \in \mathbb{N}+} A^{A^n} \cup \bigcup_{n \in \mathbb{N}+} \mathscr{P}(A^n)$. If we follow the lead Hodges ascribes to Anatoliĭ Ivanovich Mal'tsev, the existence of a language for which $\langle A, S \rangle$ serves as a duplication-free interpretation is trivial: we take S itself to be the language, its elements naming themselves autonomously.)

6 The needs of model theory

A structure comprises a domain and a set of distinguished items (individuals, operations, relations). In order to distinguish those distinguished items, we need to use labelling or indexing (or something like that). So we naturally associate labels or indices with the distinguished items in a structure. What we shouldn't do – and, as we've seen, the majority of model-theory texts do not – is to take the labelling/indexing to be a feature of the structure. (Barwise, 1977; Chang & Keisler, 1990; Marcja & Toffalori, 2003; Poizat, 2000 are the exceptions.)

When we consider homomorphisms between structures, we need to indicate which individual/operation/relation in one structure matches up with which individual/operation/relation in the other. Using a common labelling/indexing is an easy way to spell that out. So the way model-theorists proceed is entirely natural. The problem then is this: if we leave matters this way, this ties our notion of homomorphism to a shared labelling and it is this that is at odds with the "pre-theoretical", ordinary, working mathematician's notion of a homomorphism between structures. As we saw, what should be one structure turns into a family of non-isomorphic structures when presented under different labellings/indexings.

Despite the central role languages play in model theory, model-theorists are not interested in particular languages (not in the way field linguists are); they are interested in classes of languages, languages in which certain kinds of thing can be said, distinctions drawn, and the like, and the relations of (classes of) languages to their interpretations. Now, any structure (in the intuitive sense) that serves as an interpretation of a signature L, serves equally as an interpretation of any other signature in which the various grammatical categories – constants, n-place function-symbols, n-place relation-symbols – have the same cardinalities. At least some of the time, this is what model-theorists have in mind. But model-theorists cannot make do with classes of languages distinguished solely by the cardinalities of their grammatical categories. Sometimes they start with models of one language then consider interpretations of expansions of that language. Expanding a language needn't change the cardinalities of its grammatical categories. Nevertheless, a distinction needs to be made between the original language and its expansion(s). And an interpretation of an expansion is exactly that: it is not usefully to be thought of as an interpretation of the original language.

The model-theorist cannot – or, at the very least, cannot without a good deal of unedifying, pointlessly pedantic, faffing about – reduce the identification of languages to the cardinalities of the categories of their non-logical vocabulary. Such an identification *is* all that's needed in determining whether a structure can serve as an interpretation of a language and for spelling out a sensible notion of homomorphism. But in practice it won't do for all model theory's purposes. Consequently, the model theorist ends up speaking as though she does have particular languages in mind (even though she really doesn't).

References

Barwise, J. (1977). An Introduction to First-order Logic. In J. Barwise (Ed.), *Handbook of Mathematical Logic* (p. 5-46). Amsterdam: North-Holland Publishing Company.

Bell, J. L., & Slomson, A. B. (1969). *Models and Ultraproducts: An Introduction*. Amsterdam and Oxford: North-Holland Publishing Co. (Third revised printing, 1974. Reprinted Mineola NY: Dover Publications, 2006.)

Bergman, C. (2012). *Universal Algebra: Fundamentals and Selected Topics*. Boca Raton: CRC Press.

Blizard, W. D. (1989). Multiset Theory. *Notre Dame Journal of Formal Logic*, *30*(1), 36-66.

Blizard, W. D. (1991). The Development of Multiset Theory. *Modern Logic*, *1*(4), 319-352.

Chang, C. C., & Keisler, H. J. (1990). *Model Theory* (third ed.). Amsterdam: North-Holland Publishing Company. (First edition, 1973. Third edition republished, with a new preface, Mineola NY: Dover Publications, 2012.)

Cohn, P. M. (1981). *Universal Algebra* (revised ed.). Dordrecht: D. Reidel. (First published New York: Harper & Row, 1965.)

Denecke, K., & Wismath, S. (2002). *Universal Algebra and Applications in Theoretical Computer Science.* Boca Raton: Chapman & Hall/CRC Press.

Goguen, J. A., & Burstall, R. M. (1992). Institutions: Abstract Model Theory for Specification and Programming. *Journal of the Association for Computing Machinery, 39*(1), 95-146.

Grätzer, G. (1979). *Universal Algebra* (second ed.). New York: Springer. (Updated 2008. First edition Princeton/Toronto/London: D. van Nostrand Co., 1968.)

Hickman, J. L. (1980). A Note on the Concept of Multiset. *Bulletin of the Australian Mathematical Society, 22*(2), 211-217.

Hodges, W. (1993). *Model Theory.* Cambridge: Cambridge University Press.

Hodges, W. (1997). *A Shorter Model Theory.* Cambridge: Cambridge University Press.

Mal'tsev, A. I. (1936). Untersuchungen aus dem Gebeite der Mathematischen Logik. *Matematicheskiĭ Sbornik, 1*(43), 323-336. (Translated as 'Investigations in the realm of mathematical logic' in A. I. Mal'cev *The metamathematics of algebraic systems: Collected papers 1936–1967*, B. F. Wells, III (Ed. and trans.), Amsterdam: North-Holland, 1971, pp. 1-14.)

Manzano Arjona, M. (1989). *Teoría de Modelos.* Madrid: Alianza Editorial, SA.

Manzano Arjona, M. (1999). *Model Theory.* New York: Oxford University Press. (English translation of (Manzano Arjona, 1989).)

Marcja, A., & Toffalori, C. (1998). *Introduziona alla Teoria del Modelli.* Bologna: Pitagora Editrice.

Marcja, A., & Toffalori, C. (2003). *A Guide to Classical and Modern Model Theory.* Dordrecht: Kluwer Academic Publishers.

Marker, D. (2002). *Model Theory: An Introduction.* New York: Springer-Verlag.

Plotkin, B. (1994). *Universal Algebra, Algebraic Logic, and Databases.* Dordrecht: D. Reidel.

Poizat, B. (1985). *Cours de Théorie des Modèles: Une Introduction à la Logique Mathématique Contemporaine.* Villeurbanne, Lyon: B. Poizat.

Poizat, B. (2000). *A Course in Model Theory: An Introduction to Contem-*

porary Mathematical Logic. New York: Springer-Verlag. (English translation of (Poizat, 1985).)

Prestel, A. (1986). *Einführung in die Mathematische Logik und Modelltheorie*. Braunschweig: Friedrich Vieweg & Sohn.

Prestel, A., & Delzell, C. N. (2011). *Mathematical Logic and Model Theory: A Brief Introduction*. London: Springer Verlag. (English translation of (Prestel, 1986).)

Rothmaler, P. (1995). *Einführung in die Modelltheorie: Vorlesungen*. Heidelberg: Spektrum Akademischer Verlag. (Prepared by Frank Reitmaier.)

Rothmaler, P. (2000). *Introduction to Model Theory*. Amsterdam: Gordon and Breach Science Publishers. (English translation of (Rothmaler, 1995). Reprinted London: Taylor & Francis, 2002.)

Sacks, G. E. (1972). *Saturated Model Theory*. Reading, MA: W. A. Benjamin, Inc. (Reissued as a second edition with corrections, new pagination, and a new introductory paragraph in the preface, Singapore: World Scientific Publishing Co. Pte. Ltd., 2010. Page references to the 2010 edition.)

Singh, D. (1994). A Note on "The Development of Multiset Theory" [Blizard *1991*]. *Modern Logic*, *4*(4), 405-406.

Singh, D., Ibrahim, A. M., Yohanna, T., & Singh, J. N. (2007). An Overview of the Applications of Multisets. *Novi Sad Journal of Mathematics*, *37*(2), 773-792.

Tent, K., & Ziegler, M. (2012). *A Course in Model Theory*. Cambridge: Association for Symbolic Logic/Cambridge University Press.

Wechler, W. (1992). *Universal Algebra for Computer Scientists*. Berlin and Heidelberg: Springer Verlag.

Peter Milne
University of Stirling
United Kingdom
E-mail: peter.milne@stir.ac.uk

Curry's Paradox and the Inclosure Schema

MARTIN PLEITZ[1]

Abstract: To show that the set theoretic and semantic paradoxes have the same structure, Graham Priest has formulated his *Inclosure Schema* and shown that it characterizes the paradoxes of both groups. I will argue that the failure of the Inclosure Schema to also characterize *Curry's paradox* is more important than is usually realized. I will show how this failure can be addressed in a way that harmonizes with Priest's own use of the Inclosure Schema. In order to achieve this I will formulate what I call the *Curry Schema*, show that it describes curry-like paradoxes in a way similar to how the Inclosure Schema describes those logical paradoxes that establish a contradiction, and formulate a more general schema, which characterizes both inclosure paradoxes and curry-like paradoxes. I will end, however, on a more critical note by pointing out why the resulting more general diagnosis of the logical paradoxes (now including Curry's paradox) might turn out to be ambivalent with regard to Priest's project.

Keywords: Curry's paradox, logical paradoxes, paradoxes of self-reference, semantic and set theoretic paradoxes, Inclosure Schema, Principle of Uniform Solution, Graham Priest, dialetheism, paraconsistency, logical revision, metaphysical revision, vicious circle principle

1 The logical paradoxes

In this paper, I will restrict my attention to *paradoxology* – it will mainly be about the comparison of paradoxes and the question of how to sort paradoxes into categories, with only bits of diagnosis and hints at possible solutions thrown in.

The paradoxes that are of interest include the paradoxes of Cantor, Burali-Forti, Mirimanoff, Russell, Berry, König, Richard, Grelling, the Liar paradox, and the more recent addition of Yablo's paradox. Graham Priest calls these "the paradoxes of self-reference" (Priest, 1994), but as Yablo's

---

[1] I would like to thank Graham Priest, Niko Strobach, Heinrich Wansing, and all other participants of the logic colloquium on March 25th, 2014 in Münster for a helpful discussion.

paradox arguably does not make use of self-referential sentences, and as reference in the usual narrow, linguistic sense is involved only in some of these paradoxes, I will speak of them simply as *the logical paradoxes*.

Russell thought that all[2] the logical paradoxes have a common structure (cf., e.g. Russell, 1908). Ramsey, however, divided them into a set theoretic and a semantic family (Ramsey, 1925, p. 171f). Prima facie this division is justified by a criterion of *content*: The paradoxes of the set theoretic group make use of set theoretic concepts like elementhood, order type, and cardinality; the paradoxes of the semantic group make use of semantic concepts like truth, satisfaction, and reference. The division into two families has become mainstream, probably because widely accepted and highly influential proposals for a solution address only one of the two groups.

This becomes obvious when we look at the basic ingredients of the logical paradoxes. In very general terms, what happens in a logical paradox is the following:

- The claim that a certain set, predicate, sentence, etc. *exists*

- together with a certain *unrestricted bridge principle*,
  like (Naïve Set Abstraction) $\forall x(x \in \{y \mid \varphi(y)\} \leftrightarrow \varphi(x))$
  or (Tarski's Truth Schema) $\text{TRUE}(\underline{\alpha}) \leftrightarrow \alpha$,[3]

- in *classical logic* leads to contradiction and triviality.

Hence there are three broad ways to go with the goal of solving the paradoxes: We can revise metaphysics (i.e., argue against the respective existence claim), revise our set theoretic or semantic theories (i.e., restrict the respective bridge principles), or revise logic. The current mainstream (or "orthodoxy", to use a more distancing term) seems to be clear that the set theoretic paradoxes have been solved by metaphysical revision (e.g., the axiomatic set theories ZF and NBG prove that the problematic sets do not exist). And although, even today, no proposed solution to the semantic paradoxes is generally accepted, the main contenders accept the existence of the problematic expressions and must hence revise the bridge principles or logic (or both) to solve the semantic paradoxes. This is so from Tarski's hierarchy of languages, via Kripke's theory of truth, and on to current paracomplete and contextualist approaches. With a view to *both* the set theoretic and

---

[2] Minus Yablo's paradox, of course, which was not yet discovered at Russell's time.

[3] I use underlining (e.g., in '$\underline{\alpha}$') to indicate a formal name-forming device.

the semantic group of paradoxes, none of these approaches can be classified as uniform, because they all combine metaphysical revision for the set theoretic paradoxes with semantic and/or logical revision for the semantic paradoxes.

Before the division of the logical paradoxes into a set theoretic and a semantic family had been established, Russell had proposed to solve them all by revising metaphysics in the same way for both groups (using the vicious circle principle to argue against the respective existence claims; e.g., by way of the theory of types). The proposed solution of Priest and other dialetheists is equally uniform: It calls for revising logic for both groups, again in the same way: by dropping the ex falso rule, i.e., by adopting a paraconsistent logic.

2   The Principle of Uniform Solution and the Inclosure Schema

So, while orthodox approaches treat the set theoretic and the semantic paradoxes differently, both Priest's approach of logical revision and Russell's approach of metaphysical revision are *uniform*. Priest welcomes the uniformity of his dialetheist approach. He writes "it is natural to expect all the paradoxes of a single family to have a single kind of solution. Any solution that can handle only some members of the family is bound to appear somewhat one-eyed, and as not having got to grips with the fundamental issue." (Priest, 1994, p. 32). He endorses the Principle of Uniform Solution: "same kind of paradox, same kind of solution" (Priest, 1994, p. 32). Given this principle, a conclusive argument to the effect that all logical paradoxes belong to the same family would be "sufficient to sink virtually all orthodox solutions to the paradoxes" (1994, p. 32). To give such an argument, Priest takes up an observation of Russell's about the structure of the logical paradoxes (Russell, 1905, p. 142; cf. Priest, 1994, p. 27 and 2002, p. 129f) and develops it into his own Inclosure Schema (Priest, 1994, p. 28 and 2002, p. 133ff).

(**Inclosure Schema**) Let φ and ψ be predicates and δ a function. Then the following threefold condition holds:

(Existence)	$\exists \Omega (\Omega = \{x \mid \varphi(x)\} \wedge \psi(\Omega))$
(Transcendence)	$\forall x (x \subseteq \Omega \wedge \psi(x) \rightarrow \delta(x) \notin x)$
(Closure)	$\forall x (x \subseteq \Omega \wedge \psi(x) \rightarrow \delta(x) \in \Omega)$

This schema is a blueprint for paradox. Because of (Existence), 'x' can be instantiated as Ω in (Closure) and (Transcendence). Hence, $\delta(\Omega) \in \Omega$ and $\delta(\Omega) \notin \Omega$.

To show that the Inclosure Schema characterizes Russell's paradox, the Liar paradox, Berry's paradox, and Burali-Forti's paradox, Priest gives the following interpretations of φ, ψ, and δ:

	$\delta(x)$	$\psi(y)$	$\varphi(z)$	Ω
Russell	$\{w \in x \mid w \notin w\}$	$\lambda y\, y = y$	z is a set	V
the Liar	l, i.e., $\underline{l} \notin x$	y is definable	z is true	Tr
Berry	$\mu w\, w \notin x$	$y \in DN_{90}$	$z \in DN_{99}$	DN_{99}
Burali-F.	$log(x)$	$\lambda y\, y = y$	z is an ordinal	ON

Here, V is the universe of sets, Tr is the set of true sentences, DN_{99} is the set of numbers definable in 99 words, and ON is the set of all ordinals.[4] Therefore, $\delta(\Omega)$ is, for the respective interpretation, a Russell set, a Liar sentence, a Berry description, or Burali-Forti's ordinal of all ordinals. And Priest shows that these interpretations of φ, ψ, and δ satisfy the three conditions of the Inclosure Schema (Priest, 2002, pp. 129–131; 132; 134; 144f). Thus Russell's paradox, the Liar paradox, Berry's paradox, and Burali-Forti's paradox can be construed as satisfying the Inclosure Schema; and the same goes for the other logical paradoxes.

In this very brief summary, I of course disregard many details, some of them important (e.g., the role played by $\psi(y)$). I would rather relegate the unconvinced reader to Priest's own thorough treatment of these matters (Priest, 1994, 2002, pp. 128–175). I also refrain from engaging with the small but lively debate sparked by Priest's paper "The Structure of the Paradoxes of Self-Reference" (1994) and the first edition in 1995 of his book *Beyond the Limits of Thought* (2002), about whether his interpretations really capture what is essential of the particular paradoxes.[5] I have chosen to bracket the important details and the lively debate not only for reasons of space, but more importantly, for the sake of the argument. It is my aim to investigate in how far Curry's paradox presents a problem for Priest's use of his Inclosure Schema – even if not a single fault could be found with its characterization of the usual logical paradoxes.

---

[4] In the interpretation for Berry's paradox, 'μw' reads 'the least number w such that ...'. In the interpretation for Burali-Forti's paradox, 'log' is a *least ordinal greater than* operator.

[5] (Cf. Abad, 2008; Badici, 2008; Dümont & Mau, 1998; Grattan-Guinness, 1998; Priest, 2009, 2012; Weber, 2010; Zhong, 2012).

So let us take it as established that the logical paradoxes all fit the Inclosure Schema. Hence we should think of them as being all of the same kind. Together with the Principle of Uniform Solution, this result seems to provide Priest with an indirect but compelling argument for dialetheism.

3 Is Curry's paradox one of the logical paradoxes?

However, as its critics note and its friends admit, the Inclosure Schema fails to describe Curry's paradox (cf. Priest, 1994, p. 33 and 2002, p. 168f; Grattan-Guinness, 1998, p. 826f). Curry's paradox arises when we consider, for some arbitrary proposition p, a set like $\{x \mid x \in x \to p\}$ or a sentence like "If this sentence is true, then p", because we will find that these considerations seem to enable us to prove that p (Curry, 1942; Geach, 1955; cf. Beall, 2009). Curry poses a particular challenge to paraconsistent approaches to paradox, because it provides a way to triviality that does not use the ex falso rule (cf., e.g. Meyer, Routley, & Dunn, 1979).

Here are two examples of the reasoning of Curry's paradox.

(**Set Curry**)   (**Sentence Curry**)
There is a Curry set C,   There is a Curry sentence c,
i.e., $\{x \mid x \in x \to p\}$.   i.e., TRUE($\underline{c}$) $\to p$.
By (Naïve Set Abstraction),   By (Tarski's Truth Schema),
$C \in C \leftrightarrow (C \in C \to p)$.   TRUE($\underline{c}$) $\leftrightarrow$ (TRUE($\underline{c}$) $\to p$).

On each side we have thus established, for a suitable sentence γ ('$C \in C$' and 'TRUE($\underline{c}$)', respectively), that (1) $\gamma \to (\gamma \to p)$ and (2) $(\gamma \to p) \to \gamma$. By using the contraction rule on (1), we get $\gamma \to p$. Combining this with (2), two applications of the modus ponens rule will deliver p.

As p was arbitrary, we have found a way to triviality that bypasses the ex falso rule. Hence, the logical revisionist who wants to keep modus ponens must reject the contraction rule. This puts some pressure on dialetheist approaches to paradox.

So, Curry's paradox is a problem – but is it indeed one of the logical paradoxes? This question is of more than terminological or classificatory interest for those who intend to use the Principle of Uniform Solution. For if Curry's paradox and the (other) logical paradoxes are of the same kind, then it conflicts with the Principle of Uniform Solution to solve them by different kinds of logical revision – dropping ex falso for Russell, the Liar and so on, and dropping contraction for Curry.

Priest does not treat Curry's paradox as one of the logical paradoxes – or, in his own terms, he seems to claim that is is not one of the "paradoxes of self-reference" (cf. Priest, 1994, p. 33 and 1994, p. 168f).[6] I disagree. And what I will say in the following sections can be understood as a detailed argument to the conclusion that Curry's paradox should indeed be classified as one of the logical paradoxes. For now, I want to give three preliminary reasons for this classification. Firstly, Curry's paradox is based on similar existential claims as Russell's paradox and the Liar (the existence of a Curry set corresponds to the existence of a Russell set; the existence of a Curry sentence corresponds to the existence of a Liar sentence). Secondly, the Curry reasoning uses the same bridge principles, namely (Naïve Set Abstraction) or (Tarski's Truth Schema). And thirdly, Curry's paradox *overlaps* certain of the other logical paradoxes – at least in a wide range of logics.

To see this, note first that in many logical systems that have an absurd sentence '$\bot$' in their language, the following equivalence holds:

(Negation Equivalence) $\neg \alpha \Leftrightarrow \alpha \rightarrow \bot$

Note also that Curry's paradox really is a *schema* that delivers a multitude of paradoxical arguments. For each choice of the proposition p, it delivers the argument from the claim that there is the Curry set $\{x \mid x \in x \rightarrow p\}$ or from the claim that there is the formal Curry sentence $c = \text{TRUE}(\underline{c}) \rightarrow p$ to the particular conclusion p. Now, if our logic conforms to the (Negation Equivalence), we need only set $p = $ '$\bot$' to see the Curry set morph into the Russell set and the Curry sentence morph into a Liar sentence:[7]

$\{x \mid x \in x \rightarrow p\}$ $\xrightarrow{(p=\bot)}$ $\{x \mid x \notin x\}$

c, i.e., $\text{TRUE}(\underline{c}) \rightarrow p$ $\xrightarrow{(p=\bot)}$ l, i.e., $\neg\text{TRUE}(\underline{l})$

So, at least in the presence of the (Negation Equivalence), Russell's paradox is a limit case of Set Curry and the Liar is a limit case of Sentence Curry.

Priest notes this overlap between Curry and some of the (other) logical paradoxes (Priest, 1994, p. 33 and 2002, p. 168). He seems to be interested

---

[6] For Priest, Curry's paradox also does not seem to belong to his more general category of paradoxes, the "contradictions at the limits of thought". Given the arguments in the present paper for the similarity between Curry's paradox and some prime examples of Priest's contradictions at the limits of thought, I can agree only insofar as not all instances of Curry's paradox are contradictory. But they can still be *paradoxes* at the limits of thought.

[7] The metamorphosis of a Curry sentence into a Liar sentence must of course be accompanied by a corresponding metamorphosis of the term that makes the respective sentence self-referential, here from '$\underline{c}$' to '$\underline{l}$'.

only in Curry arguments with a contradictory conclusion, and observes that, in the absence of the (Negation Equivalence), these "do not involve negation" and thus "belong to a quite different family" (Priest, 2002, p. 169). But I think that he gets the direction of the overlap wrong when he states that only where the (Negation Equivalence) holds is "the curried form of each paradox [...] essentially the same as its uncurried form" (Priest, 2002, p. 169). For while not every pair of a negation and a conditional will satisfy the (Negation Equivalence), the (Negation Equivalence) can always be used to *define* some negation operator from a conditional (Restall, 2000, p. 60f). Thus there will *always* be variants of Russell's paradox and the Liar that are limit cases of a given variant of Curry's paradox.

Be that as it may – Curry's paradox will always have instances with a non-absurd conclusion, too. So the overlap is only partial; Curry is strictly more general than Russell's paradox and the Liar.

4 Mild Curry and a lesson from Church

As we have just seen, the overlap between Curry and (other) logical paradoxes is at best partial because many instances of Curry's paradox have a non-contradictory conclusion. E.g., considering the set $\{x \mid x \in x \to$ snow is white$\}$ or the sentence 'If this sentence is true, then snow is white', the Curry reasoning seems to enable us to prove that snow is white. And we could give a similar apparent proof that $1 = 1$.

What about these other Curry arguments? Are they really *milder*, because they have a non-contradictory conclusion, and in some cases, a true, or even a necessarily true conclusion? Are (as some seem to think) non-contradictory Curry arguments mostly harmless, at least on their own?

To see that even *Mild Curry* is difficult to digest, we need only generalize an observation made by Alonzo Church. He commented on the fact that Epimenides's variant of the Liar paradox alone does not allow to infer a contradiction. Church insists that it nevertheless should be "classed as a paradox" – because reflection on one Cretan saying 'All utterances by Cretans are lies' can always be turned into an apparent proof (by reductio) that there is an utterance by *another* Cretan that is not a lie (Church, 1946). The problem with this apparent proof is not its (true) conclusion in itself, but *how the conclusion is reached*. For why should the existence of one particular utterance *entail* the existence of another utterance with certain properties?

The same holds for Mild Curry. 'Snow is white' is neither a theorem of set theory nor of semantics. But it would be if there was not some fallacy somewhere in the Curry reasoning. *Any* instance of Curry's paradox, even taken in isolation, is deeply troubling, and should be understood as fully paradoxical.[8] Therefore it is an unwelcome consequence that Mild Curry – even in the presence of the (Negation Equivalence) – falls through the net of the Inclosure Schema.

5 The Curry Schema

Recall that the Inclosure Schema is a blueprint for *contradictory* paradox. An instance entails both $\delta(\Omega) \in \Omega$ and $\delta(\Omega) \notin \Omega$. Paraconsistently, this does *not* entail any p. But given modus ponens and contraction, there will *always* be a Curry argument with conclusion p. So, even in the presence of the (Negation Equivalence), most instances of Curry's paradox (namely the *mild* ones) do not fit the Inclosure Schema; they are not *inclosure paradoxes*.

What is more, the conclusion p of a particular Curry argument stands in a much more *direct* connection to the respective Curry set or Curry sentence than it would to the corresponding Russell set or Liar sentence in an argument via the ex falso rule from Russell's or the Liar paradox. (Just compare the argument that starts from considering the sentence 'If this sentence is true then snow is white' to the conclusion that snow is white with the argument that establishes the same conclusion by applying the ex falso rule after showing that a certain Liar sentence is both true and not true.)

So we can see that Curry's paradox cannot be inclosed. But it turns out that the Inclosure Schema can be curried! That is, it is possible to modify Priest's Inclosure Schema to gain a *Curry Schema* that relates to the variants of Curry's paradox in the same way as the Inclosure Schema relates to inclosure paradoxes. The modification concerns only the second condition of the Inclosure Schema, (Transcendence), which is replaced by a different

---

[8] I would like to thank Elia Zardini for pointing out to me that some instances of Curry's paradox do not fit our usual notion of a paradox as an argument from apparently true premises via apparently valid rules of logic to an apparently unacceptable conclusion. And to be sure, the proposition that snow is white (or that $1 = 1$) seems to be as acceptable as it gets. But I think we can save our usual definition of a paradox if we are willing to broaden the notion of an unacceptable conclusion. For that snow is white is surely unacceptable *as the conclusion of the Curry argument* that starts from considering the sentence 'If this sentence is true, then snow is white', because it clearly cannot be warranted by a bunch of non-empirical premises alone. So let us think of a conclusion as unacceptable either if it is false or if it intuitively cannot be warranted by the premises of the argument it is the conclusion of.

condition, which I call '(p-Spiciness)' because its parameter 'p' is a measure of how mild (or *hot*) the conclusion of the respective Curry paradox is:

(**Curry Schema**) Let φ and ψ be predicates and δ a function. Then the following threefold condition holds:

(Existence) $\quad \exists \Omega(\Omega = \{x \mid \varphi(x)\} \wedge \psi(\Omega))$
(p-Spiciness) $\quad \forall x(x \subseteq \Omega \wedge \psi(x) \to (\delta(x) \in x \to p))$
(Closure) $\quad \forall x(x \subseteq \Omega \wedge \psi(x) \to \delta(x) \in \Omega)$

This is a blueprint for a paradoxical proof of p: Because of (Existence), 'x' can be instantiated as Ω in (Closure) and (p-Spiciness). Hence, $\delta(\Omega) \in \Omega$ and $\delta(\Omega) \in \Omega \to p$. By modus ponens, p.

We can give the following interpretations of φ, ψ, and δ to show that the Curry Schema characterizes the set theoretic and the sentential variant of Curry's paradox:

	$\delta(x)$	$\psi(y)$	$\varphi(z)$	Ω
Set Curry:	$\{w \in x \mid w \in w \to p\}$	$\lambda y\, y = y$	z is a set	V
Sentence C.:	c, i.e., $\underline{c \in x \to p}$	y is definable	z is true	Tr

Given that V is the universe of sets and Tr the set of true sentences, it is clear that $\delta(\Omega)$ is a Curry Set or a Curry Sentence, respectively. And it can be shown that these interpretations of φ, ψ, and δ satisfy the three conditions of the Curry Schema – (p-Spiciness), in particular.[9]

One point is probably worth emphasizing. We got the Curry Schema from the Inclosure Schema and the recipes for Set Curry and Sentence Curry from Priest's recipes for Russell's paradox and the Liar by replacing a for-

---

[9] We want to show, e.g., that the interpretation of $\delta(x)$ as $\{y \in x \mid y \in y \to p\}$, $\psi(y)$ as $\lambda y\, y = y$, and $\varphi(z)$ as 'z is a set', which entails that $\Omega = V$ and has Set Curry as a limit case, satisfies the condition (p-Spiciness) of the Curry Schema. In the following proof the biconditional of the (Naïve Abstraction Principle) is split up into its two directions (Abs $\Leftarrow$) and (Abs $\Rightarrow$).

1	*	$\delta(x) \in x$	assumption
2	*	$\{y \in x \mid y \in y \to p\} \in x$	1, def. of δ
3	**	$\{y \in x \mid y \in y \to p\} \in \{y \in x \mid y \in y \to p\}$	assumption
4	**	$\{y \in x \mid y \in y \to p\} \in \{y \in x \mid y \in y \to p\} \to p$	3, (Abs $\Rightarrow$)
5	**	p	3, 4, modus ponens
6	*	$\{y \in x \mid y \in y \to p\} \in \{y \in x \mid y \in y \to p\} \to p$	3–5, cond. proof
7	*	$\{y \in x \mid y \in y \to p\} \in x \wedge \ldots$	
		$\{y \in x \mid y \in y \to p\} \in \{y \in x \mid y \in y \to p\} \to p$	2, 6, $\wedge$-Intro
8	*	$\{y \in x \mid y \in y \to p\} \in \{y \in x \mid y \in y \to p\}$	7, (Abs $\Leftarrow$)
9	*	p	7, 8, modus ponens
10		$\delta(x) \in x \to p$	1–9, cond. proof

mula like '$\neg\alpha$' with a formula like '$\alpha \to p$'. But that does not at all mean that the Curry Schema will work only for logical system in which the (Negation Equivalence) holds – it is strictly more general.

6 Harmonious overlap

To get clearer about their relation, let us compare Priest's Inclosure Schema with the Curry Schema. Recall:

(Inclosure Schema)   (Curry Schema)
$\exists \Omega(\Omega = \{x \mid \varphi(x)\} \wedge \psi(\Omega))$ $\exists \Omega(\Omega = \{x \mid \varphi(x)\} \wedge \psi(\Omega))$
$\forall x(x \subseteq \Omega \wedge \psi(x) \to \delta(x) \notin x)$ $\forall x(x \subseteq \Omega \wedge \psi(x) \to \ldots$
          $\ldots (\delta(x) \in x \to p))$
$\forall x(x \subseteq \Omega \wedge \psi(x) \to \delta(x) \in \Omega)$ $\forall x(x \subseteq \Omega \wedge \psi(x) \to \delta(x) \in \Omega)$

Let us also compare the interpretations of both schemata that gave us Russell's paradox and Set Curry, as well as Set Curry and the Liar. Recall:

	$\delta(x)$	$\psi(y)$	$\varphi(z)$	Ω
Russell:	$\{w \in x \mid w \notin w\}$	$\lambda y\, y = y$	z is a set	V
Set Curry:	$\{w \in x \mid w \in w \to p\}$	$\lambda y\, y = y$	z is a set	V
the Liar:	l, i.e., $\underline{l} \notin x$	y is def.	z is true	Tr
Sentence Curry:	c, i.e., $\underline{c} \in x \to p$	y is def.	z is true	Tr

What we can see here is that in the presence of the (Negation Equivalence), and for $p = \bot$, the Inclosure Schema is the *limit case* of the Curry Schema. The same goes in this situation for the particular interpretations of the two schemata: The interpretation that delivers Russell's paradox is a limit case of the interpretation that delivers Set Curry, and the interpretation that delivers the Liar paradox is a limit case of the interpretation that delivers Sentence Curry:[10]

$\{w \in x \mid w \in w \to p\} \xrightarrow{(p=\bot)} \{w \in x \mid w \notin w\}$

c, i.e., $\underline{c} \in x \to p \xrightarrow{(p=\bot)} l$, i.e., $\underline{l} \notin x$

A further point is that even in general (i.e., when we do not require that the (Negation Equivalence) holds), the two schemata and their respective interpretations look remarkably similar as far as *structure* is concerned. Priest

---

[10]Note that these are not the Curry and Russell set, nor the Curry and Liar sentence, but more general sets and sentences that depend on the parameter 'x', and that have the respective paradoxical sets and sentences as their respective limit case for $x = \Omega$.

describes what happens in an instance of the Inclosure Schema in the following way: "an irresistible force meets an unmovable object" (Priest, 1994, p. 27), where the "irresistible force" is the function δ and the "unmovable object" is Ω. Varying Priest's description we can add that in an instance of the Curry Schema, something we might call *a p-resistible force* meets an unmovable object, where the p-resistible force is the function δ and the unmovable object is, again, Ω.

7 A generalization

At this point it might look as if we had run out of similarities, and that the first differences start to emerge. For compare the following paradoxical arguments:

(Inclosure Schema)
$\exists \Omega(\Omega = \{x \mid \varphi(x)\} \wedge \psi(\Omega))$
$\forall x(x \subseteq \Omega \wedge \psi(x) \to \delta(x) \notin x)$

$\forall x(x \subseteq \Omega \wedge \psi(x) \to \delta(x) \in \Omega)$
So, $\delta(\Omega) \in \Omega$
and $\delta(\Omega) \notin \Omega$.
By ex falso, p.

(Curry Schema)
$\exists \Omega(\Omega = \{x \mid \varphi(x)\} \wedge \psi(\Omega))$
$\forall x(x \subseteq \Omega \wedge \psi(x) \to \ldots$
$\ldots (\delta(x) \in x \to p))$
$\forall x(x \subseteq \Omega \wedge \psi(x) \to \delta(x) \in \Omega)$
So, $\delta(\Omega) \in \Omega$
and $\delta(\Omega) \in \Omega \to p$.
By modus ponens, p.

Observe that here, the ex falso rule is visible in the argument on the left side, but the contraction rule is hidden in the argument on the right side. But it would surely be desirable for contraction to be visible, because (in the context of a discussion about logical revisionism) it is one of the most important logical features of the Curry reasoning.

It turns out that the contraction rule can indeed be made visible. For compare the following *modified* paradoxical arguments, which start from modifications of the Inclosure Schema and the Curry Schema, that are altered again only in their second condition – (Transcendence′) and (p-Spiciness′), as we might call them:

(**Modified Inclosure Schema**)
$\exists \Omega(\Omega = \{x \mid \varphi(x)\} \wedge \psi(\Omega))$
$\forall x(x \subseteq \Omega \wedge \psi(x) \to \ldots$
$\ldots (\delta(x) \in x \to \delta(x) \notin x))$
$\forall x(x \subseteq \Omega \wedge \psi(x) \to \delta(x) \in \Omega)$

(**Modified Curry Schema**)
$\exists \Omega(\Omega = \{x \mid \varphi(x)\} \wedge \psi(\Omega))$
$\forall x(x \subseteq \Omega \wedge \psi(x) \to \ldots$
$\ldots (\delta(x) \in x \to (\delta(x) \in x \to p)))$
$\forall x(x \subseteq \Omega \wedge \psi(x) \to \delta(x) \in \Omega)$

So, $\delta(\Omega) \in \Omega$
and $\delta(\Omega) \in \Omega \to \delta(\Omega) \notin \Omega$.
By reductio, $\delta(\Omega) \notin \Omega$.
By ex falso, p.

So, $\delta(\Omega) \in \Omega$
and $\delta(\Omega) \in \Omega \to (\delta(\Omega) \in \Omega \to p)$.
By contraction, $\delta(\Omega) \in \Omega \to p$.
By modus ponens, p.

To see that these are indeed no more than modifications of the two schemata we discussed so far, observe that in the presence of the rule of consequentia mirabilis, the Modified Inclosure Schema is equivalent to the Inclosure Schema, and in the presence of the contraction rule, the Modified Curry Schema is equivalent to the Curry Schema. The advantage of the modified schemata is that the interesting logical rule is now visible on both sides of the paradoxical reasoning that starts with the respective schema: The ex falso rule for inclosure paradoxes, and the contraction rule for Curry paradoxes. So the revisionist can easily see where to make the cut.

All three modifications of Priest's original Inclosure Schema that we have seen by now result from a change to the consequent of the quantified conditional that is its second condition, (Transcendence). This observation suggests the following generalization:

(**The Most General Schema**) Let φ and ψ be predicates and δ a function. Then the following threefold condition holds:

(Existence) $\exists \Omega(\Omega = \{x \mid \varphi(x)\} \wedge \psi(\Omega))$
(Gamma) $\forall x(x \subseteq \Omega \wedge \psi(x) \to \Gamma)$
(Closure) $\forall x(x \subseteq \Omega \wedge \psi(x) \to \delta(x) \in \Omega)$

All four schemata we discussed are special cases of the Most General Schema: Set $\Gamma =$ '$\delta(x) \notin x$' to get the Inclosure Schema, set $\Gamma =$ '$\delta(x) \in x \to p$' to get the Curry Schema, set $\Gamma =$ '$\delta(x) \in x \to \delta(x) \notin x$' to get the Modified Inclosure Schema, and set $\Gamma =$ '$\delta(x) \in x \to (\delta(x) \in x \to p)$' to get the Modified Curry Schema.

I conjecture that as long as Γ is not equivalent to '$\delta(x) \in x$', there will be a paradox. But that in itself need not put further pressure on the logical revisionist, because it is likely that a contraction-free paraconsistent logic will solve them all.

8 Discussion

In conclusion, I want to sum up what we have seen, and say something about what it might mean with regard to the diagnosis of the paradoxes and the quest for their solution.

We have seen that both Set Curry and Sentence Curry fit the Curry Schema. This is another piece of evidence for the thesis that there are deep similarities between paradoxes of the semantic and the set theoretic group, and it thus weakens the claim that these groups form two distinct families.

We have seen that the Curry Schema overlaps the Inclosure Schema, at least in the presence of the (Negation Equivalence). And we have seen that the schemata in general are similar in structure, a point that is also supported by the fact that both are special cases of the Most General Schema. This strengthens our earlier, preliminary reasons for the claim that Curry's paradox is one of the logical paradoxes.

We have seen that and how a contraction-free paraconsistent logic will solve all paradoxes that fit either the Inclosure Schema or the Curry Schema (and probably even all the paradoxes that fit the Most General Schema). But it has also become evident that it is doubtful whether this solution can count as *uniform*, because a logical revisionist could also treat inclosure paradoxes and curry-like paradoxes separately, by dropping ex falso for the former and dropping contraction for the latter.

What do these results mean for Priest's project? He can surely welcome the fact that the similarity between set theoretic and semantic paradoxes extends beyond the range of the inclosure paradoxes and into the wider realm of the curry-like paradoxes.[11] The strong support that we have found for counting Curry's paradox as one of the logical paradoxes will be less welcome to Priest, because given that the logical paradoxes form this larger family, logical revision looks much less like a uniform solution.

In fact, we might take the structural similarity between the schemata to speak not for logical but for *metaphysical* revision. In terms of the metaphor of an irresistible force (in the case of the inclosure paradoxes) or a p-resistible force (in the case of the curry-like paradoxes) crashing against the impenetrable barrier formed by an unmovable object, we can ask: Why not simply remove that barrier, and let the force run free, if this removes the need for dropping both ex falso and contraction from our logic?

In sum, the results look good and not good for Priest. In his hand, the Inclosure Schema might yet turn into a double edged sword. And that is not too surprising, because he found the material from which he forged this instrument in an observation of Russell's, who after all preferred an entirely

---

[11] Priest makes the same observation when he says that the Curry paradoxes "also cut across Ramsey's distinction" (Priest, 1994, p. 168).

different kind of uniform solution to the logical paradoxes.[12]

References

Abad, J. V. (2008). The Inclosure Scheme and the Solution to the Paradoxes of Self-Reference. *Synthese, 160*(2), 183–202.

Badici, E. (2008). The Liar Paradox and the Inclosure Schema. *Australasian Journal of Philosophy, 86*(4), 583–596.

Beall, J. (2009). Curry's Paradox. In E. N. Zalta (Ed.), *The Stanford Encyclopedia of Philosophy* (Spring 2009 ed.). http://plato.stanford.edu/archives/spr2009/entries/curry-paradox/.

Church, A. (1946). Review of "Alexandre Koyré. The Liar. Philosophy and Phenomenological Research, vol. 6 no. 3 (1946), pp. 344–362.". *The Journal of Symbolic Logic, 11*(4), 131.

Curry, H. B. (1942). The Inconsistency of Certain Formal Logics. *The Journal of Symbolic Logic, 7*(3), 115–117.

Dümont, J., & Mau, F. (1998). Are there True Contradictions? A Critical Discussion of Graham Priest's Beyond the Limits of Thought. *Journal for General Philosophy of Science, 29*, 289–299.

Geach, P. (1955). On Insolubilia. *Analysis, 15*(3), 71–72.

Grattan-Guinness, I. (1998). Structural Similarity or Structuralism? Comments on Priest's Analysis of the Paradoxes of Self-Reference. *Mind, 107*(428), 823–834.

Meyer, R. K., Routley, R., & Dunn, J. M. (1979). Curry's Paradox. *Analysis, 39*(3), 124–128.

Pleitz, M. (2015 [forthcoming]). *Logic, Language, and the Liar Paradox*. Münster: Mentis.

Priest, G. (1994). The Structure of the Paradoxes of Self Reference. *Mind, 103*(409), 25–34.

Priest, G. (2002). *Beyond the Limits of Thought*. Oxford: Clarendon.

Priest, G. (2009). Badici on Inclosures and the Liar Paradox. *Australasian Journal of Philosophy, 88*(2), 359–366.

Priest, G. (2012). Definition Inclosed: A Reply to Zhong. *Australasian Journal of Philosophy, 90*(4), 789–795.

---

[12] I myself prefer a broadly Russellian solution of metaphysical revision, requiring elementhood and reference to be *well-founded* relations, but – unlike Russell – describing the resulting hierarchical structure from within, with tools from tense logic. But that is another story; (cf. Pleitz, 2015 [forthcoming]).

Ramsey, F. P. (1925). The Foundations of Mathematics. *Proceedings of the London Mathematics Society*, *25*, 338–384.

Restall, G. (2000). *An Introduction to Substructural Logics*. London: Routledge.

Russell, B. (1905). On Some Difficulties in the Theory of Transfinite Numbers and Order Types. *Proceedings of the London Mathematical Society*, *4*, 29–53.

Russell, B. (1908). Mathematical Logic as Based on the Theory of Types. *American Journal of Mathematics*, *30*(3), 222–262.

Weber, Z. (2010). Explanation and Solution in the Inclosure Argument. *Australasian Journal of Philosophy*, *88*(2), 353–357.

Zhong, H. (2012). Definability and the Structure of the Logical Paradoxes. *Australasian Journal of Philosophy*, *90*(4), 779-788.

Martin Pleitz
Münster University
Germany
E-mail: martinpleitz@web.de

A Non-classical Logical Foundation for Naturalised Realism

EMMA RUTTKAMP-BLOEM, GIOVANNI CASINI, AND THOMAS MEYER

Abstract: In this paper, by suggesting a formal representation of science based on recent advances in logic-based Artificial Intelligence (AI), we show how three serious concerns around the realisation of traditional scientific realism (the theory/observation distinction, over-determination of theories by data, and theory revision) can be overcome such that traditional realism is given a new guise as 'naturalised'. We contend that such issues can be dealt with (in the context of scientific realism) by developing a formal representation of science based on the application of the following tools from Knowledge Representation: the family of Description Logics, an enrichment of classical logics via defeasible statements, and an application of the preferential interpretation of the approach to Belief Revision.

Keywords: scientific realism; theory-observation distinction; over-determination of theories by data; theory revision; scientific progress; description logics; preferential logics; belief revision

1 Introduction

In this paper, by suggesting a formal representation of science based on recent advances in logic-based AI, we show how three serious concerns around the realisation of traditional scientific realism can be overcome such that traditional realism is given a new guise as 'naturalised'. In the account of naturalised realism proposed by Ruttkamp-Bloem (2013), science does not advance in a teleological fashion towards 'the truth'. Instead, she suggests an epistemic definition of truth implying that the criterion for determining (realist) stances on the status of the content of science is the quality of evolutionary progressive interaction between the experimental and theoretical levels of science. The success of this process depends on the quality of available evidence. This (epistemological) analysis of science suggests a new interpretation of three well-known concerns about traditional scientific realism, which, in its turn, contributes to a firmer definition of naturalised realism as an epistemic realism vs. traditional metaphysical realism. The problems are: 1. Scientific realism should be able to deal with the messy issue of 'disentangling' empirical information and general statements and laws throughout the history of scientific investigations of a given real system, and with the added complication of the non-uniqueness of such disentanglements; 2. Theories should be robust in terms of the absorption of new empirical information in order to deal with the 'over-determination' of theories by data, akin to the definition of the equivalent empirical models of van Fraassen (1980); 3. Based on new empirical evidence, science is not prescriptive; it is not the case that there is a single 'correct' method for modifying an existing theory to incorporate new evidence. But science is also constraining; there ought to be limitations on what constitutes a possible rational modification to an existing theory.

We contend that these issues can be dealt with in a formal representation of science which is the result of applying the following tools from subareas of Knowledge Representation to the above philosophical issues in the context of scientific realism: The family of knowledge representation languages known as Description Logics (Baader, Calvanese, McGuinness, Nardi, & Patel-Schneider, 2007), an enrichment of the languages of classical logics with defeasible statements in the style of the preferential framework suggested by Kraus, Lehmann, and Magidor (1990) for propositional logic, and an application of the preferential interpretation of the approach to Belief Revision proposed by Alchourrón, Gärdenfors, and Makinson (1985), and extended to account for Iterated Belief Revision. In what follows, we first give a brief description of 'naturalised realism' as the version of realism supported by our suggested treatment of the three concerns around the possibility of actualising traditional scientific realism identified above. We then briefly discuss each of these concerns in turn, showing as we move along how in each case either Description Logics, Non-monotonic Reasoning, or Belief Revision can positively address each of these concerns. To conclude, we argue that a single coherent framework for realism could be provided which incorporates all three aspects discussed above.

2 Naturalised realism

The version of realism presented here is called 'naturalised realism' and it is based on the following twelve tenets:

1. Science is about an *independently externally existing reality* in William James's sense of experience "boiling over" (*viz.* Chang, 2012), thus in the sense of the 'resis-

A Non-classical Logical Foundation for Naturalised Realism

tance', or perhaps rather 'reaction', of nature to the application of theories as proof of the existence of the outside world; 2. Realist claims are compatible with *the nature and the history of science* in the full (naturalised) sense of reflecting the processes of science in its evaluations of the status of the content of theories; 3. The driver of scientific progress is *revision* (not retention), where retention is a special case of revision); 4. Scientific progress is *not linear or convergent* in the traditional sense; 5. The unit of appraisal for naturalised realism is a *network of systems of theories* (not single theories), where the notion of a 'system of theories' is close to the Kuhnian notion of a 'disciplinary matrix'; 6. Continuity in science is a meta-issue and is *methodological continuity*; 7. Naturalised realism is an *epistemology of science* (not a metaphysics); 8. The epistemological framework for naturalised realism is *fallibilism*; 9. The criterion for determining realist stances towards the status of the content of science is the quality of *'evolutionary progressive' interaction* between the experimental and theoretical levels of science; 10. An *epistemic account of truth as correspondence* is the best account for realism and here the account is unpacked in terms of *truth-as-method* ('method' here in the sense of the generally acceptable way in which scientists do research based on taking empirical evidence seriously, in short the Peircean 'experiental method'); 11. Relations of *'historied' reference* are epistemological trackers of what is revealed about nature as 'true' through the course of science (rather than existential claims with regards to the ontology of reality); 12. The dichotomy of the traditional scientific realism debate is collapsed into a *continuum of realist stances* towards the status of scientific theories.

The naturalised realist contends that whether realism is warranted with regards to particular scientific content depends most heavily on the kind and quality of evidence available for that content; she suggests that for the naturalised realist it makes sense to be a realist only about those aspects of scientific investigations that demonstrate actual science-world interaction, as such interaction is taken to be the source of establishing evidence for science's claims and the quality of such interaction is taken to determine the quality of evidence available for specific scientific claims at specific times.

This interaction is visible most importantly in the revisions science effects in its theories and experimental work as the result either of experimental feedback leading to theoretical adjustment or of theoretical adjustment necessitating revision at the level of experimental processes. Such revision is measured in terms of the 'evolutionary progressiveness' of theories.

A theory T is 'evolutionary progressive' at time t_n iff:

1. it is 'empirically (experimentally) adequate' according to experimental practices in the area of investigation at time t_n *in such a way that previous versions of*

theory T at time t_{n-1} *have been adapted in significant ways* in order to effect this adequacy.

2. it is 'theoretically adequate' in the sense that *theoretical descriptions made at t_{n-1} have been adapted* such that they describe or refer to (read: 'offer knowledge concerning') properties of observable and unobservable entities in the scope of the theory at time t_n.

The *naturalised realist criterion is the degree to which the definitions of reference and referential stability below are satisfied in a network of theories*, as 'refer' in their context means 'have evolutionary progressive knowledge of'. Evolutionary progressive science-world interaction is captured by *epistemic* relations of reference which define the context within which a theory could be 'true' at a given time, and which track the development of knowledge with regards to a particular event or phenomenon or target system as follows:

A term t 'refers' to a posited entity iff it satisfies a 'core causal description' (CCD) of 'identifying' properties associated with term t such that

1. the CCD in question has been adapted to fit the current experimental situation and thus describes properties currently thought to belong to the postulated entity and

2. the properties in question are such that the posited entity plays its putative causal role in virtue of these properties (i.e. these properties are the causal origin of claims associated with the putative entity) (*cf.* Psillos, 1999).

Reference for the naturalised realist is an epistemic issue and is thus viewed to be about tracking the development of knowledge claims concerning a particular target system, phenomenon or event, rather than about establishing the metaphysical existence of a real system, phenomenon or event. If it is true that the metaphysical import of successful theories would, in an ideal world, consist in their giving correct descriptions of the structure of the world, (*viz.* Ladyman & Ross, 2007), then the epistemological import of successful theories consists in their being the crystallisation of a process or method of continuous revision and sifting claims. As implied by the history of science, (*viz.* Chang, 2012; Laudan, 1981), the former option is not viable given the nature of science and therefore the naturalised realist suggests a new interpretation of 'reference' as the epistemic mechanism through which the revision and sifting of claims referred to above can be made concrete and can be tracked through the history of scientific investigation.

Returning to the definition above, 'identifying' properties are properties that can be described according to the current experimental situation – so 'core' properties are properties that have been revised, and are determined

A Non-classical Logical Foundation for Naturalised Realism

by current evidence. The noteworthy point here is that naturalised realists thus imply that what is 'core' can change, which is not a suggestion that would sit comfortably with traditional scientific realists' depiction of science. Let us now consider how the naturalised realist proposes to interpret and apply the mechanism of 'reference' in times of theory change:

Terms t and t' denote 'the same' posited entity within the same theoretical system iff

1. both t and t' each respectively satisfies a CCD of properties associated with them that has been adapted to fit the experimental situations in which the theories containing t and t' respectively have been formulated; and

2. the description of the properties in the CCD of t' has been adapted from the CCD of t; and

3. the referents of t' and t play the same causal role w.r.t. a certain set of phenomena in virtue of the properties described in their respective CCD's.

The issue is not one of identifying limiting cases of successor theories or the parts of theories that 'persevere' through theory change, but rather the possibility of finding *heuristic continuity* via revisions culminating in evolutionary progressive theories, which, *in virtue of their revision*, carry on in (heuristic/epistemic/pragmatic) continuity with their predecessor theories. In this sense, truth is assembled as science progresses through revisions and confirmations, and the content of what is *assembled* is captured or revealed by relations of reference supervening on the progressiveness of theories.

Considering the technical aspects of the definition above, it is necessary to specify that reference is reference to the 'same' posited entity *within the same theoretical system*, as the unit for naturalised realist appraisal is a network of theories, i.e. the collection of all investigations of a particular target system, phenomenon or event over time. And, in this (broader) context, obviously it may be the case that not all descriptions of the same posited entity have been adapted from previous descriptions, as there is the possibility of incompatible descriptions of the same postulated entity – e.g. Thomson, Lorentz, Bohr, Millikan, *et al.* on the properties of electrons – given that the network of investigations being evaluated may include more than one (in/compatible) theoretical system focused on the particular phenomenon or event at issue (e.g. the phlogiston system of theories vs. the oxygen system of theories). (For our purposes here, think of a system of theories as broadly a Kuhnian paradigm in the sense of a disciplinary matrix.) Thus the reference relations of every separate theoretical *'genre'* or system of theories must all be taken into account when a realist decision is made regarding the appropriate epistemic stance to adopt towards the content of knowledge concerning a particular real system, phenomenon or event at a given time.

In conclusion, note that naturalised realism is an epistemic realism which means at the most simple level simply that naturalised realists believe scientific knowledge is knowledge about the external world. (This may or may not include the belief that the entities implied to exist according to the sum of current scientific knowledge, really exist. The epistemic realist can remain agnostic to a huge degree about such metaphysical issues.) Thus *the outcome of a naturalised realist evaluation of the content of a network of theories is a particular epistemic stance* towards the content of such a network, *on a continuum of realist stances* varying from epistemic equivalents to traditional instrumentalism to such equivalents to traditional scientific realism (e.g. think of the development of knowledge concerning certain viruses). This notion of a continuum captures the sense in which naturalised realism is *naturalised*, i.e. realism traces the movements of science, as science endeavours to understand the 'movements' of real phenomena and events.

3   The empirical-theoretical distinction re-visited

Ernst Nagel (1961) offers one of the most well-known distinctions between so-called 'experimental laws' and 'proper theories'. In his sense 'experimental laws' contain only so-called observational terms, while the purpose of the formulation of 'proper theories' is to explain 'experimental laws' by the theoretical terms they introduce. This is part of the old so-called 'two-language' depiction of the positivist syntax of science of which Rudolf Carnap's (1956) *The Methodological Character of Theoretical Concepts* is the best exposition. This view, according to which science is interpreted by "relating it to an observation language (a postulated part of natural language which is devoid of theoretical terms)" (van Fraassen,1980, pp. 13–14), has now totally run its course. The debate over the distinction between theoretical and observational (or empirical in contemporary terms) rages on though, albeit in different format, and one of its most interesting playing fields is the area of scientific realism.

To see why this is so, let us briefly consider one other major contribution to the debate about the viability of the two-language distinction, namely Grover Maxwell's (1962) *The Ontological Status of Theoretical Entities* which is directly in opposition to Carnap's views and meant to be an attack on the instrumentalism of the positivist account of science. Maxwell (1962, p. 1052) starts off by quoting a statement made by the other stalwart of positivism, Ernst Nagel (1961), that the distinction between realism

A Non-classical Logical Foundation for Naturalised Realism

and instrumentalism is simply a "conflict over preferred modes of speech" (Chapter 6). Most of what Maxwell offers in support of realism turns on this remark of Nagel's, as Maxwell implies that what should be taken into account in terms of the dichotomy postulated by the positivists is that while it must be acknowledged that the 'life' of a theoretical term often starts out as a mere "heuristic crutch"(Maxwell, 1962, p. 1053), very often its life ends as an 'observation' term – e.g. the history of the discovery of 'microbes', (*viz.* Maxwell, 1962, pp. 1053–1055). Thus, against van Fraassen's definition of 'obervable' as "an unaided act of perception"(van Fraassen, 1980, p. 15), Maxwell (1962, p. 1056) argues for a view that acknowledges that "...there is, in principle, *a continuous series* [our italics] beginning with looking through a vacuum and containing these as members: looking through a windowpane, looking through glasses, looking through binoculars, looking through a low-power microscope, looking through a high-power microscope, etc., in the order given". He (ibid.) concludes from this that accepting such a 'continuous series' of candidates for 'being observable', implies that there is no "non-arbitrary line between 'observation' and 'theory' ".

The naturalised realist, while taking that 'observable' is not equated with 'existence' as Maxwell (1962, p. 1057) insists, accepts this arbitrariness given her focus on the revisable and shifting nature of evidence for current scientific theories. She proposes that Maxwell's continuous series of observability suggests an epistemic (not a metaphysical) series reflecting a continuum of epistemic (realist) stances towards the content of theories assembled at a given time. More specifically, as noted above, the outcome of a naturalised realist evaluation of the content of a network of theories is a particular epistemic stance towards the content of such a network, *on a continuum of realist stances* varying from epistemic equivalents to traditional instrumentalism to such equivalents to traditional scientific realism.

The real challenge for realism, rather than focusing on the separate parts/stances making up the continuum, actually has to do with explaining how such a continuum might work *as a continuum*, hence proposing a workable suggestion to deal with the arbitrariness of theory-observation distinctions. This is where a representation of scientific knowledge based on Description Logics can assist.

Description Logics (or DLs) are a family of knowledge representation formalisms (Baader et al., 2007), corresponding to decidable fragments of first-order logic. They are based on the notions of concepts (unary predicates in first-order logic terms) and roles (binary relations in first-order logic terms), and are mainly characterised by constructors that allow complex concepts and roles to be built from atomic ones. The DLs of interest to us are all based on an extension of the well-known DL $\mathcal{ALC}$. Concept descriptions are built from concept names using the constructors disjunction ($C \sqcup D$), conjunction ($C \sqcap D$), negation ($\neg C$), existential restriction ($\exists r.C$) and value restriction ($\forall r.C$), where C, D represent concepts and r represents a role name.

A DL knowledge base κ consists of two finite and mutually disjoint sets. A *terminology box* (or *TBox*) which introduces the *terminology*, and an *assertion box* (or *ABox*) which contains facts about particular objects in the application domain. TBox statements have the form $C \sqsubseteq D$ (*inclusions*) where C and D are (possibly complex) concept descriptions. Objects in the ABox are referred to by a finite number of *individual names* and these names may be used in two types of assertional statements: *concept assertions* of the type $C(a)$ and *role assertions* of the type $r(a, b)$, where C is a concept description, r is a role name, and a and b are individual names.

The semantics of $\mathcal{ALC}$ is the standard set-theoretic Tarskian semantics. Individual names are interpreted as elements of a domain of interest, concepts as subsets of this domain, and roles as binary relations over this domain. An interpretation I consists of a non-empty set Δ^I (the domain of I), an injective *denotation function d* which maps every individual name a to an element a^I of the domain Δ^I, a function $\cdot^I$ (the *interpretation function* of I) which maps every concept name A to a subset A^I of Δ^I, and every role name r to a subset r^I of $\Delta^I \times \Delta^I$. The interpretation function is extended to arbitrary concept descriptions as follows. Let C, D be concept descriptions and r a role name, and assume that C^I and D^I are already defined. Then

$(\neg C)^I = \Delta^I \setminus C^I, (C \sqcup D)^I = C^I \cup D^I, (C \sqcap D)^I = C^I \cap D^I$;
$(\exists r.C)^I = \{x \mid \exists y \text{ s.t. } (x, y) \in r^I \text{ and } y \in C^I\}$;
$(\forall r.C)^I = \{x \mid \forall y, (x, y) \in r^I \text{ implies } y \in C^I\}$.

An interpretation I *satisfies* $C \sqsubseteq D$ iff $C^I \subseteq D^I$. Similarly for ABox assertions, an interpretation I *satisfies* the assertion $C(a)$ iff $a^I \in C^I$, and it *satisfies* $r(a, b)$ iff $(a^I, b^I) \in r^I$. I is said to be a *model* of a DL (TBox or ABox) statement ϕ iff it satisfies the statement. I is said to be a model of a DL knowledge base κ iff it satisfies every statement in κ. A DL knowledge base κ *entails* a DL statement ϕ, written as $\kappa \models \phi$, iff every model of κ is a model of ϕ. With the formal apparatus in place, our proposal is to represent scientific theories as Description Logic knowledge bases, with the TBox of a knowledge base containing a description of the theory, and the ABox containing the accumulated empirical facts. The most relevant aspect of the structure of DL knowledge bases from the perspective of the naturalised realist is the explicit distinction drawn between statements in the TBox and the ABox. This dichotomy provides a mechanism for a formal distinction between theoretical and observational terms. The intended use of the ABox is the provision of formal empirical facts about the world, so within a DL knowledge base, the distinction between theoretical and observational terms can be formally enforced by requiring that only observational terms may be present in the ABox. The distinction between theoretical and observational terms is not *determined* by the use of DLs: it provides a framework within which the arbitrariness of the distinction between theoretical and observational terms can be formalised.

4 Over-determination of theories by data

Imagine a typical semantic account of theories, (e.g. Suppes, 1967; van Fraassen, 1980), according to which theories are depicted in terms of sets of mathematical structures that are the models of the theory in question. Some commentators on this school of thought, (e.g. Kuipers, 2001; Ruttkamp, 2002), further identify empirical reducts of the models in question as follows. Consider what it really means to formulate a model of a particular theory. A model of a theory sees to it that every predicate of the language of the theory has a definitive extension in the underlying domain of the model. Now, focusing on a particular real system at issue in the context of applying a theory, which in turn implies a specific empirical set-up in terms of the measurable quantities of that particular real system, the predicates in the mathematical model of the theory under consideration that may be termed 'empirical' predicates may be selected. This is how an empirical reduct is formulated. Recall that a 'reduct' in model-theoretic terms is created by leaving out of the language and its interpretations some of the relations and functions originally contained in these entities. This kind of structure thus has the same domain as the model in question but contains only the extensions of the 'empirical' predicates of the model. Notice that these extensions may be infinite since they still are the full extensions of the predicates in question.

The method of ('empirical') verification of each of the set of models depicting the theory (i.e. how well do each of them reflect/represent/link to the relevant system in the real world?), is decided by the specific nature of the specific model in question, *as well as* by the nature of the specific real system in question. Hence if the phenomena in some real system and the experimental data concerned with those phenomena are logically reconstructed in terms of a mathematical structure – call it an 'empirical' or 'data' model – then the relation of empirical adequacy becomes a relation which is an isomorphism from the empirical model into an empirical reduct of a relevant model of the theory in question.[1] To summarise: In the empirical reduct are interpreted only the terms called 'empirical' in the particular relevant context of application or empirical situation. Think of this substructure of the model at issue as representing the set of all atomic sentences expressible in the particular empirical terminology true in the model. An empirical model – still a mathematical structure – can be represented as a finite subset of these sentences, and contains empirical data formulated in the relevant language of the theory.

But what is the status of the models or empirical reducts or empirical models not chosen at a specific time, then? The knowledge or information about the particular empirical model(s) in question that they carry, certainly still *is* knowledge, is it not? Well, yes and no. What is needed is a formal mechanism by which to depict choices, the motivations for choices, and the change of both of these, should there be a change of context within which a theory is being applied. A decision is taken to work with a certain model or empirical reduct at a certain time, but, on the one hand, a different choice/decision based on new information can be made at any time, or on the other hand, what are deemed to be 'empirical predicates' can change according to new knowledge and thus the formulation of empirical reducts can change, (*cf.* Ruttkamp, 2005). For instance, the general theory of relativity was formulated by Einstein (and Hilbert) in 1915. For almost a century now physicists have been constructing literally dozens of different types of models – all models of precisely the same theory – to fit both experimental and observational data about the space-time structure of the real universe *and* certain paradigmatic preferences. This illustrates (perhaps in a different way from what he (Kuhn, 1996) intended), Kuhn's point that neither the content of science nor any system in reality should be claimed to be "uniquely exemplified" by scientific theories from the viewpoint of studies of "finished scientific achievements". In this sense the 'open-endedness' of science can be taken to imply that the terms of an already established the-

---

[1] Recall that according to van Fraassen (1976, p. 631) a theory is empirically adequate if "all appearances are isomorphic to empirical substructures in at least one of its models".

ory are 'about' an on-going potential of entities in some system of reality to relate to some objects and relations in *any* model of that theory. The actualization of this potential requires human action in the sense of finding and finally articulating 'satisfying' relations between aspects of systems in reality and certain empirical aspects (reducts) of models of the theory at issue. It is in the process of establishing such relations that one becomes aware of 'empirical proliferation' in the sense of 'over-determination of theories by data'. In a sense this is the reverse of the traditional scenario of the underdetermination of theories by data.[2]

Scientific knowledge is amendable and even defeasible, because of its contingent and particularized links with the reality it describes (and explains). In general scientific theories, depicted as syntactic (linguistic) entities that need to be interpreted to be given semantic meaning and reference, are not able to uniquely capture their semantic content. In terms of theory application, taking now both the syntactic and semantic accounts of theories into account, within a model-theoretic context, two sets of relations are conducive to empirical proliferation: the set of relations between the terms in some theory and their extensions in its various models; and the set of relations between the terms of models (or of only one model) – via an empirical reduct (or empirical reducts) of that (those) model(s) – and the objects and relations of some real system (or systems) conceptualized in one or many empirical models. Retaining the notion of scientific theories as linguistic expressions at the 'top' level of science solves the problems regarding the justification of the existence of many (conceptual) models as interpretations of any one theory by the simple (formal) fact of the incompleteness of formal languages. Thus the possibility of a given scientific theory being interpreted in more than one mathematical model (structure) is natural in a very basic sense in model-theoretic terms. The second proliferation of relations between models and their empirical reducts and between these and empirical models may also turn out to be less counterintuitive than might be supposed at first glance, if it can be shown that the possibility of articulating rational reasons for choosing one relation among the others at any given time is *not* jeopardized under such circumstances.

In what follows we claim in particular that an application of *defeasible* reasoning to situations of over-determination of theories by models and data may enable formalising this second kind of complexity in terms of a particu-

---

[2]The context of this discussion is empirical equivalence in van Fraassen's sense (1980, p. 67).

lar kind of preferential ranking of these models (*Cf.* Ruttkamp, 2005). In this context, let us consider the possibility of introducing into the wide empirical equivalence debate, concentrated on issues concerning over-determination of theories by data, the non-monotonic mechanism of defeasible reasoning, refined into a model-theoretic non-monotonic logic based on the preferential logics of Lehmann and Magidor (1992) offering a formal method to rank models. To enable defeasible reasoning, the language of classical propositional logic is enriched with a notion of *defeasible implication*, represented with the connective $\mid\sim$, with statements of the form $\alpha \mid\sim \beta$ informally intended to represent the information that β normally follows (defeasibly) from α. The intention is to exploit the ability to represent defeasible information in order to deal with the over-determination of theories.

For example suppose that our theory contains the information that mammalian and avian red blood cells are vertebrate red blood cells (MRBC $\rightarrow$ VRBC, ARBC $\rightarrow$ VRBC), that vertebrate red blood cells have a nucleus (VRBC $\rightarrow$ hasN), but that mammalian red blood cells don't (MRBC $\rightarrow$ ¬hasN). Notice that mammalian red blood cells, by virtue of being vertebrate red blood cells, are also said to have a nucleus (MRBC $\rightarrow$ hasN). So a consequence of the theory is that mammalian red blood cells have, and don't have a nucleus – a classical case of over-determination.

The proposal we are advancing here is that a language with defeasible implication allows for a more appropriate and accurate expression of scientific theories. In the case of the example above a more accurate theory would be one in which it is stated that mammalian and avian red blood cells are vertebrate red blood cells (MRBC $\rightarrow$ VRBC, ARBC $\rightarrow$ VRBC), as before, but that vertebrate red blood cells *normally*, but not necessarily, have a nucleus (VRBC $\mid\sim$ hasN), and that mammalian red blood cells normally don't (MRBC $\mid\sim$ ¬hasN). Furthermore, an appropriate notion of entailment for such a richer language would correctly identify mammalian red blood cells as *exceptional* vertebrate red blood cells, therefore not conforming to the default property of vertebrate red blood cells having a nucleus, and in doing so, avoid any over-determination.

It turns out that suitable model-theoretic notions of entailment can be obtained by employing a form of preferential semantics. Recall that the semantics of propositional logic is based on the set of propositional valuations V, where a valuation v is a function which assigns a truth value (true or false) to every propositional atom from which the language is generated. The preferential semantics for defeasible reasoning proposed by Lehmann and colleagues (Kraus et al., 1990; Lehmann & Magidor, 1992) involves

A Non-classical Logical Foundation for Naturalised Realism

the establishment of an ordering $\prec$ over valuations,[3] with valuations lower down in the ordering viewed as more plausible. That is, $v \prec w$ means that we consider the situation described by the valuation v as more plausible than the situation described by the valuation w; a *preferential model* $\mathcal{M}$ consists of a set of valuations and an ordering, on which we can impose the satisfaction of specific properties. Preferential models can be used to define the semantics of classical propositional formulas, as well as defeasible implications: let $\hat{\alpha}$ be the set of the valuations in $\mathcal{M}$ that satisfy α, $\mathcal{M}$ satisfies $\alpha \mathrel{|\!\sim} \beta$ iff $\min_{\prec}(\hat{\alpha}) \subseteq \hat{\beta}$, that is, all the most plausible valuations satisfying α satisfy also β. Lehmann and Magidor (1992) propose that *rational* forms of entailment for defeasible reasoning be captured in terms of a specific class of models. Given a knowledge base κ expressed in a propositional language enriched with defeasible implication, their proposal is that the set of formulas entailed by κ ought to be exactly those that are satisfied in one of the ranked models satisfying all elements of κ. Since, for most knowledge bases κ, there would be more than one ranked model satisfying it, there may well be more than one appropriate form of entailment as well. A detailed investigation into different forms of entailment is beyond the scope of this paper, but the point is that defeasible reasoning, and more specifically, the introduction of defeasible implication based on a preferential model-theoretic semantics, can provide a successful formal mechanism for managing the problem of over-determination in logic-based representation of scientific theories.

5 Revision

Returning now to naturalised realism, is it a case of anything goes? All of science is about revision after all, so are naturalised realists, realists about all of science? The issue for the naturalised realist is not simply noting that theories are revised in the course of science, it is *about determining what scientists learn from revision*, as the naturalised realist is specifically interested in revision that is 'evolutionary progressive'. Thus the naturalised realist suggests that scientific progress is the result of the available network of knowledge claims becoming more refined as scientists learn from their 'mistakes' and *adapt* their theories such that they *show how their theories can accommodate revision* (note that maintaining the *status quo* when no revision is necessary in the face of new evidence, is recognised as a special

---

[3] Strictly speaking, an ordering over states, where states are mapped to valuations.

case of 'revision'). Continuity lies in the method of science – thus in revision, in the epistemic sense that the quality of knowledge claims depends for ever on the quality of evidence available at the time. In this sense perhaps theories that have been most revised (e.g. phlogiston) are theories with the most realist potential.

This is not in contradiction with the fact that no such thing as phlogiston exists, because naturalised realism expands the scope of realism from being tightly (almost exclusively) focused on metaphysical issues to allowing an emphasis on the epistemological aspects of theory content evaluation. A theory that can absorb revision (thus which is evolutionary progressive) is much stronger than one that is true in all possible worlds, because an evolutionary progressive theory contains much more knowledge of the system (or phenomenon or event) it describes than a tautology does, in the sense that it reflects knowledge of what should be left out of descriptions (based on current evidence), while a theory that is always true does not reflect such knowledge. Naturalised realism is first and foremost an epistemology of science, not a metaphysics of science.

All investigations of a phenomenon or event over time inform our current knowledge of it. The *unit of (naturalised) realist appraisal is networks of (systems of) theories*, not single theories. The more revision there is, the better movements within such a network of theories can be tracked and interpreted and the better the strength of the current realist stance towards knowledge of the event or phenomenon at issue can be evaluated. We suggest *belief revision* as a method through which to demonstrate these aspects of a naturalised realist definition of progress in science in terms of revision rather than retention.[4]

Generally, belief revision is concerned with the rational management of changes to an agent's information about the world it inhabits. More specifically, it aims to describe how an agent ought to change its mind when confronted with new, and possibly conflicting, information to be admitted into its knowledge about the world. To incorporate belief revision into the methodology put forward by the naturalised realist, it is best to consider the approach to revision advanced by Alchourrón, Gärdenfors, and Makinson, the so-called AGM approach (1985). Technically, AGM belief revision is applicable to any logic with a standard Tarskian semantics for which the

---

[4] A related issue is the case of theory change, e.g. Kepler to Newton, Newton to Einstein, or the 18th and 19th century ether theories of light to electromagnetism. Coping with such examples demands a more specific extension of vocabulary (languages) which is somewhat outside the scope of this paper, but which we address in a series of forthcoming works.

A Non-classical Logical Foundation for Naturalised Realism

Compactness Theorem holds, but we shall make matters concrete here by focusing on propositional logic. A knowledge base in this context is represented as a set κ of propositional formulas. For the naturalised realist κ would therefore represent a scientific theory. With the knowledge base κ fixed, a revision operator $\circ$ takes as input a propositional formula to be incorporated into the current knowledge base and produces a new knowledge base (a set of propositional formulas).

The most important aspect of AGM revision is that it does not prescribe a unique method for revising a knowledge base. Instead, it provides the following set of postulates to which any *rational* revision operator $\circ$ should conform:

(**K1**) $\kappa \circ \alpha = Cn(\kappa \circ \alpha)$;
(**K2**) $\alpha \in Cn(\kappa \circ \alpha)$;
(**K3**) $Cn(\kappa \circ \alpha) \subseteq Cn(\kappa \cup \{\alpha\})$;
(**K4**) If $\kappa \cup \{\alpha\} \not\models \bot$ then $\kappa \circ \alpha = Cn(\kappa \cup \{\alpha\})$;
(**K5**) If $\kappa \circ \alpha \models \bot$ then $\alpha \models \bot$;
(**K6**) If $\alpha \equiv \beta$ then $Cn(\kappa \circ \alpha) = Cn(\kappa \circ \beta)$;
(**K7**) $Cn(\kappa \circ (\alpha \wedge \beta)) \subseteq Cn(\kappa \circ \alpha \cup \{\beta\})$;
(**K8**) If $(\kappa \circ \alpha) \cup \{\beta\} \not\models \bot$
then $Cn((\kappa \circ \alpha) \cup \{\beta\}) \subseteq Cn(\kappa \circ (\alpha \wedge \beta))$

The significance of these postulates have been discussed at length (e.g., by Gärdenfors (1988)), and we won't do so here. For the naturalised realist this approach to revision is important because revision need not be unique, thereby allowing for the development of systems of theories, all conforming to the eight AGM revision postulates. At the same time, revision is not naïvely open-ended either. The constraints imposed by the eight postulates are suitably restrictive, ensuring ample justification for the maintenance of a system of theories. AGM revision provides a formal framework for developing a system of theories, but since it only describes a mechanism for a single revision step, it does not address the issue of the maintenance of a *network* of systems of theories. For that it is necessary to consider the area of *iterated belief revision* which seeks to describe the dynamics of a sequence of belief revision steps. A detailed description of iterated revision is beyond the scope of this paper. It will suffice to remark that the most successful approach (Darwiche & Pearl, 1997) involves an extension to AGM revision, and allows for the construction of a network of results in line with the naturalised realist stance.

6 Conclusion

We have argued that DL can possibly shed new light on the formal representation of the empirical-theoretical distinction, that Non-monotonic Logic in the form of defeasible implication can be employed to manage the overdetermination of scientific theories, and that Belief Revision can be used as a mechanism for representing the naturalised realist approach to progress in science. A single unifying framework for doing so would involve the incorporation of Non-monotonic logic and Belief Revision into Description Logics. In recent years, significant progress has been made in this regard, with Description Logics being extended to include notions of *defeasible subsumption* (Giordano, Gliozzi, Olivetti, & Pozzato, 2013) and Belief Revision (Meyer, Lee, & Booth, 2005). What is still missing for a truly integrated framework is a system in which Belief Revision can be applied to Description Logics that are enriched with defeasible subsumption. We are currently investigating this.

References

Alchourrón, C., Gärdenfors, P., & Makinson, D. (1985). On the Logic of Theory Change: Partial Meet Contraction and Revision Functions. *Journal of Symbolic Logic, 50*, 510–530.

Baader, F., Calvanese, D., McGuinness, D., Nardi, D., & Patel-Schneider, P. (2007). *The DL Handbook* (2nd ed.). Cambridge: Cambridge University Press.

Carnap, R. (1956). The Methodological Character of Theoretical Concepts. *Minnesota Studies in the Philosophy of Science, 1*, 38–76.

Chang, H. (2012). *Is Water H2O? Evidence, Realism and Pluralism*. Berlin: Springer.

Darwiche, A., & Pearl, J. (1997). On the Logic of Iterated Belief Revision. *Artificial Intelligence, 89*(1-2), 1–29.

Gärdenfors, P. (1988). *Knowledge in Flux: Modeling the Dynamics of Epistemic States*. Cambridge: MIT Press.

Giordano, L., Gliozzi, V., Olivetti, N., & Pozzato, G. L. (2013). Rational Closure for Description Logics of Typicality. In *IAF-13* (pp. 1–9).

Kraus, S., Lehmann, D., & Magidor, M. (1990). Nonmonotonic Reasoning, Preferential Models and Cumulative Logics. *Artificial Intelligence, 44*, 167–207.

Kuipers, T. A. F. (2001). Epistemological Positions in the Light of Truth Approximation. *Proceedings of 20th World Congress of Philosophy, 2001*, 79–88.

Ladyman, J., & Ross, D. (2007). *Every Thing Must Go: Metaphysics Naturalized*. Oxford: Oxford University Press.

Laudan, L. (1981). A Confutation of Convergent Realism. *Philosophy of Science, 48*(1), 19–49.

Lehmann, D., & Magidor, M. (1992). What Does a Conditional Knowledge Base Entail? *Artificial Intelligence, 55*, 1–60.

Maxwell, G. (1962). The Ontological Status of Theoretical Entities. *Minnesota Studies in the Philosophy of Science, 3*, 181–192.

Meyer, T., Lee, K., & Booth, R. (2005). Knowledge Integration for Description Logics. In *Proceedings of AAAI-05* (pp. 645–650).

Nagel, E. (1961). *The Structure of Science: Problems in the Logic of Scientific Explanation*. San Diego: Harcourt, Brace & World.

Psillos, S. (1999). *Scientific Realism: How Science Tracks Truth*. London: Routledge.

Ruttkamp, E. (2002). *A Model-theoretic Realist Interpretation of Science*. Berlin: Springer.

Ruttkamp, E. (2005). Overdetermination of Theories by Empirical Models: A Realist Interpretation of Empirical Choices. *Poznan Studies in the Philosophy of the Sciences and the Humanities, 84*(1), 409–436.

Ruttkamp-Bloem, E. (2013). Re-enchanting Realism in Debate with Kyle Stanford. *Journal for General Philosophy of Science, 44*(1), 201–224.

Suppes, P. (1967). What Is a Scientific Theory? In S. Morgenbesser (Ed.), *Philosophy of Science Today* (pp. 55–67). New York: Basic Books.

van Fraassen, B. (1976). To Save the Phenomena. *Journal of Philosophy, 73*, 623–632.

van Fraassen, B. (1980). *The Scientific Image*. Oxford: Oxford University Press.

Emma Ruttkamp-Bloem, Giovanni Casini, and Thomas Meyer

Emma Ruttkamp-Bloem
University of Pretoria
South Africa
E-mail: emma.ruttkamp-bloem@up.ac.za

Giovanni Casini
University of Pretoria & CAIR (CSIR-Meraka), University of Luxemburg
South Africa, Luxemburg
E-mail: giovanni.casini@gmail.com

Thomas Meyer
University of Cape Town & CAIR (CSIR-Meraka)
South Africa
E-mail: tmeyer@cs.uct.ac.za

Action Frames for Weak Relevant Logics

IGOR SEDLÁR[1]

Abstract: The article introduces extended models for the propositional dynamic logic **PDL**. In extended models, valuation assigns to every state a set of atomic formulas and a **PDL** program. The program is informally construed as an action preferred by a contextually fixed agent. **PDL** is then extended by introducing a conditional connective expressing partial correctness claims. The main contribution of the article is the observation that the partial correctness conditional is in fact a substructural implication. It is shown that a variant of the minimal distributive substructural logic **DB** is a fragment of **PDLS**, an extension of **PDL** that contains the partial correctness conditional and fusion. It is also shown that the main results can be extended to a specific class of positive substructural logics.

Keywords: Action, propositional dynamic logic, partial correctness, substructural logics

1 Introduction

Valuation in propositional modal models is usually defined as a function that assigns sets of states to atomic formulas. Equivalently, valuation can be seen as a partial description of states, since it lists the atomic formulas that 'hold' in the respective states. A natural extension of this is to re-define valuation in such a manner that it provides *more information* about states. This article introduces such *extended models* for the propositional dynamic logic **PDL**, see (Fischer & Ladner, 1979; Harel, Kozen, & Tiuryn, 2000). In these models, valuation assigns to every state a set of atomic formulas *and* a **PDL**-program. Informally, the program is seen as an action *pre-*

---

[1] The author acknowledges the support of the grant VEGA 1/0221/14 of the Scientific Grant Agency of the Ministry of Education, Science, Research and Sport of the Slovak Republic.

ferred by an agent. The language of **PDL** is then extended by a conditional connective '$\to$', expressing global partial correctness claims. On the truth definition employed there, $\phi \to \psi$ holds in state x of a model iff the action preferred in x is partially correct with respect to input (precondition) ϕ and goal (postcondition) ψ throughout the model.

The main contribution of this article is the observation that the partial correctness conditional is a substructural implication. This provides a link between substructural logic and dynamic logic. More specifically, we show that **DB**, a variant of the basic distributive substructural logic, is sound and complete with respect to extended models for **PDL**. As a result, it is shown that **DB** is a fragment of **PDLS**, a specific extension of **PDL** that contains the partial correctness conditional and fusion. Our proof of this result utilises a representation of the ternary Routley-Meyer accessibility relation of substructural models, see (Restall, 2000), by the binary relation corresponding to the action preferred at particular states. It is also shown that the main results can be extended to a specific class of positive substructural logics.

We note that our work is related to Dunn's investigations into the links between relevant and action logics, see (Bimbó & Dunn, 2005; Dunn, 2001, 2003) and to some informal interpretations of the ternary accessibility relation in Routley-Meyer frames, see (Mares, 1996, 2004; Restall, 1995; Slaney & Meyer, 1997) and especially (Beall et al., 2012). However, our approach is specific in that it uses action semantics to provide models for substructural logics (not the other way around) and defines the ternary relation in terms of simpler notions explicitly. Links between dynamic and substructural logics have been studied also by Kozen and Tiuryn (2003). Another related approach is epistemic logic with non-rigid agent names and varying agent domains, see (Grove, 1995; Grove & Halpern, 1993). Our framework can be seen as a fragment of the first-order version of non-rigid epistemic logic.

The article is organised as follows. Section 2 introduces the main logics to be discussed in this article, **PDL** and **DB**. Extended models for **PDL** are introduced and independently motivated in Section 3. Section 4 explains that extended models allow to interpret a conditional connective that corresponds to partial correctness claims describing specific actions. Section 5 introduces the logic **PDLS** and establishes the main technical result of the article: **DB** is a fragment of **PDLS**. Section 6 extends the main result to stronger substructural logics and specific extensions of **PDLS**. Section 7 concludes the article and outlines one interesting topic of future research.

2 Preliminaries: PDL and DB

This section briefly introduces the logics to be discussed. We build on (Harel et al., 2000) and (Restall, 2000), but our formulation of **DB** is slightly adjusted.

2.1 PDL

Let π be a countable set of *atomic programs* and Φ a countable set of *atomic formulas* (disjoint from π). The sets of *formulas* and *programs* of the 'dynamic' language $\mathcal{L}(\pi, \Phi)$ are defined by mutual induction as follows:

$$\phi ::= p \mid \neg\phi \mid \phi \wedge \phi \mid \phi \vee \phi \mid \phi \supset \phi \mid [\alpha]\phi$$
$$\alpha ::= a \mid \alpha \cup \alpha \mid \alpha;\alpha \mid \alpha^* \mid \phi?$$

where $p \in \Phi$ and $a \in \pi$. Φ shall be the set of atomic formulas for every language considered in this article. Accordingly, $\mathcal{L}(\pi, \Phi)$ shall be referred to as $\mathcal{L}$.

Program operators '$\cup$', ';', '$*$' and '?' stand for indeterministic choice, sequential composition, indeterministic iteration and test, respectively. Set $Prog(\pi)$ is the set of programs over π. '$[\alpha]\phi$' is read 'It is necessary that after executing α, ϕ is true'.

Formulas of the form

$$\phi \supset [\alpha]\psi \tag{1}$$

express that program α is *partially correct* with respect to input ϕ and output ψ: whenever the program is started in a state satisfying ϕ, then if it halts, it does so in a state satisfying ψ. Precursors of **PDL** such as the Floyd-Hoare Logic were devised specially for the purpose of proving partial correctness of specific programs.

A model for $\mathcal{L}$ (a 'dynamic model') is a structure

$$M = \langle S, R, v \rangle \tag{2}$$

where S is a non-empty set of 'states', v is a function from Φ to subsets of S and R is a function from $Prog(\pi)$ to binary relations on S such that

- $R(\alpha \cup \beta) = R(\alpha) \cup R(\beta)$,
- $R(\alpha;\beta) = R(\alpha) \circ R(\beta) = \{\langle x, y \rangle \mid \exists z.(R(\alpha)xz \wedge R(\beta)zy)\}$,
- $R(\alpha^*) = R(\alpha)^*$, i.e. $R(\alpha^*)$ is the reflexive transitive closure of $R(\alpha)$,

- $R(\phi?) = Id(S) \cap \{x \mid M, x \models \phi\}$, where $Id(S)$ is the identity relation on S and the truth relation $M, x \models \phi$ is defined below.

Truth conditions for Boolean ϕ in pointed models (M, x) are as usual. Moreover:

- $M, x \models [\alpha]\phi$ iff $M, y \models \phi$ for all y such that $R(\alpha)xy$.

Validity in M is defined in the usual way as truth in every state of M (Notation: $M \models \phi$). **PDL** is the set of dynamic formulas valid in every dynamic model.

2.2 DB

The sets of *formulas* of the 'positive substructural language' $\mathcal{L}_s$ and the 'positive substructural consecution language' $\mathcal{L}_s^{\supset}$ (referred to hereafter as the 'consecution language') are defined as follows:

$$\phi ::= p \mid \phi \wedge \phi \mid \phi \vee \phi \mid \phi \to \phi \mid \phi \otimes \phi$$
$$\phi' ::= \phi \mid \phi \supset \phi$$

where $p \in \Phi$. The connective '$\to$' is the substructural implication and '$\otimes$' is fusion.

A *flat model* for $\mathcal{L}_s^{\supset}$ (a 'flat substructural model' or simply a 'flat model') is a structure

$$\mathcal{M}_s = \langle \mathcal{P}, \mathcal{R}, \nu \rangle \qquad (3)$$

where $\mathcal{P}$ is a non-empty set of 'points', ν is a function from Φ to subsets of $\mathcal{P}$ and $\mathcal{R}$ is a *ternary* relation on $\mathcal{P}$. The truth conditions of substructural formulas in flat models are as follows:

- $\mathcal{M}, x \models p$ iff $x \in \nu(p)$,
- $\mathcal{M}, x \models \phi \wedge \psi$ iff $\mathcal{M}, x \models \phi$ and $\mathcal{M}, x \models \psi$,
- $\mathcal{M}, x \models \phi \vee \psi$ iff $\mathcal{M}, x \models \phi$ or $\mathcal{M}, x \models \psi$,
- $\mathcal{M}, x \models \phi \supset \psi$ iff $\mathcal{M}, x \not\models \phi$ or $\mathcal{M}, x \models \psi$,
- $\mathcal{M}, x \models \phi \to \psi$ iff for all y, z, if $\mathcal{M}, y \models \phi$ and $\mathcal{R}xyz$, then $\mathcal{M}, z \models \psi$,
- $\mathcal{M}, x \models \phi \otimes \psi$ iff there are y, z such that $\mathcal{R}yzx$ and $\mathcal{M}, y \models \phi$ and $\mathcal{M}, z \models \psi$.

ϕ is valid in $\mathcal{M}_s$ iff $\mathcal{M}, x \models \phi$ for all $x \in \mathcal{P}$. **DB** is the set of consecution formulas valid in every flat model.

Note that the distributive laws

$$\phi \wedge (\psi \vee \chi) \supset (\phi \wedge \psi) \vee (\phi \wedge \chi) \tag{4}$$
$$\phi \vee (\psi \wedge \chi) \supset (\phi \vee \psi) \wedge (\phi \vee \chi) \tag{5}$$

and their converses are clearly valid in **DB**.

Our logic **DB** corresponds to the positive fragment of **DBSub** without truth and falsity constants and the converse implication '←', see (Restall, 2000). Note that $\phi_1 \odot \phi_2$ ($\odot \in \{\supset, \to, \otimes\}$) is a consecution formula only if ϕ_i do not contain occurrences of '⊃'. This formulation of the language replaces 'consecutions' $X \vdash \phi$, where

$$X ::= \phi \mid X; X \mid X, X \tag{6}$$

by material implications $f(X) \supset \phi$, where $f(X)$ is the result of replacing every occurrence of ';' in X by an occurrence of '⊗' and every occurrence of ',' by an occurrence of '∧' (Semantically, $\phi; \psi$ is equivalent to $\phi \otimes \psi$, ϕ, ψ to $\phi \wedge \psi$ and a consecution $\phi \vdash \psi$ is valid in a model iff there is no point x such that ϕ holds at x and ψ does not.) Hence, the original formulation of **DB** as a set of consecutions is replaced by a formulation of **DB** as a set of formulas. Note that every $\phi \in$ **DB** is a material implication.

Our flat models are a special case of 'standard' relational models, where $\mathcal{P}$ is a poset $(P, \leq)$. The poset is 'flattened' here as $\leq$ is replaced by $=$. The use of flat models is allowed by the fact that **DB** is sound and complete with respect to flat models, see (Restall, 2000, Chapter 13.3).

Proposition 1 **DB** *has the finite model property.*

Proof. A simple corollary of Theorem 14.11 of (Restall, 2000). □

3 Extended dynamic models

This section explains the ideas motivating extended valuations and extended dynamic models. We concentrate on extending models for **PDL**, but it is clear that the technique may be applied to models of any multi-modal logic.

Assume that we have a robot Rob. Let $p \in \Phi$ stand for some possible atomic features of the environments Rob is built to operate in. Let $a \in \pi$

stand for some atomic instructions Rob is designed to carry out. Now every dynamic model (2) can be seen as specifying a set of relevant possible environments (S) and transitions on these environments given by the available actions built on π (given by R). Importantly, the valuation v provides a *partial description* of the possible environments: it lists for every $x \in S$ the atomic features of the environment x.

Now assume that we have a user Ann who interacts with Rob to achieve specific goals. One interesting feature of the Rob/Ann interactions are Ann's *preferences*. In particular, we can see Ann as preferring Rob to perform specific actions $\alpha \in Prog(\pi)$. Note that Ann's preferences can be seen as part of Rob's environment. One way of formalising this is to *extend* the notion of valuation to describe Ann's preferences (in the various possible environments $x \in S$), in addition to the atomic facts represented by $p \in \Phi$.

Definition 1 (Extended dynamic models) *An extended dynamic model is a structure*

$$\mathfrak{M} = \langle S, R, V \rangle. \tag{7}$$

where S and R are as in (2), and $V : S \to (Prog(\pi) \times 2^{\Phi})$. $V(x)$ will also be denoted as $\langle \alpha_x, \Phi_x \rangle$. Truth conditions for dynamic formulas and validity are defined as before, with the exception that

- $\mathfrak{M}, x \models p$ *iff* $p \in \Phi_x$

It is clear that $\phi \in \mathbf{PDL}$ iff ϕ is valid in every $\mathfrak{M}$, i.e. extended models are 'proper' models for **PDL**.

Again, α_x can be seen as the action Ann prefers Rob to perform in x.[2] It is clear that although extended models are defined as models for $\mathcal{L}$, interesting new operators can be introduced and interpreted in them. One simple example follows.

Example 1 We can extend $\mathcal{L}$ by a unary operator A and read $A\phi$ as 'the preferred action yields ϕ'. The corresponding truth condition would be

- $\mathfrak{M}, x \models A\phi$ iff for all $y \in S$, if $R(\alpha_x)xy$, then $\mathfrak{M}, y \models \phi$.

This extended language would be expressive enough to describe how Rob's actions can influence Ann's preferences. For example

$$p \supset [\alpha]Aq \tag{8}$$

---

[2] Or, in general, α_x can be seen as the action preferred in x by some contextually fixed agent. Our informal discussions will be confined to the Rob/Ann context throughout the article.

means that if p is the case, then Rob's performing α necessarily results in a state where Ann's preferred action (whatever it is) yields q.

Note that we could select an action $a_i \in \pi$ and read it in terms of Ann's doxastic accessibility, i.e. translate $R(a_i)xy$ as 'Everything Ann believes in x holds in y'. Hence, $[a_i]\phi$ could be read 'Ann (implicitly) believes that ϕ' and written as '$B\phi$'. A situation in which Ann falsely believes that her preferred action yields p, for example, could then be formalised by

$$\neg Ap \wedge BAp. \tag{9}$$

However, we shall not investigate this extension here.

4 The partial correctness conditional

The dynamic language with the partial correctness conditional $\mathcal{L}_c$ ('correctness language') extends $\mathcal{L}$ by a binary operator '$\to$', with the proviso that ϕ? is a program only if ϕ contains no occurrence of '$\to$'.

Definition 2 ($\to$ in extended models)   *We add to the $\mathcal{L}$-truth conditions in extended models the following:*

- $\mathfrak{M}, x \models \phi \to \psi$ *iff for all* $y, z \in S$, *if* $\mathfrak{M}, y \models \phi$ *and* $R(\alpha_x)yz$, *then* $\mathfrak{M}, z \models \psi$.

Validity is defined in the usual way.

Hence, $\phi \to \psi$ holds at x iff the following holds: if the action preferred at x is performed at a state satisfying ϕ, then every possible output state satisfies ψ. This condition is familiar, as the next theorem shows.

Theorem 1   *The following are equivalent (with a slight abuse of notation):*

- $\mathfrak{M}, x \models \phi \to \psi$,
- $\mathfrak{M} \models \phi \supset [\alpha_x]\psi$.

Proof. Firstly, assume that (a) $\mathfrak{M}, x \models \phi \to \psi$ and (b) $\mathfrak{M} \not\models \phi \supset [\alpha_x]\psi$. (b) implies that there is y such that $\mathfrak{M}, y \models \phi$ and $\mathfrak{M}, y \not\models [\alpha_x]\psi$. The latter means that there is z such that $R(\alpha_x)yz$ and $\mathfrak{M}, z \not\models \psi$. But (a) entails that this is impossible.

Secondly, assume that (c) $\mathfrak{M}, x \not\models \phi \to \psi$ and (d) $\mathfrak{M} \models \phi \supset [\alpha_x]\psi$. (c) entails that there are y, z such that $\mathfrak{M}, y \models \phi$, $R(\alpha_x)yz$ and $\mathfrak{M}, z \not\models \psi$. But (d) entails that this is impossible. □

Theorem 1 vindicates our calling '$\to$' a partial correctness conditional: $\phi \to \psi$ holds at x iff the action preferred at x is (globally) partially correct with respect to input (precondition) ϕ and output (postcondition) ψ.

Providing an axiomatisation of **PDLC**, the set of correctness formulas valid in every extended dynamic model, is a natural open problem. Instead of solving the problem, we turn to investigating the links between extended dynamic models and substructural logic.

5 DB in extended models

The correctness language with fusion $\mathcal{L}_{ds}$ (the 'dynamic substructural language') extends $\mathcal{L}_c$ by the fusion connective '$\otimes$', with the proviso that ϕ? is a program only if ϕ contains no occurrence of '$\to$' or '$\otimes$'.

Definition 3 ($\otimes$ in extended models) *Formulas in the $\mathcal{L}_c$-fragment of $\mathcal{L}_{ds}$ are interpreted in extended models as usual. Moreover,*

- $\mathfrak{M}, x \models \phi \otimes \psi$ *iff there are* $y, z \in S$ *such that* $R(\alpha_y)zx$ *and* $\mathfrak{M}, y \models \phi$ *and* $\mathfrak{M}, z \models \psi$.

Validity is defined in the usual way. **PDLS** *is the set of dynamic substrucutral formulas valid in every extended dynamic model.*

$\phi \otimes \psi$ holds at x iff x can be obtained by performing an action that is preferred at some state that satisfies ϕ in a state that satisfies ψ. Note that the consecution language $\mathcal{L}_s^{\supset}$ is a fragment of $\mathcal{L}_{ds}$. The $\mathcal{L}_s^{\supset}$-fragment of **PDLS** is the set of consecution formulas valid in every extended dynamic model.

Theorem 2 *If $\phi \in$ DB, then $\phi \in$ PDLS.*

Proof. We show that if $\phi \in \mathcal{L}_s^{\supset}$ and $\phi \notin$ **PDLS**, then $\phi \notin$ **DB**. Assume that $\mathfrak{M} \not\models \phi$ for some $\mathfrak{M} = \langle S, R, V \rangle$. Define $\mathfrak{M}_s = \langle S, \mathcal{R}^V, \nu^V \rangle$ as follows:

- $\mathcal{R}^V$ is a ternary relation on S such that $\mathcal{R}^V xyz$ iff $R(\alpha_x)yz$ (in $\mathfrak{M}$).

- ν^V is a function from Φ to subsets of S such that $x \in \nu^V(p)$ iff $p \in \Phi_x$ (in $\mathfrak{M}$).

It is plain that $\mathfrak{M}_s$ is a flat model. It suffices to show that for all $x \in S$ and $\phi \in \mathcal{L}_s^{\supset}$, $\mathfrak{M}, x \models \phi$ iff $\mathfrak{M}_s, x \models \phi$. But this is established easily by induction on the complexity of ϕ. $\square$

Theorem 3 $\phi \in \mathcal{L}_s^\supset$ and $\phi \in$ **PDLS**, then $\phi \in$ **DB**.

Proof. We show that if $\phi \in \mathcal{L}_s^\supset$ and $\phi \notin$ **DB**, then $\phi \notin$ **PDLS**. Assume that $\phi \notin$ **DB**. By Proposition 1, there is a finite $\mathcal{M} = \langle \mathcal{P}, \mathcal{R}, \nu \rangle$, such that $\mathcal{M} \not\models \phi$. Hence, there is an injective $\tau : \mathcal{P} \to \pi$. Let us have any $\mathcal{M}^\tau = \langle \mathcal{P}, R^\tau, V^\tau \rangle$ that satisfies the following:

- R^τ is a function from $Prog(\pi)$ to binary relations on $\mathcal{P}$ such that for all $a \in \pi$: if $a = \tau(x)$ for some $x \in \mathcal{P}$, then $R^\tau(a)yz$ iff $\mathcal{R}xyz$. Moreover, R^τ complies with the requirements concerning '∪', ';', '*' and '?', explained in Section 2.1 (Note that it is 'always' possible to define R^τ so that the requirements are satisfied).

- $V^\tau(x) = \langle \tau(x), \{p \mid x \in \nu(p)\} \rangle$.

It is plain that every such $\mathcal{M}^\tau$ is an extended model and that at least one such $\mathfrak{M}^\tau$ exists. It suffices to show that for every $\phi \in \mathcal{L}_s^\supset$ and $x \in \mathcal{P}$, $\mathcal{M}, x \models \phi$ iff $\mathcal{M}^\tau, x \models \phi$. But this is established easily by induction on the complexity of ϕ. □

Corollary 1 **DB** *is the* $\mathcal{L}_s^\supset$-*fragment of* **PDLS**.

The results of this section establish a link between dynamic logic and substructural logic: a variant of the basic distributive substructural logic is a fragment of a dynamic logic where the substructural conditional '→' corresponds to partial correctness claims describing specific actions.

6 Extensions of the main result

It is natural to ask if similar results can be obtained for other substructural logics. This section provides some observations.

An *extension* of **PDLS** is a set of $\mathcal{L}_{ds}$-formulas valid in a specific class of extended dynamic models. We will show that every substructural logic over the consecution language $\mathcal{L}_s^\supset$ that satisfies certain conditions is a $\mathcal{L}_s^\supset$-fragment of some extension of **PDLS**.

Firstly, let us reiterate that R and V in every $\mathfrak{M}$ induce a ternary relation $\mathcal{R}^V$ such that

$$\mathcal{R}^V xyz \iff R(\alpha_x)yz. \tag{10}$$

(This observation was used in the proof of Theorem 2.) Secondly, we list several familiar 'frame-conditions' that are used in defining models for specific extensions of **DB**, see (Restall, 2000, Ch. 11) (for all $x, y, z, w \in \mathcal{P}$):

$$\exists v.(\mathcal{R}xyv \wedge \mathcal{R}vzw) \Rightarrow \exists v.(\mathcal{R}yzv \wedge \mathcal{R}xvw) \quad (11)$$
$$\exists v.(Ryzv \wedge Rxvw) \Rightarrow \exists v.(\mathcal{R}xyv \wedge \mathcal{R}vzw) \quad (12)$$
$$\mathcal{R}xyz \Rightarrow \mathcal{R}yxz \quad (13)$$
$$\mathcal{R}xxx \quad (14)$$

(11) is usually called the 'associativity condition', (12) the 'converse associativity condition', (13) the 'weak commutativity condition' and (14) the 'weak contraction condition'.

Lemma 1 *Given* (10), *the following claims hold:*

1. $\mathcal{R}^V$ *satisfies* (11) *iff (for all $x, y, z \in S$) $R(\alpha_x)yz$ implies $R(\alpha_z) \subseteq R(\alpha_y; \alpha_x)$ ($\mathfrak{M}$ that comply with this condition will be called 'associative'),*

2. $\mathcal{R}^V$ *satisfies* (12) *iff (for all $x, y, z, w \in S$) $R(\alpha_y; \alpha_x)zw$ implies that there is u such that $R(\alpha_x)yu$ and $R(\alpha_u)zw$ (conversely associative models),*

3. $\mathcal{R}^V$ *satisfies* (13) *iff (for all $x, y, z \in S$) $R(\alpha_x)yz$ implies $R(\alpha_y)xz$ (weakly commutative models),*

4. $\mathcal{R}^V$ *satisfies* (14) *iff (for all $x \in S$) $R(\alpha_x)xx$ (weakly contractive models).*

Proof. We shall prove the right-to-left direction of the first claim. Proofs of the other claims are similar. Given (10), (11) entails that if there is v such that $R(\alpha_x)yv$ and $R(\alpha_v)zw$, then there is u such that $R(\alpha_y)zu$ and $R(\alpha_x)uw$. In other words, then $R(\alpha_y; \alpha_x)zw$. But then, by first-order logic,

$$\forall v. (R(\alpha_x)yv \supset (R(\alpha_v)zw \supset R(\alpha_y; \alpha_x)zw)). \quad (15)$$

□

Let the following extensions of **PDLS** be defined as sets of dynamic substructural formulas valid in the set of extended dynamic models indicated in the brackets: **PDLS**$_B$ (associative models); **PDLS**$_{B^c}$ (conversely associative models); **PDLS**$_{CI}$ (weakly commutative models); **PDLS**$_{WI}$ (weakly contractive models).

Theorem 4 *For all $\phi \in \mathcal{L}_s^\supset$:*

1. *If ϕ is valid in every flat substructural frame that satisfies* (11), *then $\phi \in$* **PDLS**$_B$.

2. *If ϕ is valid in every flat substructural frame that satisfies* (12), *then $\phi \in$* **PDLS**$_{B^c}$.

3. *If ϕ is valid in every flat substructural frame that satisfies* (13), *then $\phi \in$* **PDLS**$_{CI}$.

4. *If ϕ is valid in every flat substructural frame that satisfies* (14), *then $\phi \in$* **PDLS**$_{WI}$.

Proof. Assume that $\phi \notin$ **PDLS**$_B$. Then there is an associative model $\mathfrak{M} \not\models \phi$. Define $\mathfrak{M}_s$ as in the proof of Theorem 2. By Lemma 1, $\mathfrak{M}_s$ is a flat substructural model that satisfies (11). It is easy to show by induction on the complexity of $\phi \in \mathcal{L}_s^\supset$ that (for all $x \in S$) $\mathfrak{M}, x \models \phi$ iff $\mathfrak{M}_s, x \models \phi$. Proofs of the other claims are similar. □

We shall say that a substructural logic **Log** (over $\mathcal{L}_s^\supset$) has the *countable flat model property* iff $\phi \notin$ **Log** implies that there is a *countable* flat model $\mathcal{M} \not\models \phi$.

Theorem 5 *Any substructural logic (over $\mathcal{L}_s^\supset$) that is sound and complete over a class of frames specified by (some) of the frame properties* (11)–(14) *and has the countable flat model property is a $\mathcal{L}_s^\supset$-fragment of some extension of* **PDLS**.

Proof. Given Theorem 4, it is sufficient to prove that if $\mathcal{M} \not\models \phi$ for a countable flat model $\mathcal{M} = \langle \mathcal{P}, \mathcal{R}, \nu \rangle$ that satisfies some of the conditions (11)–(14), then there is a corresponding[3] extended dynamic model $\mathfrak{M} \not\models \phi$. If $\mathcal{M}$ is countable, then there is an injection $\tau : \mathcal{P} \to \pi$ and we may proceed as in the proof of Theorem 3. □

7 Conclusion

We have shown that an independently motivated extension **PDLS** of **PDL** is interestingly linked to substructural logic. The substructural conditional

---
[3] Associative extended dynamic models correspond to (11), associative and (at the same time) weakly commutative models correspond to the conjunction of (11) and (13), etc.

'→' within the logic **DB** can be seen as expressing partial correctness claims concerning specific 'preferred' actions. The main technical result is a soundness and completeness proof for **DB** with respect to extended dynamic models, which entails that **DB** is a specific language-fragment of the dynamic logic **PDLS**. We have shown that our results extend to a specific class of stronger substructural logics.

Perhaps the most interesting topic for further research is to extend the results of this article to logics with negation. Substructural negation is a *modal* operator, usually interpreted in terms of a 'compatibility' relation C in substructural models: $\sim\phi$ holds in x iff ϕ is false in every y compatible with x, that is Cxy, see (Restall, 2000). Technically, the compatibility relation could be 'incorporated' into extended models quite easily: just pick any $a_i \in \pi$ and read $R(a_i)$ as the compatibility relation. In other words, one could define '$\sim$' thus:

$$\sim\phi ::= [a_i]\neg\phi. \qquad (16)$$

As shown in Example 1, such an operator could be read informally in terms of Anne's implicit belief. Hence, the substructural $\sim\phi$ could be read as 'Anne believes that ϕ is false', or, equivalently, 'Anne believes that $\neg\phi$'. Further investigations into this are left for another occasion.

References

Beall, J., Brady, R., Dunn, J. M., Hazen, A. P., Mares, E., Meyer, R. K., ... Sylvan, R. (2012). On the Ternary Relation and Conditionality. *Journal of Philosophical Logic, 41*, 595–612.

Bimbó, K., & Dunn, J. M. (2005). Relational Semantics for Kleene Logic and Action Logic. *Notre Dame Journal of Formal Logic, 46*, 381–513.

Dunn, J. M. (2001). Ternary Relational Semantics and Beyond: Programs as Data and Programs as Instructions. *Logical Studies, 7*, 282–301.

Dunn, J. M. (2003). Ternary Semantics for Dynamic Logic, Hoare Logic, Action Logic, etc. In *Smirnov's Readings* (pp. 70–71). Moscow: Institute of Logic, Russian Academy of Sciences.

Fischer, M. J., & Ladner, R. E. (1979). Propositional Dynamic Logic of Regular Programs. *Journal of Computer and System Sciences, 18*, 194–211.

Grove, A. J. (1995). Naming and Identity in Epistemic Logic, Part II: A First-order Logic for Naming. *Artificial Intelligence, 74*, 311–350.

Grove, A. J., & Halpern, J. Y. (1993). Naming and Identity in Epistemic Logics, Part I: The Propositional Case. *Journal of Logic and Computation*, *3*, 345–378.

Harel, D., Kozen, D., & Tiuryn, J. (2000). *Dynamic Logic*. Cambridge: MIT Press.

Kozen, D., & Tiuryn, J. (2003). Substructural Logic and Partial Correctness. *ACM Transactions on Computational Logic*, *4*, 355–378.

Mares, E. D. (1996). Relevant Logic and the Theory of Information. *Synthese*, *109*, 345–360.

Mares, E. D. (2004). *Relevant Logic: A Philosophical Interpretation*. Cambridge: Cambridge University Press.

Restall, G. (1995). Information Flow and Relevant Logics. In J. Seligman & D. Westerståhl (Eds.), *Logic, Language and Computation: The 1994 Moraga Proceedings* (pp. 463–477). Stanford: CSLI Publications.

Restall, G. (2000). *An Introduction to Substructural Logics*. London: Routledge.

Slaney, J., & Meyer, R. (1997). Logic for Two: The Semantics of Distributive Substructural Logics. In D. M. Gabbay, R. Kruse, A. Nonnengart, & H. J. Ohlbach (Eds.), *Qualitative and Quantitative Practical Reasoning* (pp. 554–567). Berlin: Springer.

Igor Sedlár
Comenius University in Bratislava
Slovakia
E-mail: `igor.sedlar@uniba.sk`

On Strong Fragments of Peano Arithmetic

VÍTĚZSLAV ŠVEJDAR

Abstract: We analyze the classical proof of Paris and Kirby, showing that Σ_{n+1}-collection is not provable using Σ_n-induction. We then mention another principle that is also violated in the model of Paris and Kirby and that might be weaker than Σ_{n+1}-collection: there is no Σ_{n+1} definable bounded one-one function.

Keywords: induction, collection, fragments of PA

1 Robinson and Peano arithmetics, induction

Robinson arithmetic Q is a weak axiomatic theory of natural numbers. Its language contains the symbols $+$ and $\cdot$ for addition and multiplication of natural numbers, the symbol S for the successor function (adding the number one), and the constant 0. The original version of Robinson arithmetic, as defined in (Tarski, Mostowski, & Robinson, 1953), has seven simple axioms Q1–Q7, where for example the axioms Q4 and Q5 are $\forall x(x + 0 = x)$ and $\forall x \forall y (x + S(y) = S(x+y))$. Because of the bounded quantifiers (to be discussed soon), it nowadays seems more practical to add the symbols $\leq$ and $<$ for unstrict and strict ordering to the language of Q, and formulate two additional axioms about these symbols:

Q8: $\quad \forall x \forall y (x \leq y \equiv \exists v (v + x = y))$,
Q9: $\quad \forall x \forall y (x < y \equiv \exists v (S(v) + x = y))$.

The resulting version of Q, with nine axioms, is conservative over the original version defined in (Tarski et al., 1953).

Despite the fact that the two sample axioms Q4 and Q5 mentioned above look like a recursive definition of uniquely defined addition, expected properties of addition, like associativity and commutativity, are not provable in Q. The same is true about multiplication. Expected properties of operations can be proved in *Peano arithmetic* PA, which is a theory obtained

Vítězslav Švejdar

from Q by adding the *induction schema*: the sentence

Ind: $\quad \forall \underline{y}(\varphi(0,\underline{y}) \,\&\, \forall x(\varphi(x,\underline{y}) \to \varphi(\mathrm{S}(x),\underline{y})) \to \forall x \varphi(x,\underline{y}))$,

where $\underline{y}$ stands for $y_1,..,y_k$, is an axiom for any choice of a formula φ and a variable x. The variable x can be called *induction variable*, or *variable of induction*, the remaining variables $y_1,..,y_k$ are *parameters*. It is understood that all free variables of φ are among $x, y_1, .., y_k$. When using Ind to show that $\forall x\varphi(x,\underline{y})$, we say that $\forall x\varphi(x,\underline{y})$ is proved *by induction on x for (fixed) parameters $y_1,..,y_k$*. As an example or an exercise, the reader may want to write down proofs of the sentences $\forall x(0 + x = x)$, $\forall x \forall y (\mathrm{S}(x) + y = \mathrm{S}(x+y))$ and $\forall x \forall y(x + y = y + x)$. Peano arithmetic is incomplete because Gödel incompleteness theorems apply to it. However, it is a strong theory: natural examples of independent sentences are difficult to find. Peano arithmetic can also prove the schema

B: $\quad \forall \underline{y} \forall x (\forall u {<} x\, \exists v \varphi(u,v,\underline{y}) \to \exists z \forall u {<} x\, \exists v {<} z\, \varphi(u,v,\underline{y}))$,

where again $y_1,..,y_k$ are parameters. This schema is called *collection schema*, or *boundedness schema*. It prevents the existence of an arithmetically definable relation that is *unbounded on some x*, where a relation unbounded on x is an "impossible" relation the characteristic function of which is suggested in Fig. 1: for each z there are columns $u < x$ that only contain zeros in all lines $v < z$, but, what might not be apparent in Fig. 1, in every column $u < x$ there are some ones.

2 Hierarchy of formulas and bounded induction

To prove the expected properties of operations, like the commutativity of addition mentioned above, it is often sufficient to use the induction schema for very simple or even open (quantifier free) formula φ. The study of fragments of PA is motivated by the following question: in which situations, if ever, does one need induction (or collection) for a complicated φ, with alternating quantifiers?

Let $\forall v {<} z\, \varphi$ and $\exists v {<} z\, \varphi$ be shorthands for $\forall v(v < z \to \varphi)$ and $\exists v(v < z\, \&\, \varphi)$ respectively, and similarly for $\forall v {\leq} z\, \varphi$ and $\exists v {\leq} z\, \varphi$. The expressions $\forall v {<} z$, $\exists v {<} z$, $\forall v {\leq} z$, and $\forall v {\leq} z$, where v and z are different variables, are called *bounded quantifiers*. A formula is a *bounded formula*, or a Δ_0-*formula*, if all quantifiers in it are bounded. A formula is Σ_n if it has

On Strong Fragments of Peano Arithmetic

⋮	⋮	⋮	⋮	⋮	⋮	⋮	⋮	⋮	⋮	⋮	⋮	⋮	⋮
0	0	0	0	0	0	0	0	0	0	0	0	1	0
0	0	0	0	0	0	0	0	0	0	0	0	0	0
0	0	0	0	0	0	0	0	0	0	0	0	0	0
0	0	0	0	0	0	0	0	0	0	0	0	0	0
0	0	0	0	0	0	0	0	0	0	0	1	0	0
0	0	0	0	0	0	0	0	0	0	0	0	0	0
0	1	0	0	0	0	0	0	0	0	0	0	0	0
0	0	0	0	0	0	0	0	0	0	0	0	0	0
0	1	0	0	0	0	0	0	0	0	0	0	0	0
0	0	0	0	0	0	0	0	0	0	0	0	0	0
0	0	0	0	0	0	0	0	0	0	0	0	0	0
0	0	0	0	0	0	0	0	0	0	0	0	0	0
0	0	0	0	0	0	0	0	1	0	0	0	0	0
0	0	0	0	0	0	0	0	1	0	0	0	0	0
0	0	0	0	0	0	0	0	0	0	0	0	0	0
0	...				...						...		x

Figure 1: An unbounded relation on x

the form $\exists v_1 \forall v_2 \exists .. v_n \varphi$ where φ is bounded. Thus a Σ_n-*formula* has a prefix of n alternating quantifiers the first of which is existential, followed by a Δ_0-formula. Symmetrically, a Π_n-*formula* has the form $\forall v_1 \exists v_2 \forall .. v_n \varphi$ where φ is Δ_0-formula.

The class of all relations definable by Δ_0-formulas in the structure **N** of natural numbers is an easily defined (proper) subclass of the class of all recursive relations. It is quite inclusive in the sense that all r.e. relations can be obtained from Δ_0 relations by the projection operation (i.e. by existential quantification). It is also inclusive in the sense that it contains many important sets like the divisibility relation, the set of all primes, and the set of all powers of two. It also contains the graph of the function $n \mapsto 2^n$; that is, there is a Δ_0-formula $\varepsilon(x, y)$ such that $m = 2^n$ if and only if $\mathbf{N} \models \varepsilon(\overline{n}, \overline{m})$, where $\overline{n}$ and $\overline{m}$ are the numerals for n and m, i.e. the closed terms $S(S(.. (0)..)$, having the corresponding number of occurrences of the symbol S. This fact has a not so trivial proof, see (Pudlák, 1983) or (Bennet, 1962).

$I\Delta_0$ is a theory like Peano arithmetic, but with the induction schema restricted to Δ_0-formulas. The *theory* $I\Delta_0$ is also called *bounded arithmetic*,

Vítězslav Švejdar

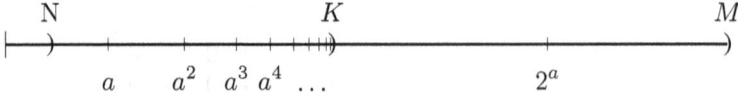

Figure 2: A simple construction of a model of IΔ_0

but this name is sometimes also used for similar theories that can be even weaker and that can have a somewhat different language.

In IΔ_0, i.e. using induction for Δ_0-formulas only, one can prove various facts about addition and multiplication: both operations are associative and commutative, multiplication is distributive over addition, for each pair x and y there is a unique z such that $z + x = y$ or $z + y = x$. It is also possible to develop coding of finite sets and sequences. Thus writing $x \in w$ has a good sense in IΔ_0. We also have

$$\text{I}\Delta_0 \vdash \varepsilon(0,\overline{1}) \ \& \ \forall x \forall y \forall z(\varepsilon(x,y) \to \varepsilon(x+\overline{1},z) \equiv z = \overline{2} \cdot y) \qquad (1)$$

where ε is the formula mentioned above, that defines the graph of the exponentiation function. Also, one can prove in IΔ_0 that every two formulas ε satisfying (1) are equivalent; thus exponentiation with base 2 has a unique meaning in IΔ_0. However, $\forall x \exists y \varepsilon(x,y)$ is *not* provable in IΔ_0. That is, the function $x \mapsto 2^x$ is a uniquely defined but possibly partial function. Let Exp be the sentence $\forall x \exists y \varepsilon(x,y)$, and let I$\Delta_0$+Exp be I$\Delta_0$ enhanced with the sentence Exp as an additional axiom. We take IΔ_0+Exp as the *base theory*, although we know that this choice is somewhat arbitrary. For some purposes, like coding of logical syntax, a weaker axiom would be sufficient, while in some other situations a stronger additional axiom would be needed. In IΔ_0+Exp, coding of finite sets and sequences works as expected; in particular, there are sets and sequences of arbitrary (finite) lengths. The structure **N** of natural numbers is a model (the *standard model*) of the theories PA, IΔ_0, and IΔ_0+Exp. It is also a model of all the remaining theories we will consider. However, one should not forget that all these theories have non-standard models as well. So for example, when we speak, inside a theory, about the length of a (finite) sequence, it is good to see a non-standard element of some model behind that length. We also have Δ_0-comprehension in IΔ_0+Exp: for each Δ_0-formula $\varphi(v)$, possibly with parameters, IΔ_0+Exp $\vdash \forall z \exists w \forall v(v \in w \equiv v < z \ \& \ \varphi(v))$. That is, for each z there exists a number w that codes the set $\{ v \ ; \ v < z \ \& \ \varphi(v) \}$.

Consider a model $M \models$ PA like in Fig. 2, with domain M, the standard part N and a non-standard element a. Put $K = \{b \in M\,;\, \exists m{\in}\mathrm{N}(b < a^m)\}$. K is evidently closed under $+$ and $\cdot$. Thus the structure $\boldsymbol{K}$ with domain K and operations induced from $\boldsymbol{M}$ is a substructure of $\boldsymbol{M}$. One can show by induction on complexity of $\varphi(x_1,..,x_k) \in \Delta_0$ that if $b_1,..,b_k \in K$ then $\boldsymbol{K} \models \varphi(\underline{b}) \equiv \boldsymbol{M} \models \varphi(\underline{b})$. From this, absoluteness of Δ_0-formulas, it follows that $\boldsymbol{K} \models \mathrm{I}\Delta_0$. For each standard m, there exists a standard k such that $2^x > x^m$ for all $x > k$. For example, if $m = 3$ then $2^x > x^m$ for all $x > 9$. If $m = 4$ then $2^x > x^m$ for all $x > 16$. Since a is non-standard, all elements a^2, a^3, a^4, ... are less than 2^a. It follows that $2^a \notin K$. From this we have that 2^x does not exist in $\boldsymbol{K}$, i.e. $\boldsymbol{K} \not\models \exists y \varepsilon(a,y)$. Thus we have $\boldsymbol{K} \not\models \mathrm{Exp}$.

3 The hierarchy of strong fragments of Peano arithmetic

Let IΓ, for $\Gamma = \Sigma_n$ or $\Gamma = \Pi_n$, be the theory obtained from IΔ_0 by adding the schema IΓ, induction restricted to Γ-formulas. Thus IΣ_n and IΠ_n is a notation for both an axiomatic schema and an axiomatic theory. IΣ_0 and IΠ_0 are the same theories as IΔ_0. For the formula $\varepsilon(x, y)$ defining the graph of the function $x \mapsto 2^x$, one can show by induction on x that $\forall x \exists y \varepsilon(x,y)$. Since $\exists y \varepsilon(x,y)$ is a Σ_1-formula, the sentence $\forall x \exists y \varepsilon(x,y)$, i.e. the sentence Exp, is an example of a sentence provable in IΣ_1 but not in IΔ_0. Using dummy quantifiers, it is easy to show that a Σ_n- or a Π_n-formula is (already in predicate logic, and thus in all theories considered here), equivalent to a Σ_m-formula for each $m > n$, and also to a Π_m-formula for each $m > n$. Thus the sentence Exp, a useful axiom possibly added to IΔ_0, brings nothing new if added to IΣ_n for $n > 0$ or to IΠ_n for $n > 1$.

Besides induction and collection, we can consider yet another axiom schema, the *least number principle*

L: $\forall \underline{y}(\exists x \varphi(x, \underline{y}) \to \exists x(\varphi(x, \underline{y}) \,\&\, \forall v{<}x\, \neg\varphi(v, \underline{y})))$.

Let LΓ and BΓ be the theories defined similarly as IΓ, i.e. the theories obtained by adding the schema L or B restricted to Γ-formulas, to the theory IΔ_0. Again, LΣ_n, LΠ_n, BΣ_n, and BΠ_n denote both an axiom schema and an axiomatic theory. We will see that BΠ_0 is not provable in IΔ_0+Exp. However, it is not difficult to check that BΠ_0 is valid in the model $\boldsymbol{K}$ from Fig. 2.

Assume that $\varphi(u,v,s)$ is a Π_n-formula, with possible parameters that are not shown. Assume further that $\exists s\,\varphi(u,v,s)$ defines an unbounded relation on x, i.e. that $\forall u\!<\!x\,\exists v\exists s\,\varphi(u,v,s)$, but $\neg\exists z\forall u\!<\!x\,\exists v\!<\!z\,\exists s\,\varphi(u,v,s)$. Then $\forall u\!<\!x\,\exists w\exists v\!<\!w\,\exists s\!<\!w\,\varphi(u,v,s)$. This is because if $\exists v\exists s\,\varphi(u,v,s)$ then one can take arbitrary $w > \max\{v,s\}$ to have $\exists v\!<\!w\,\exists s\!<\!w\,\varphi(u,v,s)$. Further

$$\neg\exists z\forall u\!<\!x\,\exists w\!<\!z\,\exists v\!<\!w\,\exists s\!<\!w\,\varphi(u,v,s).$$

Indeed, from $\forall u\!<\!x\,\exists w\!<\!z\,\exists v\!<\!w\,\exists s\!<\!w\,\varphi(u,v,s)$ we could evidently conclude that $\forall u\!<\!x\,\exists v\!<\!z\,\exists s\,\varphi(u,v,s)$, which would contradict our assumption. Thus we see that if $\exists s\,\varphi(u,v,s)$ defines an unbounded relation on x then $\exists v\!<\!w\,\exists s\!<\!w\,\varphi(u,v,s)$ defines an unbounded relation (in x and w) on the same x. The whole argument shows that $\mathrm{B}\Pi_n$ implies $\mathrm{B}\Sigma_{n+1}$ provided $\exists v\!<\!w\,\exists s\!<\!w\,\varphi(u,v,s)$ is a Π_n-formula. Which it is, as it follows from (a) of the following theorem.

Theorem 1 *The theory $\mathrm{B}\Pi_n$ proves the following.*
(a) Σ_{n+1}- and Π_{n+1}-formulas are closed under bounded quantification. That is, a formula obtained from Σ_{n+1}- or Π_{n+1}-formula by bounded quantification is in $\mathrm{B}\Pi_n$ equivalent to some Σ_{n+1}- or Π_{n+1}-formula respectively.
(b) Σ_{n+2}-formulas are closed under $\exists$, Π_{n+2}-formulas are closed under $\forall$.
(c) Both Σ_{n+2}- and Π_{n+2}-formulas are closed under conjunctions and disjunctions.

It is trivial that Σ_0- and Π_0-formulas are closed under bounded quantification and Boolean connectives—recall that $\Sigma_0 = \Pi_0 = \Delta_0$. It is also true that Σ_1-formulas are closed under $\exists$ and Π_1-formulas are closed under $\forall$, and that both Σ_1- and Π_1-formulas are closed under conjunctions and disjunctions. These facts are not listed in the previous theorem because their proofs do not require the collection schema (do not require anything above $\mathrm{I}\Delta_0$).

Theorem 2 *(a) $\mathrm{B}\Sigma_{n+1}$ and $\mathrm{B}\Pi_n$ are equivalent theories.*
(b) All $\mathrm{I}\Sigma_n$, $\mathrm{I}\Pi_n$, $\mathrm{L}\Sigma_n$, and $\mathrm{L}\Pi_n$ are equivalent theories.
(c) $\mathrm{I}\Sigma_{n+1}$ proves $\mathrm{B}\Pi_n$, and $\mathrm{B}\Pi_n$ proves $\mathrm{I}\Sigma_n$.

We already gave the proof of (a). The remaining proofs are similarly natural. Theorems 1 and 2 contain the basic relationships between strong fragments of PA and are well known. Details can be found in (Paris &

Kirby, 1978) and (Hájek & Pudlák, 1993). With Theorem 2 in mind, we have the following hierarchy of theories:

$$I\Delta_0+\text{Exp} \Leftarrow B\Pi_0+\text{Exp} \Leftarrow I\Sigma_1 \Leftarrow B\Pi_1 \Leftarrow \ldots \Leftarrow \text{PA}$$
$$\Updownarrow \qquad \Updownarrow \qquad \Updownarrow$$
$$B\Sigma_1+\text{Exp} \qquad I\Pi_1 \qquad B\Sigma_2$$
$$\Updownarrow$$
$$L\Sigma_1$$
$$\Updownarrow$$
$$L\Pi_1$$

A formula is $\Sigma_0(\Sigma_n)$ if it is obtained from Σ_n formulas using Boolean connectives and bounded quantification. It is known that $I\Sigma_n$ for $n \geq 1$ proves $\Sigma_0(\Sigma_n)$-comprehension: for each $\Sigma_0(\Sigma_n)$-formula $\varphi(v)$ the theory $I\Sigma_n$ proves that for each z there exists a number w that codes the set $\{v\,;\,v < z\ \&\ \varphi(v)\,\}$. It is clear that $\Sigma_0(\Delta_0) = \Delta_0$. We have already noted that $I\Delta_0+\text{Exp}$ proves Δ_0-comprehension.

A *universal Σ_{n+1}-formula* is a formula $\exists v \gamma(z,x,v)$ such that $\gamma \in \Pi_n$ and for any Σ_{n+1}-formula $\psi(x)$ there exists a number $e \in \mathbb{N}$ such that $\psi(x)$ is in $I\Sigma_1$ equivalent to $\exists v \gamma(\overline{e},x,v)$. Universal Σ_{n+1}-formulas exist for each $n \geq 1$. Using universal Σ_{n+1}-formulas, it is possible to show that all theories $I\Sigma_n$ and $B\Pi_n$ for $n \geq 1$ are finitely axiomatizable.

4 The hierarchy of fragments does not collapse

We now rephrase the proof of (Paris & Kirby, 1978) that $B\Pi_n$ is not provable in $I\Sigma_n$. We will use the notion of sparse relation. We say that a *relation defined by a formula $\theta(u,v)$ is sparse with respect to t* if $\forall u \exists v \theta(u,v)$ but, for each s, the set $\{[u,v]\,;\,u < s\ \&\ v < s\ \&\ \theta(u,v)\,\}$ has less than t elements. In other words, if a relation is sparse then every square $s \times s$ contains less than t elements (pairs) of that relation. The definition of a sparse relation only makes sense in the presence of comprehension for the formula θ: only then the number of pairs in a square $s \times s$ is uniquely determined. Thus let the notion of sparse relation be made more precise as follows: in $I\Delta_0+\text{Exp}$, we speak about sparse relations defined by formulas $\theta \in \Delta_0$, while in $I\Sigma_n$ for $n \geq 1$ we speak about sparse relations defined by formulas $\theta \in \Sigma_0(\Sigma_n)$. The following theorem says that a sparse relation is

Vítězslav Švejdar

something even more weird than a relation unbounded on x. The lemma is easily proved using Theorem 2(a): a $\Sigma_0(\Sigma_n)$-formula is Σ_{n+1} in $I\Sigma_n$.

Theorem 3 *In $I\Sigma_n$+Exp, if a relation defined by $\theta \in \Sigma_0(\Sigma_n)$ is sparse w.r.t. t then it is unbounded on any $x \geq t$. Thus $B\Pi_n$+Exp proves that there is no $\Sigma_0(\Sigma_n)$ sparse relation with respect to any t.*

Let n be given and let M be a model of PA containing non-standard Σ_{n+1}-definable elements, where an element $a \in M$ is Σ_{n+1}-definable if it is the only element of M satisfying certain Σ_{n+1}-formula. To obtain M, we can use Gödel incompleteness theorems: a witness for a false Σ_1-sentence (if 1st Gödel incompleteness theorem is used) or the least proof of contradiction in formalized PA (if 2nd incompleteness theorem us used) is Δ_0-definable and non-standard. Thus we can begin with the same M regardless of n. Let $K = \{\, a \in M \,;\, a \text{ is } \Sigma_{n+1}\text{-definable in } M \,\}$.

First, we claim that K contains all standard elements of M as well as some non-standard elements, and is closed under addition and multiplication. Indeed, if $\psi_1(x) \in \Sigma_{n+1}$ defines a and $\psi_2(x) \in \Sigma_{n+1}$ defines b then the formula

$$\exists u \exists v (\psi_1(u) \ \& \ \psi_2(v) \ \& \ x = u+v) \qquad (2)$$

is satisfied by exactly one element of the model M, the sum $a+b$, and so it defines the element $a+b$. The argument for $a \cdot b$ is analogical. The formula (2) is Σ_{n+1} because M is a model of full PA (in which Σ_{n+1}-formulas are closed under $\exists$ and $\&$). Thus K, i.e. the set K with operations inherited from M, is a substructure of M. A difference to the model in Fig. 2 is that K might be not downwards closed.

Second, we claim that if $\varphi(x, y_1, \ldots, y_k)$ is a Σ_{n+1}-formula, $a_1, \ldots, a_k$ elements of K and $M \models \exists x \varphi(x, \underline{a})$ then there exists a $b \in K$ such that $M \models \varphi(b, \underline{a})$. Indeed, let $\varphi(x, \underline{y})$ be $\exists v \theta(v, x, \underline{y})$ where $\theta \in \Pi_n$, and let $\psi_1(x), \ldots, \psi_k(x)$ be Σ_{n+1}-formulas that define the elements $a_1, \ldots, a_k$ respectively. Let $[v, x] \mapsto (v, x)$ be the pairing function, i.e. the function defined as $(v, x) = \frac{1}{2}(v+x)(v+x+\overline{1}) + v$, and let l and r be the two inverse functions to the pairing functions, i.e. the functions satisfying the equations $(l(w), r(w)) = w$, $l((v, x)) = v$, and $r((v, x)) = x$ for each w, v, and x. Then

$$\exists u_1 \ldots \exists u_k \exists w (\psi_1(u_1) \ \& \ \ldots \ \& \ \psi_k(u_k) \ \& \ \theta(l(w), r(w), \underline{u}) \ \& \\ \& \ \forall z {<} w \, \neg \theta(l(z), r(z), \underline{u}) \ \& \ x = l(w))$$

is a Σ_{n+1} formula satisfied by exactly one element of M, the left part $l(w)$ of the least w satisfying $\theta(l(w), r(w), \underline{u})$ with respect to the uniquely defined elements $a_1, \ldots, a_k$. Recall that the trick of working with the left part of the least w satisfying $\theta(l(w), r(w), \underline{u})$ rather than the least x satisfying $\exists v\theta(v, x, \underline{u})$ is taken from recursion theory. The least x satisfying $\exists v\theta(v, x, \underline{u})$ is not Σ_{n+1}-definable.

Third, a usual argument from model theory shows that Σ_{n+1}-formulas are absolute between K and M: if $\varphi(\underline{y}) \in \Sigma_{n+1}$ and $a_1, \ldots, a_k \in K$ then $K \models \varphi(\underline{a}) \Leftrightarrow M \models \varphi(\underline{a})$. We can describe this by saying that K is a Σ_{n+1}-*elementary substructure of* M and denote as $K \preceq_{n+1} M$.

Fourth, from $K \preceq_{n+1} M$ it follows that any Π_{n+2} sentence valid in M is also valid in K. Since all axioms of $I\Sigma_n$ are Π_{n+2}, we have $K \models I\Sigma_n$.

Fifth, it remains to show what is violated in K. Let $\exists v\gamma(z, x, v)$ where $\gamma \in \Pi_n$ be a universal Σ_{n+1}-formula. Since every $a \in K$ is Σ_{n+1}-definable in M and every Σ_{n+1} formula $\psi(x)$ is equivalent to $\exists v\gamma(\overline{e}, x, v)$ for some $e \in \mathbb{N}$, we see that for every $a \in K$ there exists an $e \in \mathbb{N}$ such that a is the only element x of M satisfying $\exists v\gamma(\overline{e}, x, v)$ in M. Let a non-standard $t \in K$ be fixed. Since Σ_{n+1}-formulas are absolute between K and M, we have

$$K \models \forall x \exists z < t \forall u (\exists v \gamma(z, u, v) \equiv u = x), \tag{3}$$

which says that, in K, every number x is the only number u (in the entire universe) satisfying $\exists v\gamma(z, u, v)$ for a suitable $z < t$. Work in K and consider the following relation between z, x, and w:

$$\begin{aligned}
& x \leq w \ \& \ \forall w' < w \forall x' < w \neg\gamma(z, x', w') \ \& \\
& \quad \& \ \forall x' \leq w (x' \neq x \to \neg\gamma(z, x', w)) \ \& \\
& \quad \& \ \exists w' \leq w \gamma(z, x, w').
\end{aligned} \tag{4}$$

Let the formula (4) be denoted $\delta(z, x, w)$. To understand its meaning, think of z as fixed, imagine a characteristic function of a binary relation like in Fig. 1, with columns x or x' and lines w', and think of w as a size of a square. Then the three lines in (4) say that (i) the square $(w+1) \times (w+1)$ contains zeros everywhere except possibly in the right and upper borders, i.e. in column w or in line w, (ii) there are zeros in all columns $x' \neq x$ (even in line w), but (iii), there are some ones in column x. Put otherwise, either $x = w$ and the column x is the only column containing some ones, or $x < w$ and the coordinates of the only one in the entire square are $[x, w]$. Evidently, there exists at most one pair $[x, w]$ such that $\delta(z, x, w)$. For such a pair it is the case that $x \leq w$. Consider now all pairs $[x, w]$ such that $\exists z< t\,\delta(z, x, w)$, now with t fixed as said above. Then $\exists z< t\,\delta(z, x, w)$ can be seen as a union of t binary relations each of which is at most single-element. Thus the relation $\exists z< t\,\delta(z, x, w)$ satisfies the condition from the definition of sparse relation with respect to $t+1$: every square $s \times s$ contains less than $t + 1$ elements of this relation. Let x be given and consider z like in (3), satisfying $z < t$ and $\forall u (\exists v \gamma(z, u, v) \equiv u = x)$. If w is the least number such that $\exists w' \leq w\, \gamma(z, x, w')$ then $\delta(z, x, w)$. This shows that $\forall x \exists w \exists z < t\,\delta(z, x, w)$, and so $\exists z < t\,\delta(z, x, w)$ satisfies also the remaining condition from the definition of sparse relation.

We can summarize that $\exists z< t\,\delta(z, x, w)$ is a sparse relation; hence by Theorem 3, $B\Pi_n$ is violated in K.

Theorem 4 *The following conditions are equivalent over $I\Sigma_n + \text{Exp}$.*
(i) There exists a $\Sigma_0(\Sigma_n)$-definable sparse relation.
(ii) There exists a Σ_{n+1}-definable one-one function which is bounded.
(iii) There exists a Σ_{n+1}-definable one-one function the range of which is the interval $\{ y\, ;\, y < t\, \}$ for some t.

We omit the proofs of this theorem, but we give some remarks and hints. if $\varphi(x, y, v) \in \Sigma_0(\Sigma_n)$ defines a one-one function bounded by z then the relation defined by

$$\exists w \exists y \leq w \exists v \leq w\, (\varphi(x, y, v)\ \&\ \forall y' < w\, \forall v' < w\, \neg \varphi(x, y', v'))$$

is $\Sigma_0(\Sigma_n)$ and sparse with respect to any $t > z$. The proof of (ii) $\Rightarrow$ (iii) is due to Jeff Paris and is probably unpublished. Actually, the notion of sparse relation is extracted from Paris's proof of (ii) $\Rightarrow$ (iii).

5 Final remarks

There are several variants of the pigeon hole principle (PHP), the most common being that there is no one-one function from $t + 1$-element set to t-element set. The negation of the condition in (ii) of Theorem 4 says that there is no Σ_{n+1}-definable one-one function *from the entire universe* to some t-element set. This principle, i.e. the negation of the condition (ii) of Theorem 4, can be called a *weak PHP principle* and denoted $\text{WPHP}(\Sigma_{n+1})$. I conjecture (and have some ideas how to prove) that this principle is weaker than $B\Pi_n$, i.e. that $I\Sigma_n + \text{WPHP}(\Sigma_{n+1})$ does not prove $B\Pi_n$. There may also be an open problem connected with $\text{WPHP}(\Sigma_{n+1})$: I do not know whether $I\Delta_0 + \text{WPHP}(\Sigma_{n+1})$ proves $I\Sigma_n$.

References

Bennet, J. H. (1962). *On Spectra.* Dissertation, Princeton University, Princeton, NJ.

Hájek, P., & Pudlák, P. (1993). *Metamathematics of First Order Arithmetic.* Berlin: Springer.

Paris, J. B., & Kirby, L. A. S. (1978). Σ_n-collection Schemas in Arithmetic. In A. Macintyre, L. Pacholski, & J. Paris (Eds.), *Logic Colloquium '77* (pp. 199–209). Amsterdam: North-Holland.

Pudlák, P. (1983). A Definition of Exponentiation by a Bounded Arithmetical Formula. *Comm. Math. Univ. Carolinae, 24*(4), 667–671.

Tarski, A., Mostowski, A., & Robinson, R. M. (1953). *Undecidable Theories.* Amsterdam: North-Holland.

Vítězslav Švejdar
Charles University in Prague
The Czech Republic
E-mail: `vitezslav.svejdar@cuni.cz`

Towards a New Theory of Historical Counterfactuals

JACEK WAWER AND LESZEK WROŃSKI

Abstract: We investigate the semantics of historical counterfactuals in indeterministic contexts. We claim that "plain" and "necessitated" counterfactuals differ in meaning. To substantiate this claim, we propose a new semantic treatment of historical counterfactuals in the *Branching Time* framework. We supplement our semantics with supervaluationist postsemantics, thanks to which we can explain away the intuitions which seem to talk in favor of the identification of "would" with "would necessarily."

Keywords: Conditionals, counterfactuals, modality, branching time, David Lewis, Robert Stalnaker

1 Introduction

There is a recurring idea in the semantics of counterfactuals summarized well by the following statement: "all counterfactual conditionals express necessitation" (Pollock, 1976, p. 34). It was shared by Lewis (1986) for whom, as Stalnaker (1981, p. 93) neatly puts it, "the antecedents of conditionals act like necessity operators on their consequents." Recently, a similar claim was made by Leitgeb (2012, p. 36) who takes

(A) If the match were struck, it would light.

and

(B) If the match were struck, it would necessarily light.

to be synonyms.

We do not find this semantic identification convincing. Not only we do not think that "necessitated" reading of "would" is the default one, but we are even uncertain if there is any non-artificial reading of "would" that has the meaning of "necessarily would."

Indeterministic contexts support our intuition. Consider first a fair coin that I am about to toss and a pair of sentences:

(C) The coin will land heads.

(D) The coin will necessarily land heads.

While we are strongly inclined to reject (D) (on the grounds that the coin is fair), our reaction towards (C) is much more hesitant; we can even say "Who knows, maybe." In terms of degrees of belief, our degree of belief in (C) is much higher (about 0.5) than our degree of belief in (D) (about 0), which itself should suggest that we do not naturally take them to be synonymous. Now, suppose that I could have but did not toss the coin and consider a pair of conditionals:

(E) If I had flipped the coin, it would have landed heads.

(F) If I had flipped the coin, it would have necessarily landed heads.

We claim that our intuitive reactions towards (E) and (F) resemble those towards (C) and (D) respectively. Similarly, our degree of belief in (E) (about 0.5) is much higher than our degree of belief in (F) (about 0). We think that just as (C) and (D) clearly differ in meaning, so do (E) and (F). Observations of this sort support the view that "would" should *not* be be identified with "necessarily would."

To validate our intuition we develop, against the background of the theory of branching time (section 2), a novel semantic treatment of counterfactuals (sections 3 and 4). We combine our semantics with supervaluational postsemantics (section 5) and use this rich conceptual apparatus to explain the subtle affinity of plain and necessitated counterfactuals (section 6).

We choose the branching-time setting since it suits the discussion of indeterminism especially well. This setting has already been used for analysis of counterfactual future contingents (Placek & Müller, 2007; Thomason & Gupta, 1980). However, our semantic approach is, as far as we know, a novel one. We will minimize the discussion of the formal part of the theory to the necessary minimum, leaving the full description of the formalities involved for a different paper.

2 Branching time

Branching time is a natural way of representing how possibilities evolve in time. The framework aims at explicating the idea that some eventualities which are possible at some earlier time cease to be possible at later

Towards a New Theory of Historical Counterfactuals

times. It is also meant to capture the intuition that the future, contrary to the past, is open to multiple possible realizations. Additionally, it incorporates the observation that what is possible and what is settled is very much circumstance-dependent: what is possible at a given moment depends on what the world is like at this very moment (such situation-dependent necessity was sometimes called "accidental necessity" in the scholastic terminology).

All these intuitions are meant to be captured by a pictorial representation of temporal possibilities in the shape of a tree which branches upwards, but never downwards. Each point at a tree represents a possible state of the world – these possible states are called "(possible) moments". At a given moment, each "branch" growing out of a given moment represents its possible future continuation while the "trunk" of the moment represents its unique possible past. Each maximal line throughout the tree represents a full possible course of events – a "(possible) history". ("Moment" and "history" will be used as technical terms from now on.)

Here is an example of a simple branching-time structure used as a depiction of the coin-tossing story we introduced above:

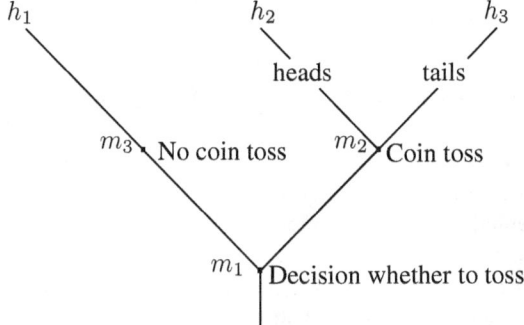

Figure 1: Coin toss

Formally put, a branching time (BT) structure is a partially ordered set (i.e. the ordering relation is reflexive, weakly antisymmetric, and transitive) satisfying the extra condition of no backward branching (or simply backwards-linearity):

- $\forall m, m_1, m_2 [((m_1 < m) \wedge (m_2 < m)) \rightarrow ((m_1 < m_2) \vee (m_2 < m_1) \vee (m_1 = m_2))]$.

Elements of the set represent possible moments (and will be called "moments"), and the ordering represents the relation of *earlier-possibly later*. A "history" is a maximal linearly ordered subset of the set.

To give a formal account of our idea, we need an extra notion of "co-presence" relation ($\sim$) defined on possible moments. Two moments stand in this relation if and only if they happen at the same time, possibly in different histories. Formally, ($\sim$) is an equivalence relation on possible moments, the standard definition of which can found in (Belnap, Perloff, & Xu, 2001, pp. 194–195).

Branching time structures are meant to interpret the language in which we talk about the future, past, and temporal possibility; it is often used as a model for the propositional language with temporal operators "It will be the case that" (F) and "It has been the case" (P) and a modal operator of historical possibility ($\Diamond$). In the standard Ockhamist semantics, the truth value of a sentence depends on both a moment and a history at which it is evaluated. (Sentences are evaluated at moment-history pairs m/h such that $m \in h$.) For example, the sentence $F\phi$ is true at a moment-history pair m/h if and only if there is a later moment m' in history h at which ϕ is true. A modal sentence $\Diamond\phi$ is true at moment-history pair m/h if and only if there is a history h' "passing through" m (i.e. min h) such that ϕ is true at m/h'.

3 Semantics of historical counterfactuals

Since we want to use BT structures to model the behavior of counterfactuals, we need to introduce an additional two-argument operator $>$. The basic idea behind the semantic interpretation is rather standard for the possible world setting: the truth value of a counterfactual at a moment-history pair, depends on whether the consequent is true at the "closest"[1] moment-history pair at which the antecedent is true. The BT structure provides a very natural account of "closeness" of possibilities. From the perspective of a moment-history pair m/h, the closest alternative m'/h' is the one that (a) makes the antecedent true; (b) the history h' "branched off" the history h as recently as possible to make the antecedent true; and (c) the moment m' is co-present with the moment m. Thus, to evaluate the counterfactual, we are looking for the history which shares as much past as possible with the actual history; in other words, the history which deviates from the actual history as recently as possible to make the antecedent true now.

---

[1] Yes, of course there might be ties. Hold on, we will deal with this.

Towards a New Theory of Historical Counterfactuals

To give a formal account of this idea we follow the lead of Stalnaker (1968) and add a selection function parameter of truth. A selection function s takes as its argument a pair consisting of sentence ϕ and a moment-history pair m/h, while delivering as its value the "closest" moment-history pair m'/h' such that:

1. ϕ is true at m'/h' ($m'/h' \vDash \phi$),

2. $m \sim m'$, i.e. m is co-present with m',

3. $\forall m'' \forall h'' ((m'' \sim m \land (h \cap h' \subset h \cap h'')) \to m''/h'' \nvDash \phi)$.

For the sake of simplicity, we ignore the possibility of ever-closer ϕ-histories and counterfactuals with historically impossible antecedents (if you require a stance on the latter issue, we suggest you follow the lead of Thomason and Gupta (1980, p. 74) and declare all such counterfactuals as true).

We also ignore, due to the limitation of space, the so-called "late departure" problem and counterfactual future contingents like "If I had bet tails, I would have won" which are notoriously difficult to tackle in the simple BT setting (examples of this sort are discussed in Placek & Müller, 2007; Thomason & Gupta, 1980).

It is important to note that conditions 1–3 above under-determine the choice of a selection function. This is clearly visible in indeterministic contexts. Take the coin toss model above (Figure 1 two pages earlier): the heads-history h_2 branches off the history h_1 at the same moment as the tails-history h_3. Therefore, both these histories are equally viable candidates for values of a selection function for the argument $\langle coin\ toss, m_3, h_1 \rangle$. It is a very important feature of our semantics which we are happy to embrace.

With the selection function at our disposal, we can define the truth conditions of a counterfactual connective $>$:

Definition 1 (Counterfactual) $\quad m/h, s \vDash \phi > \psi$ iff $s(\phi, m/h), s \vDash \psi$.

So, a counterfactual is true at a triple *moment-history-selection function* $m/h, s$ iff the selection function s when given the antecedent of the counterfactual and the m/h pair as its argument gives back a *moment-history* pair at which the consequent is true. Observe that the truth of a sentence is relative to a selection function while the definition of the selection function appeals to the notion of truth. To avoid circularity, we define $\vDash$ and the set of all selection functions bottom-up by use of double induction. The curious reader can consult the rigorous definition stated in Appendix B.

Already at this point we have achieved some interesting goals. Our semantics generates some validities we desire to preserve. For example, the counterfactual law of excluded middle $(A > B) \vee (A > \neg B)$ is valid and what we believe to be the natural interaction of counterfactual with negation is preserved, i.e. $\neg(A > B) \leftrightarrow (A > \neg B)$ is true (unless A is a historical impossibility).

4 Counterfactuals and historical modality

The notion of historical necessity is inherently local. However, introduction of a counterfactual connective indicates "the transition to a tense logic in which what is true at moments co-present with i can be relevant to what is true at i" (Thomason & Gupta, 1980, p. 78). This significantly influences the behavior of historical possibility, especially when the whole counterfactual is inside the scope of the possibility operator. In such cases, to determine what is necessary and what is possible at a given moment, we need to take into account what must and what could have happened at different moments, situated in alternative scenarios. For example, to determine whether "It is possible that if yesterday I had met a person in need, I would have helped" is true, I need to know the possible outcomes of the encounter I could have had the day before.

To take these alternative scenarios into account, we need to modify the behavior of the possibility operator, extending the domain of quantification in its semantics. So far, to determine what is possible, we had to take into account the histories passing through the moment of evaluation only, but as soon as counterfactuals are introduced, we need to somehow consider the counterfactual possibilities. We propose to do it in the following manner:

Definition 2 (Possibility) $m/h, s \vDash \Diamond \phi$ *iff* $\exists_{s'} \exists_{h'} m/h', s' \vDash \phi$.

Thus, $\Diamond \phi$ is true at a moment m, history h passing through that moment and according to the selection function s iff there is a history passing through that moment which makes ϕ true according to some selection selection function s'. (We preserve the natural duality of modalities: $\Box = \neg \Diamond \neg$.)

Thanks to this definition we can semantically distinguish "plain would" counterfactuals from "necessitated would" counterfactuals. The coin toss model from Figure 1 can serve as a paradigm example: on the one hand, at

m_3/h_1 the sentence "If I had tossed the coin it would have landed heads" (coin toss $>$ F heads) is either true or false *depending on the choice* of the selection function.[2] On the other hand, the sentences "If I had tossed the coin it would have been necessary that it would land heads" (coin toss $>$ $\Box$ F heads) and "Necessarily, if I had tossed the coin it would have landed heads" ($\Box$(coin toss $>$ F heads)) are both false *independently* of the choice of the selection function. We will give a more thorough interpretation of this phenomenon in the next section.

Simultaneous introduction of possibility and counterfactual operators to the language opens a research perspective regarding their interactions. The pioneering research in this domain has been conducted by Thomason and Gupta (1980). Our definitions significantly differ and simplify those of our predecessors, nonetheless, we were able to restate some of their results. The conclusions of our preliminary research can be found in appendix C.

5 Semantics and postsemantics

To further explore the behavior of plain and modal counterfactuals we use the conceptual apparatus of John MacFarlane (2003, 2014), who distinguishes truth-at-an-index ($\models$) from truth-at-a-context ($\Vdash$). The first notion of truth is largely a technical tool of a given semantic theory. It is used to guarantee that the semantic theory validates intuitive tautologies and rules of inference. It is also meant to guarantee the natural interaction between different logical connectives. Moreover, it is used to explicate the notion of truth-at-a-context. This last notion is more closely related to the ordinary practice of speech. At some points (e.g. 2014, p. 208) MacFarlane even suggests that a sentence's truth-at-a-context is sufficient and necessary for accuracy of an assertion of that sentence in that context. MacFarlane calls the theory of truth-at-an-index "semantics" and the theory of truth-at-a-context "postsemantics". For our simple application, the context might be reduced to the moment on the tree where the sentence is used; in turn, the index reduces to the *moment-history-selection function* triple.

Following Kaplan (1989) and Belnap et al. (2001) we understand the notion of the context as metaphysically loaded (it represents some aspects of the world at the moment of utterance), therefore taking the notion of truth-at-a-context to be metaphysically relevant as well. Specifically, we assume

---

[2]Perhaps the antecedent in the formal representation of the sentence should include the P operator; we ignore this issue as it is not important for the current task.

that the sentence is true-at-a-context if and only if the context is sufficient to ground the truth of that sentence at that context. We do not believe there is anything in the context of use of the counterfactual in virtue of which the counterfactual is true; at the same token we think that there is nothing in virtue of which the negation of the counterfactual is true. Consequently, we propose a postsemantic theory according to which neither a counterfactual future contingent nor its negation is true at a context at which it is used. It turns out that given the semantic definitions we proposed, it is sufficient to slightly generalize supervaluationism of Thomason (1970) to get the appropriate result.

Definition 3 (Truth-at-a-context)

$$m \Vdash \phi \text{ iff } \forall_s \forall_h \ m/h, s \vDash \phi.$$

That is, a counterfactual is true-at-a-context only if the choice of a "maximally close" counterfactual scenario (i.e. of a selection function) makes no difference since all the maximally close alternatives make the consequent true. The counterfactual is false-at-a-context if and only if its negation is true-at-a-context. This definition of postsemantics is very close in spirit to the theory of counterfactuals advocated in (Stalnaker, 1981).

Such a postsemantics generates the desired result: Every counterfactual future contingent is neither true nor false at the context of its use. Take again the coin toss model from Figure 1, p. 3 as an example. At the context m_3 neither "If I had tossed the coin, it would have landed heads" nor "If I had tossed the coin, it would not have landed heads"[3] is true:

$$m_3 \nVdash \text{coin-toss} > F \text{ heads} \ \& \ m_3 \nVdash \text{coin-toss} > \neg F \text{ heads}.$$

It is so because the truth-at-an-index of each of these counterfactuals crucially depends on the choice of selection function. At the same time, the postsemantics guarantees that the disjunction of the two is true at context m_3:

$$m_3 \Vdash (\text{coin-toss} > F \text{ heads}) \vee (\text{coin-toss} > \neg F \text{ heads}).$$

Interestingly, $A > B$ and $\Box(A > B)$ are true at the very same contexts, but they might be false at different contexts! This observation brings us to the last section of the paper.

---

[3] Remember that in our semantics "If I had tossed the coin it would not have landed heads" is equivalent to "It is not the case that if I had tossed the coin it would have landed heads."

6 Giving necessitarians their due

In this part, we try to explain away the lure of the idea that all counterfactuals should be read in the strong, necessitated sense. One of the most distinguished proponents of this idea is David Lewis (1986) who motivates this approach by the observation that to reject a counterfactual like "Had I tossed the coin, it would have landed heads," it is often enough to say "No! If you had tossed the coin, it might have landed tails." (p. 8). Since it is sufficient to use a might-not-counterfactual to deny a would-counterfactual, it is tempting to hold that a "would" has something of a "must" in it. Since we try to resist this temptation, we need to explain away Lewis' observation. We believe (in line with the views of Stalnaker, 1981) that the phenomenon is postsemantic rather than semantic in nature and that it should be explained by the criteria of correctness of assertions rather than by the semantics of counterfactuals. We can utilize the notion of truth-at-a-context to this purpose if we accept the following norm:

Definition 4 (Truth norm of assertion) *An act of assertion is correct in a context only if the sentence asserted is true at that context.*

This norm of assertion together with supervaluational postsemantics naturally generates correctness gaps. Since counterfactual future contingents are true at no contexts, they can never be correctly asserted. For example, neither "If I had tossed the coin, it would have landed heads," nor "If I had tossed the coin, it would not have landed heads" can be correctly asserted. Consequently, one can correctly assert a counterfactual only if it is settled. To show that a counterfactual is not true at a given context it suffices to establish that it might false. Together with the truth norm of assertion, it follows that such a counterfactual cannot be correctly asserted. This fact can be summarized by the following postsemantic theorem:

$$m \not\Vdash \psi > \phi \text{ iff } m \Vdash \Diamond(\psi > \neg\phi).$$

Therefore, it is enough to establish possibility of falsity of a counterfactual to undermine the right to assert it. It is the reason why we can deny "If I had tossed the coin, it would have landed heads" with "If I had tossed the coin, it might not have landed heads."

This postsemantic fact also explains why it is never correct to assert "If I had tossed the coin it would have landed heads, but if I had tossed it, it might not have landed heads." The reason is not that the sentence is self-contradictory, as Lewis would have us believe, but because the sentence is true in no context and as such can never be correctly asserted; that is, for any m in any model:

$$m \not\Vdash (\phi > \psi) \wedge \Diamond(\phi > \neg\psi).$$

Therefore, the intuition that "would" has a necessitating force is explained by a general linguistic mechanism (truth norm of assertion) rather than by the specific modal strengthening of the meaning of "would". Observe that the very same mechanism is at play in different kind of contexts. For example: "I were at the party, but possibly I weren't" sounds just as strange, but we do not conclude that the sentence is self-contradictory.

We noted in Section 4 that plain and necessitated counterfactuals can be semantically distinguished. This fact has a postsemantic consequence: even though would- and would-necessarily-counterfactuals are true at the very same contexts, they are false at different contexts (i.e. their negations are true at different contexts); that is, there is a model with a moment m such that:

$$m \Vdash \neg\Box(\phi > \psi) \text{ and } m \not\Vdash \neg(\phi > \psi).$$

Any indeterministic example can attest to that. Let us get back to the coin tossing example. The sentence, "If I had tossed the coin, it would necessarily have landed heads," is false, while the sentence, "If I had tossed the coin, it would have landed heads," is neither true nor false. We think that it follows the linguistic intuition rather well. It also suggests that the semantic identification of "would" and "necessarily-would" is not justified.

Lastly, let us come back to the issue of the relationship between counterfactuals like (A) and (B) (p. 1). Again, we see a postsemantic rather than semantic connection. There are two ways of describing our point.

First, it is sometimes maintained (see e.g. Stalnaker, 1981) that counterfactuals whose natural language formulation includes a word like "necessarily" in the consequent have the logical form in which the corresponding modal operator stands in front of the whole formula, so e.g. "If the match were struck, it would necessarily light" has the logical form of $\Box(A > B)$. If we take this route, notice that in our framework the following rules hold:

$$\frac{m \Vdash (\phi > \psi)}{m \Vdash \Box(\phi > \psi)} \qquad \frac{m \Vdash \Box(\phi > \psi)}{m \Vdash \phi > \psi}$$

Second, if we insist that the aforementioned sentence has the logical form of $A > \Box B$, then the issue is more subtle, but the connection of the two is still close enough. The sentence $\phi > \psi$ does follow from $\phi > \Box\psi$ and the converse entailment holds under an assumption (satisfied by most everyday counterfactuals, including those used in Leitgeb's example) that the antecedent of the counterfactual is not a future contingent. We can express these two observations in the form of the following rules:

$$\frac{m \Vdash (\phi > \psi) \land (\phi > \Box\phi)}{m \Vdash \phi > \Box\psi} \qquad \frac{m \Vdash \phi > \Box\psi}{m \Vdash \phi > \psi}$$

The bottom line is that sentences used in Leitgeb's examples – counterfactuals and necessitated counterfactuals – are not synonymous, but their affirmations are assertible in exactly the same contexts, which explains why they look like synonyms.

A Acknowledgments

We would like to thank Tomasz Placek, Thomas Müller, Michael De, Antje Rumberg, the audiences of the ∃ntia et Nomin∀ and Logica 2014 conferences, as well as members of the Izydora Dąmbska Research Group for feedback.

The part of the project realized by Jacek Wawer has been financed by means of the Polish National Science Center granted by the decision number DEC-2013/11/N/HS1/04805. Jacek Wawer also gratefully acknowledges the financial support of the Foundation for Polish Science received in years 2014–2015 in form of the scholarship START no. 122.2014.

B The details of the truth definition

Introductory comments:

- Assume M/H is the set of all moment-history pairs of a given BT model such that $m \in h$; when we write $m/h \sim m'/h'$ we mean "m is co-present with m'".

- Assume the issue of everywhere false antecedents is dealt with in some suitable way; e.g. declare that all such counterfactuals are true.

- Assume the standard definition of Ockhamist truth $\vDash_{BT}$.

Definition 5 (Alphabet) *The alphabet of our language consists of:*

- *countably many propositional variables;*
- *parentheses;*
- *connectives:* ¬, ∧, ◊, F, P, >.

Definition 6 (Formula)

1. *Every propositional variable is a formula;*
2. *If ϕ, ψ are formulas, then $\neg\phi$, $\phi \wedge \psi$, $F\phi$, $P\phi$, $\Diamond\phi$, $\phi > \psi$ are;*
3. *Nothing else is a formula.*

The class off all formulas is denoted by $\mathcal{F}$; the class of formulas which do not contain $>$ is denoted by $\mathcal{F}_0$.

Definition 7 (Degree of a formula) *The degree of a formula ϕ is denoted by $deg(\phi)$ and is defined as follows:*

- *For $\phi \in \mathcal{F}_0$, $deg(\phi) = 0$;*
- *For $\phi \in \{\neg\psi, F\psi, P\psi, \Diamond\psi\}$, $deg(\phi) = deg(\psi)$;*
- *For $\phi = \psi \wedge \chi$, $deg(\phi) = max(deg(\psi), deg(\chi))$;*
- *For $\phi = \psi > \chi$, $deg(\phi) = max(deg(\psi), deg(\chi)) + 1$.*

The symbol $\mathcal{F}_n$ denotes the set of all formulas of degree up to – and including – n.

B.1 Base step

- $S_0 := \{s : \mathcal{F}_0 \times M/H \to M/H \mid \forall m/h \in M/H, \phi \in \mathcal{F}_0$
  - $s(\phi, m/h) \vDash_{BT} \phi$;
  - $s(\phi, m/h) \sim m/h$;
  - $\forall m''/h'' \in M((m'' \sim m) \wedge (h \cap h' \subset h \cap h'') \Rightarrow m''/h'' \nvDash_{BT} \phi)$, where h' is the history in $s(\phi, m/h)\}$.

The parameters for $\vDash_0$ are moment-history pairs from M and selection functions from S_0; the full definition of $\vDash_0$ is as follows, for $\phi \in \mathcal{F}_0$:

Definition 8 (0-truth)
$m/h, s \vDash_0 \phi$ iff $m/h \vDash_{BT} \phi$

Towards a New Theory of Historical Counterfactuals

B.2 Inductive step

In what follows, $s|_n$ is the restriction of the selection function s to formulas in $\mathcal{F}_n$.

Definition 9 (Class of all n-selection functions S_n) *An n-selection function is a function $s : \mathcal{F}_n \times M/H \mapsto M/H$ such that:*

1. $s|_{n-1} \in S_{n-1}$

2. *for any $\phi \in \mathcal{F}_n$ and $m/h \in M/H$:*

   - $s(\phi, m/h), s|_{n-1} \vDash_n \phi;$
   - $s(\phi, m/h) \sim m/h;$
   - $\forall_{m''/h'' \in M/H}((m'' \sim m) \wedge (h \cap h' \subset h \cap h'') \Rightarrow m''/h'', s|_{n-1} \nvDash_n \phi)$ *where h' is the history in $s(\phi, m/h)$}*

We can now proceed with the "meat" of the induction step. Assume that $\vDash_{n-1}$ and S_{n-1} are given and that $\phi \in \mathcal{F}_n$. The parameters for $\vDash_n$ are moment-history pairs and and selection functions from S_{n-1}.

Definition 10 (n-truth)

- *for $\phi \in \mathcal{F}_{n-1}$:*
  $m/h, s \vDash_n \phi$ *iff* $m/h, s \vDash_{n-1} \phi;$

- *for $\phi, \psi \in \mathcal{F}_{n-1}$:*
  $m/h, s \vDash_n \phi > \psi$ *iff* $s(\phi, m/h), s \vDash_{n-1} \psi;$

- *for $\phi, \psi \in \mathcal{F}_n \setminus \mathcal{F}_{n-1}$:*

  - $m/h, s \vDash_n \neg\phi$ *iff* $m/h, s \nvDash_n \phi$
  - $m/h, s \vDash_n \psi \wedge \phi$ *iff* $m/h, s \vDash_n \psi$ *and* $m/h, s \vDash_n \phi$
  - $m/h, s \vDash_n \mathrm{F}\phi$ *iff* $\exists_{m'>m} m'/h, s \vDash_n \phi$
  - $m/h, s \vDash_n \mathrm{P}\phi$ *iff* $\exists_{m'<m} m'/h, s \vDash_n \phi$
  - $m/h, s \vDash_n \Diamond\phi$ *iff* $\exists s' \in S_{n-1} \exists h' \in H\ m/h', s' \vDash_n \phi$

B.3 Final definitions

We will now construct the "general" selection functions, that is, functions which are not defined for formulas up to some particular degree, but for arbitrary formulas. Each member of the set S^+ defined below is a "proto-selection function", that is, a set of selection functions containing exactly a single function of each degree such that for any two of those one is an extension of the other.

Definition 11 (Set S^+ of all proto-selection functions)
$S^+ := \{ s^+ \subset \bigcup_{n \in \mathbb{N}} S_n \mid$

- $\forall_n \exists!_s$ such that $s \in S_n$ and $s \in s^+$
- $\forall_{s',s'',m}$ if $s', s'' \in s^+$, and $s' \in S_m$, and $s'' \in S_{m+1}$, then $s''|_m = s'\}$

The required selection functions are simply unions of proto-selection functions:

Definition 12 (Set S of all selection functions) $\quad S := \{s \mid s = \bigcup s^+,$ for some $s^+ \in S^+\}$

By relativization of truth to a choice of a selection function $s \in S$ we arrive at our ultimate definition of truth (we omit as obvious the parts concerning classical and temporal connectives):

Definition 13 (Truth)

- $m/h, s \vDash p$ iff $m/h \vDash_{BT} p$
- ...
- $m/h, s \vDash \phi > \psi$ iff $s(\phi, m/h), s \vDash \psi$
- $m/h, s \vDash \Diamond \phi$ iff $\exists_{s' \in S} \exists_{h' \in H} \, m/h', s' \vDash \phi$

C Reasoning with modalities and counterfactuals

Let us present some results of our explorations of the interplay of historical counterfactuals with historical necessity. To begin with, consider the pair of sentences:

Towards a New Theory of Historical Counterfactuals

- It is possible that if I had met a person in need yesterday, I would have helped.

- If I had met a person in need yesterday, it would have been possible that I would help.

These two seem to be synonymous; they seem to be true (or false) together and for the very same reason: our readiness (or lack thereof) to help.

Examples like these suggest that, as far as historical possibility is concerned, it makes no difference whether the modal operator takes a wide or a narrow scope with respect to the counterfactual; this would indicate the validity of the following rules of inference:

$$\text{(narrow-to-wide)} \quad \frac{\phi > \Diamond\psi}{\Diamond(\phi > \psi)} \qquad \text{(wide-to-narrow)} \quad \frac{\Diamond(\phi > \psi)}{\phi > \Diamond\psi}$$

If these rules were valid, it would open a new perspective on the debate regarding the scope ambiguity of might-counterfactuals. Prima facie, the sentence like "Had I flipped the coin, it might have landed heads" can be understood either as $\phi > \Diamond\psi$ or as $\Diamond(\phi > \psi)$. Given the validity of the rules, the scope ambiguity would make no difference, at least as long the historical modalities were concerned. With our semantic definitions, we can establish that in most cases, such two sentences are indeed equivalent, but in full generality, neither of the two inferences are valid. The following two examples show the invalidity of the two above rules under our semantics.

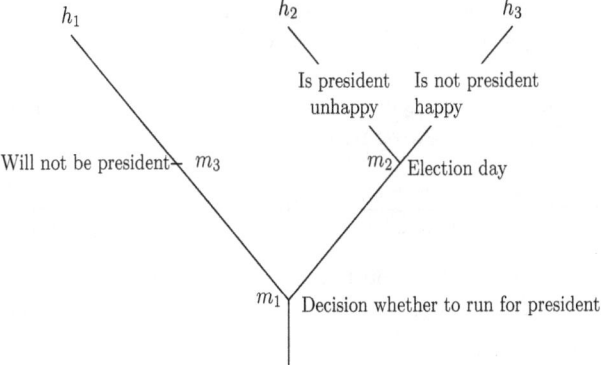

Figure 2: Potential election

Jacek Wawer and Leszek Wroński

The first case, illustrating the failure of the narrow-to-wide rule, uses an antecedent which is counterfactually contingent; that is, it is a sentence in the future tense which is not settled at the moment-history pair we "jump to" using the selection function. Consider Figure 2; the model obviously uses a lot of idealization but is just an illustration of a technical point. Suppose we had made the decision not to run for president; m_3 occurs on the date of the election in which, in h_1, we do not take part in the election. Consider the sentence, "If I were to become president, I might be happy" (F president > ◇F happy).

For an arbitrary selection function s, we have that $s(\text{F president}, m_3/h_1) = m_2/h_2$, where the sentence "◇F happy" is true (due to the existence of h_3, where we turn out to be not elected, but happy). However, keeping the same s, we see that at $m_3/h_1, s$ the sentence "◇(F president > F happy)" is false: there is no selection function which would make us jump to moment-history pair in which both "F president" and "F happy" hold. Thus, it is not possible that if I were to be elected, I would be unhappy. Therefore the narrow-to-wide principle fails in full generality; however, it holds when we restrict our attention to counterfactuals with antecedents speaking about the settled past (or, in general, antecedents ϕ such that $\phi > \Box\phi$).

The converse inference fails for an unrelated reason. Consider a model which illustrates the following scenario (Figure 3). There are two rigged coins in a box, a double-headed coin and a double-tailed coin. At moment m_0, I can draw and later toss one of the coins from the box. In history h_1, I draw a double-headed coin, in history h_3, I draw a double-tailed one and in history h_2, which is the actual one, I decide not to draw any coin at all. Let us take a moment m_2 in history h_2 and a selection function s such that $s(\text{toss}, m_2/h_2) = m_3/h_3$. Then we have that $m_2/h_2, s \vDash \Diamond(\text{toss} > \text{F(heads)})$, but $m_2/h_2, s \nvDash \text{toss} > \Diamond\text{F(heads)}$.

We have stated the rules in terms of ◇, but they can be equally well stated in terms of □ as the following two equivalences hold:

$$\frac{\phi > \Diamond\psi}{\Diamond(\phi > \psi)} \text{ iff } \frac{\Box(\phi > \psi)}{\phi > \Box\psi} \quad \text{and} \quad \frac{\Diamond(\phi > \psi)}{\phi > \Diamond\psi} \text{ iff } \frac{\phi > \Box\psi}{\Box(\phi > \psi)}$$

Since we cannot get the most general rules of inference we pursued, we can try the "second best". Consider the following strengthening of the so called "Edelberg inferences" advocated by Thomason and Gupta (1980):

(Edelberg) $\dfrac{\Box(\phi > \psi)}{\phi > \Box(\phi \to \psi)}$ (Weak Edelberg) $\dfrac{\phi > \Box\phi, \Box(\phi > \psi)}{\phi > \Box\psi}$

Towards a New Theory of Historical Counterfactuals

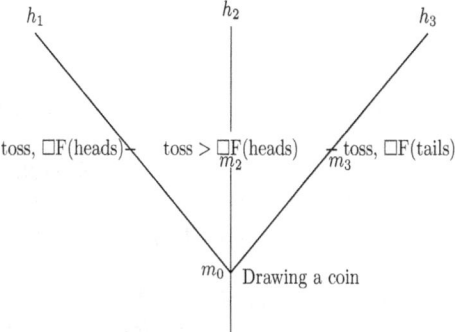

Figure 3: Two rigged coins

To validate these, the authors replaced the notion of history with a highly complex concept of a "future choice function", and then impose as many as 9 conditions on selection functions. Interestingly, the three simple requirements on selection functions we proposed above, together with our definition of possibility, are sufficient to validate both these inferences (we omit the formal details). We take it to be an advantage of our theory over the theory of Thomason and Gupta.

References

Belnap, N., Perloff, M., & Xu, M. (2001). *Facing the Future: Agents and Choices in Our Indeterministic World*. Oxford: Oxford University Press.

Kaplan, D. (1989). Afterthoughts. In J. Almong, J. Perry, & H. Wettstein (Eds.), *Themes from Kaplan* (pp. 565–614). Oxford: Oxford University Press.

Leitgeb, H. (2012). A Probabilistic Semantics for Counterfactuals. Part A. *The Review of Symbolic Logic*, *5*, 26–84.

Lewis, D. (1986). *Philosophical Papers* (Vol. II). Oxford: Oxford University Press.

MacFarlane, J. (2003). Future Contingents and relative Truth. *The Philosophical Quarterly*, *53*(212), 321–336.

MacFarlane, J. (2014). *Assessment Sensitibity: Relative Truth and Its Aplications*. Oxford: Clarendon Press.

Placek, T., & Müller, T. (2007). Counterfactuals and Historical Possibility. *Synthese, 154*, 173–197.
Pollock, J. (1976). *Subjunctive Reasoning* (Vol. 28) (No. 113). Reidel.
Stalnaker, R. (1968). A Theory of Conditionals. In N. Rescher (Ed.), *Studies in Logical Theory* (pp. 23–42). Blackwell.
Stalnaker, R. (1981). A defense of conditional excluded middle. In R. S. W. Harper & G. Pearce (Eds.), *Ifs* (pp. 87–104). Dordrecht.
Thomason, R. H. (1970). Indeterminist Time and Truth-value Gaps. *Theoria, 36*, 264–281.
Thomason, R. H., & Gupta, A. (1980). A Theory of Conditionals in the Context of Branching Time. *The Philosophical Review, 89*(1), 65–90.

Jacek Wawer
Jagiellonian University, Kraków
Poland
E-mail: jacek.wawer@fulbrightmail.org

Leszek Wroński
Jagiellonian University, Kraków
Poland
E-mail: leszek.wronski@uj.edu.pl

Clarifying Sense: Frege's Commentary on His Symbolism

ALEXANDER YATES[1]

Abstract: Frege was a foundationalist. But in the early parts of Grundgesetze, he gives arguments for some of his basic laws. Why? According to Heck, Frege is arguing for the truth of sentences, and trying to provide something like a soundness proof. According to Ricketts and Weiner, Frege is not at all engaged in genuine semantics, but is merely elucidating his system. Both of these interpretations fail to make sense of the way in which Frege phrases his arguments. A middle way is needed, and I argue that this is the sense-clarification interpretation – Frege is eliciting the self-evidence of basic laws by clarifying their sense.

Keywords: Frege, foundationalism, sense, *Grundgesetze*

1 Introduction

In preliminary exposition of his concept-script in *Grundgesetze der Arithmetik*, Frege states what the referents of his logical symbols are, and then uses these stipulations to informally argue for the truth of several of his axioms. If we take these passages at face value, Frege appears to be giving arguments for some of his basic logical laws. That he does so is perplexing – Frege believed that the basic laws of *Grundgesetze* are justificatory bedrock, fundamental logical laws which are themselves self-evident and unprovable, in which case any purported justifications of them would be circular. So why did Frege give such arguments for basic laws, and how are we to view Frege's aim in his introduction of his logical system?

The *elucidatory reading*, held by Thomas Ricketts and Joan Weiner among others, denies that the introductory passages ought to be understood as genuine semantic argumentation. Instead they are mere elucidations – loose ways of speaking which help familiarize readers with Frege's system, which may be simply dropped afterwards, and which could not, given

---

[1] Thank you to Peter Milne and Peter Sullivan for their insightful feedback, and to the attendees of Logica and of the St Andrews postgraduate Friday Seminar for many helpful comments.

some of Frege's other commitments, be made into something more formal. At the opposite extreme, the *semantic reading*, held by Richard Heck, sees Frege's arguments as foreshadowing later semantic methods – their primary purpose is to establish the correctness of parts of a symbolic system, rather than to argue directly for the laws that this system is fashioned to reflect. In this paper, I will offer a third alternative, the *sense–clarification reading*. Expanding upon ideas due to Tyler Burge, this reading aims to retain key insights of the semantic and elucidatory interpretations, while also remedying where they fall short, and also emphasizing that the purposes of the arguments in question have a relatively circumscribed role in the overall goals of *Grundgesetze*.

Firstly, I show that Frege's foundationalism is a commitment which remained constant throughout his career, and thus that any interpretation of *Grundgesetze* must say something about it. Next, I examine the semantic and elucidatory interpretations, with a focus on rebutting arguments from proponents that their respective interpretations best make sense of how Frege phrases his apparent arguments. I then present the sense-clarification interpretation, and argue that it makes more sense of the way in which Frege presents his basic laws, and that it fits well with his comments elsewhere concerning the requirements formal systems must meet. I finish with a discussion of the respects in which my view differs from that of Burge, the sense-clarification interpretation's closest competitor.

2 Frege's foundationalism

Frege was, from beginning to end, a *Euclidean foundationalist* about justification – he believed that thoughts stand in an objective justificatory relation, and thus that those laws which constitute the foundations needed to have certain special epistemic properties. There's much debate over how central a commitment this is, and how it constrains his choice of axioms. The semantic and elucidatory interpretations are alike in a lack of emphasis on Frege's Euclideanism. Heck thinks that we can't make good interpretive sense of a Fregean notion of objective basicness required of axioms for two reasons – firstly, Frege was quite aware that there are multiple ways to select axioms, and secondly, he says very little about how we should determine what counts as a primitive truth (Heck, 2007, p. 39–41). Ricketts, on the other hand, says little about basic laws partly because he thinks that inferences, not laws, are Frege's main concern (Ricketts, 1986b, p. 74). Since the tension I'm exam-

Clarifying Sense: Frege's Commentary on His Symbolism

ining is a non-issue if Frege's Euclideanism isn't to be taken seriously, I'll now argue that Euclidean foundationalism was an abiding commitment that lasted throughout Frege's career.

As early as the *Begriffsschrift*, Frege insists:

> It seems natural to derive the more complex of these judgments from simpler ones, not in order to make them more certain, which would be unnecessary in most cases, but in order to make manifest the relations of the judgments to one another. (Frege, 1879/1977, §13)

For this to make sense, we must suppose that the complexity of a proposition is, for Frege, an extra-systemic notion, a property that the proposition has independently of particular axiomatic systems in which it might figure. Similar comments appear in §3 of *Grundlagen*, where Frege discusses analyticity and the a priori in foundationalist terms:

> When a proposition is called a posteriori or analytic in my sense... it is a judgement about the ultimate ground upon which rests the justification for holding it to be true. (1884/1960, §3)

> But if, on the contrary, its proof can be derived exclusively from general laws, which themselves neither need nor admit of proof, then the truth is a priori. (1884/1960, §3)

Frege speaks of proofs as bottoming out in statements with some particular characteristic – in this case, unprovability. Frege must take "provable" in a restrictive and justificatory sense, as distinct from mere derivability – to prove a proposition A from a collection propositions Γ is for A to be justified on the basis of Γ. This is clear from context. We must suppose that there is an objective, system-independent conception of unprovability.

The comments cited above were, of course, written many years before Frege's introduction of the sense/reference distinction. But despite the changes which resulted from its introduction, his confidence in the foundational structure of logical truths appears to have persisted. In *Grundgesetze*:

> It cannot be required that everything be proven, as this is impossible; but it can be demanded that all propositions appealed to without proof are explicitly declared as such, so that it can be clearly recognized on what the whole structure rests. One must strive to reduce the number of these fundamental laws as far as

possible by proving everything that is provable. (1893/2013a, VI)

The reasoning used above applies here as well – we must suppose that Frege means something extra-systemic by "provable". This passage is significant because it makes it clear that unprovability is a criterion which constrains one's choice of axioms – we go about proving things, and once we reach bedrock, we call the propositions which constitute that bedrock axioms. Frege is quite explicit about this in "On the Foundations of Geometry: First Series":

> Traditionally, what is called an axiom is a thought whose truth is certain without, however, being provable by a chain of inferences. The laws of logic, too, are of this nature. (1903/1984b, p. 319)

And in his later "Logic in Mathematics":

> Science demands that we prove whatever is susceptible of proof and that we do not rest until we come up against something unprovable. (1914/1979b, pp. 204–205)

Frege's conception of the foundational structure of mathematical justification is not unusual, but his take on it is. Euclidean foundationalism demands more of axioms than mere successes in generating a body of truths – they must, in addition, have certain epistemic properties. As for what these properties are, Frege says notoriously little. Several criteria show up in his writings, however; axioms must be true (1903/1984b, p. 319), certain (1903/1984b, p. 319), unprovable (1884/1960, §3, 1903/1984b, p. 319), self-evident (1903/2013b, §60)[2] and independently recognizable as true (1899–1906/1979c, p. 168).

The upshot of all of this is that one must acknowledge Frege's foundationalist stance as a commitment that lasted throughout his career – it shows up in his early, mid, and later writings, often in introductory passages explaining how he conceives of his technical work. Unprovability is

---

[2] Frege says that "any expression which is not completely self-evident must actually be proven" (1903/2013b, §60). Since genuine axioms are unprovable, they must be self-evident. The context here is a discussion of arithmetic – thus, the logical laws which Frege believed arithmetical truths reduced to are self-evident.

an objective, extra-systemic notion for Frege – logical truths occur in a rational order of justification, connected by inferences, with more basic truths at the bottom. But this rational ordering does not *determine* the choice of axioms, as Frege knew there are different combinations of basic laws which would allow one to derive the same truths. Given how little Frege says about the process of discovering and selecting axioms, an account of it must be, at best, a reconstruction. However, the fact that Frege evidently held his choice of axioms to be constrained without being determined counts strongly in favor of Burge's picture of Frege's foundationalism (1998/2005a). According to Burge, it's implicit in Frege that there is a set of objectively basic laws which collectively overdetermine the set of all logical truths. That is, there are more "unprovable" (in Frege's objective, justificatory sense), self-evident propositions than are needed as axioms, since we can derive some basic truths from others. That there are inferential connections of this sort between basic laws does not contravene their basicness, for deriving a basic law φ from a set of basic laws Γ will not be a justification of φ – genuine proofs are something like valid derivations which respect to the objective rational ordering of propositions, where more fundamental truths are proven on the basis of ones that are less so.

3 Frege's commentary on his symbolism

After some preliminary comments on judgment and the unsaturated nature of functions, Frege introduces his horizontal stroke (§5), negation sign (§6), and, a bit later on, his conditional stroke:

> I introduce the function with two arguments $\begin{smallmatrix} \top \xi \\ \zeta \end{smallmatrix}$ by means of the specification that its value shall be the False if the True is taken as the ζ-argument, while any object that is not the True is taken as the ξ-argument; that in all other cases the value of the function shall be the True. (1893/2013a, §12)

Although the introduction above is similar to the standard truth-table explanations of propositional connectives[3], it makes no explicit use of semantic ascent. It contains no names for sentences or thoughts, nor does it make

---

[3]This similarity, though undeniable, mustn't be exaggerated. Frege's functions are total, taking any object whatsoever as input, and thus are distinct from the truth functions with which we are so familiar.

use of "refers to" – strictly speaking, it is an explicit introduction only of a function, not of the sign that this function denotes. The same is true of Frege's introduction of negation and the horizontal stroke. But Frege's introductory passages are far from consistent on this point, and in the next two introductions, the semantic ascent is explicit:

> Identity: "$\Gamma = \Delta$" refers to the True, if Γ is the same as Δ; in all other cases it is to refer to the False. (§7)
>
> Universal Quantifiers: Let "$\overset{a}{\smile} \Phi(\mathfrak{a})$" refer to the True if the value of the function $\Phi(\zeta)$ is the True for every argument, and otherwise the False. (§8)[4]

Frege then goes on give what are, prima facie, arguments for basic laws:

> Basic Law I: According to §12, ⊤⊤ Γ would be the False only if Γ and Δ were the True while Γ was not the True. This is impossible. Therefore, ⊢⊤ a.
>
> Basic Law II: $\overset{a}{\smile} \Phi(\mathfrak{a})$ is the True only if the value of the corresponding function $\Phi(\xi)$ has to be the True for every argument. So, $\Phi(\Gamma)$ likewise has to be the true. From this it follows that ⊤ $\Phi(\Gamma)$ is always the True, whatever function with one argument $\Phi(\xi)$ may be. (§20)

In these arguments, semantic ascent is entirely absent. This shouldn't be seen as willful abstention on Frege's part, however, as the preceding discussion is shot through with semantic talk. §17, for example, contains an argument that "$\overset{a}{\smile}_\Gamma \Phi(\mathfrak{a})$" always refers to the same value as "⊤ $\Phi(\mathfrak{a})$", for any proper name "Γ". This argument bobs back and forth between talk of functions and of signs – in part, he says:

> ...Compare this with "⊤ $\Phi(\mathfrak{a})$". The latter formula refers to the False if Γ is the True and $\overset{a}{\smile} \Phi\mathfrak{a}$ is the False. But this is the

---

[4] Frege says that this explanation is, as it stands, incomplete, since we need to specify precisely the function which is in question, and give some rules concerning how we ought to replace bound letters with free variables. (§8)

case if for some argument the value of the function — $\Phi(\xi)$ is the False. In all other cases $\vdash\!\!\!\!-\overset{\mathfrak{a}}{\smile}\Phi(\mathfrak{a})$ is the true.

Frege's arguments for the validity of *inferences*, when given, make no use of semantic ascent either, though the *conclusion* that a certain inference pattern is permissible speaks of signs. He doesn't give arguments for all of his inference rules – he simply states the law of universal generalization, for example. Nor does he argue for all of his basic laws – he sometimes just cites the section where the semantic explication of the terms occur, and leaves it at that (e.g. §20). What we should take from this is that although it is prima facie plausible that Frege is engaged in genuine semantic talk, he is not doing so systematically. He flips back and forth between talking about symbols and talking about their Bedeutung, and also doesn't sharply separate between theoretical and metatheoretical terms (Cook, 2013, A-3).

4 The semantic and elucidatory interpretations

According to the semantic interpretation of the introductory passages of *Grundgesetze*, Frege is engaged in semantics – he stipulates the referents of complex expressions involving the logical connectives, and argues for the truth of sentences on this basis. Lack of explicit semantic ascent doesn't show lack of semantic intent. For instance, Frege states modus ponens in terms of what sorts of propositions (i.e. sentences which express thoughts) can be inferred from which others, and his argument for the correctness of this mode of inference relies solely on his explanation of the conditional function given in §12, which says nothing explicitly of symbols. There's surely something right in this – Frege's discussion in §§1–32 is certainly semantic in *some* sense, and §31 in particular, as I shall argue later, cannot be seen as anything other than an attempt at proving something concerning the semantic properties of Frege's symbolism.

Frege had to adopt a meta-theoretical perspective in some sense. His insistence on making sure nothing "slips in unnoticed" into arguments requires that his proofs be gap-free (1879/1977, p. 5), and thus that it be mechanically checkable whether a purported derivation accords to the logical rules. Showing this checkability requires that one have variables that range over formula. Heck claims more than this though – he goes as far as to compare the sophistication of Frege's semantics to the semantics of Tarski (Heck, 2007, p. 27). Heck points out that the portions of *Grundgesetze* in which

Frege argues that his basic laws are true, and that the transitions between them which he allows are truth-preserving, suffice to prove the soundness[5] of the Basic Law V-free fragment of Frege's system. While this may well be the case, Frege did not see it that way – he never formulated notions of soundness or completeness for formal systems.

Heck goes on to argue that Frege's philosophical commitments in fact *required* him to provide a soundness proof. Heck's case hinges on his reading of Frege's critique of formalist theories of arithmetic, found in their most detailed form in §§86–137 of Volume II. As Frege sees it, formalists such as Heine and Thomae have a faulty notion of content – they content themselves with shuffling symbols according to arbitrary rules and only avoid the hard work of constraining these rules at the cost of shunting the task off to the various sciences applying the rules in question. Frege demands instead that any account of arithmetic should make sense of applicability, and that our rules "find their grounding in the reference of signs" (Frege, 1903/2013b, §92). Heck takes this demand to imply that Frege owed us something like a proof of the correctness of his logical system if he was to avoid the sins with which he himself charges the formalists (Heck, 2007, pp. 47–48).

Contra Heck, I don't think that Frege's goal of grounding his rules for statements containing number-words in the reference of signs, and demonstrating that such grounding holds, requires anything like a soundness proof. The key mistake Heck makes is failing to properly appreciate the importance, for Frege, of the objective justificatory structure in which thoughts stand. For Frege, arithmetical statements are emphatically *not* logically basic ones – he argues in *Grundlagen* that we would be amiss to take them as unprovable (1884/1960, §§5–8). Also, in the system found in *Grundgesetze*, numbers are not primitive terms – they are *defined*. Putting these two facts together, we are faced with a natural alternative to Heck's reading – to ground arithmetical laws in the reference of signs *just is* to show that they are derivable from the basic laws, after terms containing numerals have been replaced with their definitions in terms of extensions. Contra Heck, I don't think that Frege's goal of grounding his rules for statements containing number-words in the reference of signs, and demonstrating that such grounding holds, requires anything like a soundness proof. The key mistake Heck makes is failing to properly appreciate the importance, for Frege, of the objective justificatory structure in which thoughts stand. For

---

[5] I'm not, of course, using "soundness" in the model-theoretic sense – when Heck says Frege tries to prove soundness, he just means that he tries to prove that his formula are true and his inference rules truth-preserving (2007, p. 28).

Clarifying Sense: Frege's Commentary on His Symbolism

Frege, arithmetical statements are emphatically *not* logically basic ones – he argues in *Grundlagen* that we would be amiss to take them as unprovable (1884/1960, §§5–8). Also, in the system found in *Grundgesetze*, numbers are not primitive terms – they are *defined*. Putting these two facts together, we are faced with a natural alternative to Heck's reading – to ground arithmetical laws in the reference of signs *just is* to show that they are derivable from the basic laws, after terms containing numerals have been replaced with their definitions in terms of extensions. Contra Heck, I don't think that Frege's goal of grounding his rules for statements containing number-words in the reference of signs, and demonstrating that such grounding holds, requires anything like a soundness proof. The key mistake Heck makes is failing to properly appreciate the importance, for Frege, of the objective justificatory structure in which thoughts stand. For Frege, arithmetical statements are emphatically *not* logically basic ones – he argues in *Grundlagen* that we would be amiss to take them as unprovable (1884/1960, §§5–8). Also, in the system found in *Grundgesetze*, numbers are not primitive terms – they are *defined*. Putting these two facts together, we are faced with a natural alternative to Heck's reading – to ground arithmetical laws in the reference of signs *just is* to show that they are derivable from the basic laws, after terms containing numerals have been replaced with their definitions in terms of extensions.

If Frege really was providing something like a correctness proof, Heck needs to explain why Frege doesn't systematically employ semantic ascent. Heck thinks that Frege is merely subject to a use/mention confusion, and that he would have employed uniform semantic ascent had he been more careful (2007, p. 43). This can't be quite right. First of all, Frege was meticulous when it came to distinguishing between symbols and their meanings (1903/2013b, §98) – he believed that an unclear distinction between the two was a primary cause of many of the formalists' perceived intellectual sins (1903/2013b, §§96–97). Secondly, Frege's habit of employing semantic ascent only sporadically is not an instance of use/mention confusion, at least in the usual sense – he always studiously uses quotes when speaking of an expression. Although his habit of switching between talk of function-names and talk of functions is far from systematic, he never makes the mistake of saying that an object falls in the extension of "F", or that F refers to a concept, for example. The only potential for confusion on the readers part is the fact that Frege uses "Δ", "Γ", "Θ" both as names for objects and as meta-systemic variables for expressions (Cook, 2013, A-3).

The elucidatory interpretation lies on the opposite extreme – Frege's perspective not only made proving soundness pointless, but it prevented him from doing semantics of any sort. Frege's talk of symbols, reference, and functions are all mere elucidatory devices – they help familiarize one with his symbolism, and once one has gained fluency with the concept-script, semantic talk plays no further role (Ricketts, 2010, p. 195). One motivation for this reading is that it follows from Frege's "universalist" conception of logic – Ricketts thinks that Frege's view of logic as an all-encompassing framework allows for no vantage point from which we can meaningfully talk about logic itself (Ricketts, 1986b, pp. 136–137). Weiner is of a similar mind – she thinks it's important that Frege eschewed a semantic ascent in his discussion of basic laws (2005, pp. 346–348).

The arguments in §29– §31 don't readily bear this interpretation. In §29, Frege gives recursive clauses for what it means for a symbol to have a referent. The clauses are quite explicitly semantic, and in §31 are used in an intricate argument that every sentence in the Begriffsschrift has a unique referent – an incontrovertibly semantic claim. Had Frege's proof succeeded, then an immediate corollary would have been that every sentence expressible in his symbolism refers to a unique truth value. As this implies consistency, Frege's proof was, in the end, flawed. But successful or not, the conclusion of the proof is explicitly semantic. Heck's main point is a good one, – the arguments in §§29–32 look for all the world like a meta-theoretical proof of exactly the sort proponents of the elucidatory interpretation vehemently deny Frege could have understood.

5 The sense-clarification interpretation

The elucidatory and semantic interpretations are, in most respects, diametrically opposed to one another. As Ricketts would have it, Frege's comments in the introduction enable us to grasp, in a novel notation, that which we must already know to be able to think at all, whereas Heck thinks that these arguments demonstrate something new – either the logicality of the basic laws, or the soundness of the symbolic system which aims to capture them. Both, however, make the mistake of opting for interpretations that dictate too rigidly the way in which Frege ought to have spoken about his logic. If the semantic interpretation is correct, then Frege should have employed uniform semantic ascent throughout. If the elucidatory interpretation is correct, then, as Ricketts himself admits (1986a, p. 3) there's a deep tension between how Frege conceives of logic, and how he talks about it.

Clarifying Sense: Frege's Commentary on His Symbolism

There are two stages to my argument that Frege is eliciting self-evidence by clarifying sense. First, I look at passages in Frege's corpus where Frege explicitly states his demand that sense be clarified. Secondly, I proceed to explain why this gives an elegant reading of the *Grundgesetze* passages in particular. I'll conclude with some comments concerning the circumscribed role of self-evidence in *Grundgesetze*, and an explanation of how my account of sense-clarification differs from that of Burge.

5.1 Frege's concern with sense

An abiding element in Frege's writings is his repeated admonitions against the danger of reasoning with empty symbols. The system found in *Grundgesetze* is a *calculus ratiocinator*, with mechanically checkable deductions, but there is no *point* to these deductions if not all symbols involved are properly endowed with content – we would find ourselves playing a mere game with words (1903/2013b, §§89–95). This gives rise to Frege's requirement that every symbol, and combination of symbols, in his formal system be meaningful. In "Function and Concept", he says it is "demanded by scientific rigor" that we have provisos which prevent the formation of meaningless compound expressions (Frege, 1891/1984a, p. 19).

These comments were written close to publication of *Grundgesetze*, so Frege would have had the requirement firmly in mind while writing his magnum opus. Strange, then, that *Grundgesetze* says little about sense, while it says a great deal directly about reference, such as when Frege provides stipulations governing the reference of complex formula containing his various function symbols. But, of course, this puzzle is no real puzzle, because for Frege, "it is via a sense, and only via a sense that a proper name is related to an object" (1892–1895/1979a, pp. 124–125). We cannot fail to involve ourselves with sense – whenever we say something about the referent of a name, simple or complex, this will tell us something about sense, even if only that the name has one and that it determines that referent. Thus, although virtually all of Frege's explicit discussion of meaning in *Grundgesetze* concerns reference, sense is always in the background. For instance, in §32, after giving his (flawed) proof that every name has a referent, he immediately concludes that every sentence expresses a thought.

In Frege's discussion of definitions and basic terms found in his 1914 "Logic and Mathematics", he says quite explicitly that

> The effect of the logical analysis of which we spoke will then be precisely this – to articulate the sense clearly. ... Before

the work of construction is begun, the building stones have to be carefully prepared so us to be usable; i.e. the words, signs, expressions, which are to be used, must have a clear sense, so far as a sense is not to be conferred on them in the system itself by means of a constructive definition. (1914/1979b, p. 211)

There are two things to take from all this. Firstly that Frege was so concerned with articulating sense, and with assuring us that all formulas in his system express distinct thoughts, lends plausibility to the sense-clarification interpretation. Secondly, the relative lack of direct discussion of sense does not count at all against the sense-clarification interpretation. In Frege's way of seeing things, we involve ourselves with sense any time we speak of reference.

5.2 Making sense of Frege's arguments

Now let's consider how this interpretation makes sense of the ways in which Frege argues for his basic laws. A problematic element which the elucidatory and semantic interpretations have in common is that they both place restrictions on the sort of language Frege ought to have employed or avoided in his discussion of basic laws. The semantic interpretation says that Frege ought to have uniformly utilized semantic ascent in his discussion of his basic laws, and that his failure to do so is a use/mention confusion (Heck, 2007, p. 43), while proponents of the elucidatory interpretation, Weiner in particular, draw far-reaching conclusions from its absence (2005).

Frege's introduction to his symbolic script and the system based upon it is informal, but far from sloppy. It contains a welter of finely structured arguments and comments on particulars of derivation, utilizing varied means of expression. Frege slides seamlessly between talk of functions and talk of function symbols. One of the points in favor of the sense-clarification interpretation is that it makes sense of these varied means of expression. Frege's goal was to elicit the self-evidence of basic laws by clarifying their sense, which is accomplished by leading one to reflect on the compositional structure of the thoughts. And this is something which can be accomplished by using or mentioning symbols. Consider, again, the argument for Basic Law I, which I cited above. My contention is that it is of little note that Frege says "is the true" here, rather than "refers to the true", as in the following imagined variation:

Clarifying Sense: Frege's Commentary on His Symbolism

<u>Modified Argument</u>: According to §12, "⊢⊤ Γ " would refer to
⊤ Δ
Γ
the False only if "Γ" and "Δ" referred to the True while "Γ" did not refer to the True. This is impossible. Therefore, ⊢⊤ a.
⊤ b
a

Heck thinks that Frege ought to have given the modified argument, rather than the original one. Weiner think's it's important that Frege did not give the modified argument. Contra Weiner, the lack of semantic ascent in the introductory passages ought not to be seen as deliberate or crucially important, because Frege cites these supposedly non-semantic passages when arguing for semantic conclusions. As part of his §31 argument, he needs to show, as a base case, that "— ξ" refers to something no matter what name is substituted for "ξ". Frege says, with no further comment or argument, that "this follows immediately from our explanations" (§31). The explanation that he's referring to could only be his introduction of "— " in §5. But this introduction makes no use of semantic ascent – he tells us when the function — ξ *is the true*, rather than when "— ξ" *refers to the True* (Heck, 2007). Thus, §5 is clearly semantic in some sense, even if it doesn't use a truth-predicate. Introducing a symbol, and securing its sense, may be done either with or without semantic ascent. Heck thinks that Frege ought to have given the modified argument, rather than the original one. Weiner think's it's important that Frege did not give the modified argument. Contra Weiner, the lack of semantic ascent in the introductory passages ought not to be seen as deliberate or crucially important, because Frege cites these supposedly non-semantic passages when arguing for semantic conclusions. As part of his §31 argument, he needs to show, as a base case, that "— ξ" refers to something no matter what name is substituted for "ξ". Frege says, with no further comment or argument, that "this follows immediately from our explanations" (§31). The explanation that he's referring to could only be his introduction of "— " in §5. But this introduction makes no use of semantic ascent – he tells us when the function — ξ *is the true*, rather than when "— ξ" *refers to the True* (Heck, 2007). Thus, §5 is clearly semantic in some sense, even if it doesn't use a truth-predicate. Introducing a symbol, and securing its sense, may be done either with or without semantic ascent.

The sense-clarification reading also makes sense of the diversity in Frege's treatment of basic laws. He gives brief arguments for Basic Laws I, II, III, and IV in (§18, §20), but gives no arguments whatsoever for Basic Laws

V and VI, which respectively govern his smooth-breathing notation and his backslash operator. It's not hard to speculate why – I–IV are just those laws which make use of what correspond roughly to propositional connectives, the sense of which can be clarified by drawing attention to the way in which the truth-value of the whole depends on the values of its parts. V and VI, on the other hand, are simple identity statements. For VI, one just sees, on the basis of the introduction of the operator, that this identity is true. Basic Law V is, of course, a trickier case – Frege does not directly stipulate what sort of objects terms formed with the smooth-breathing operator refer to, but instead gives a stipulation to determine the reference of identity statements in which such terms appear on both sides. Frege later (in the appendix to Volume II of *Grundgesetze* discussing Russell's paradox) says that he always realized it lacked the same degree of obviousness as the other laws (1903/2013b, p. 253). This suggests that he may have simply refrained from giving an argument for V because he was unsure what to say about it. However, if it is a true, basic logical principle, then it seems that Frege would have thought of it in terms similar to VI – there is simply nothing more to say about it, because there is nothing more to be done to clarify it.

Another virtue of the sense-clarification interpretation is that while it denies that Frege's intent in the introductory passages would have been carried out more correctly (or at least rendered more explicit) had Frege systematically used semantic ascent, it need not – and does not – hold that Frege never involves himself in semantic talk, or even that he abstains from semantic arguments. Frege *could* have used a truth-predicate when arguing for the truth of the formula expressing his basic laws without violating any of his own substantive philosophical commitments – he simply didn't, and, due to his view of the epistemic status of basic laws, would have seen little point in doing so. Frege is doing semantics of a sort in §31, however, and it is totally consistent with the sense-clarification interpretation to see it as such.

5.3 The role of self-evidence

I've given a rough picture of what the sense-clarification interpretation is. Now, it's worth taking a few moments to emphasize what it isn't. In particular, a few words need to be mentioned concerning self-evidence, the role of which shouldn't be exaggerated. I've argued that Frege is engaged in sense-clarification in his arguments for basic laws. I've also suggested that Frege's aim in doing so is to elicit the self-evidence of basic laws. I endorse both,

Clarifying Sense: Frege's Commentary on His Symbolism

but am more certain of the truth of the first, and it's important to realize the independence of the two claims. There are several reasons one might think that Frege wanted to clarify the sense of his basic laws:

1. In order to elicit our recognition of the logicality of basic laws.

2. In order to elicit our recognition of the self-evidence of basic laws.

3. In order to elicit the self-evidence of basic laws.

4. In order to get us to recognize the basic laws as true.

2 and 3 are importantly different. Getting us to recognize *that* a truth is self-evident (2) would be pointless – as Burge has emphasized (Burge, 1998/2005a, p. 340), self-evidence is not a reason for taking something to be true, but rather a label given to truths which are recognizable as such once one understands them properly. Seen in this way, (3) comes to much the same thing as (4) – if clarifying the sense of a proposition suffices to elicit our recognition of it as true, then that proposition just is self-evident. Finally, (1) and (4) are not mutually exclusive – it's conceivable that Frege's apparent arguments play a double role of both eliciting self-evidence, and providing some sort of evidence for logicality.

Frege is quite clear that self-evidence is not a distinctive mark of the logical – we must always ask whether the self-evidence (einleuchtet) of a thought ought to be attributed to logicality, or rather to some sort of mathematical intuition (1884/1960, §90). He also says in a letter to Huntington that "nothing must be left to mere self-evidence, for its nature and laws are unknown" (Frege, 1902/1980, p. 57). Moreover, self-evidence is not a ground for taking something to be true, but a property of propositions whereby they may be recognized as true independently of other truths, provided one has the requisite level of understanding (Burge, 1998/2005a, p. 340). Frege's lack of emphasis on, and circumscribed role of, self-evidence does not count against the present interpretation. The arguments for basic laws form a very small part of Frege's explication of his concept-script – the vast majority is devoted to expounding inference rules, with the arguments for basic laws concentrated in §18 and §20. Thus, that Frege aimed to elicit self-evidence in these passages does not give self-evidence an untenably central place in Frege's overall goals.

5.4 Differences from Burge's view

It's important to distinguish the version of the sense-clarification interpretation being expounded here from its closest competitor – the interpretation of Tyler Burge, from which the current account draws much inspiration. I take Burge to have argued decisively for the following points – firstly, that Frege conceived of logical truths as standing in a system-independent rational order with objectively basic and unprovable truths at the bottom, and secondly that in the introduction to *Grundgesetze*, Frege is eliciting the self-evidence of basic-laws, rather than justifying them, or arguing for the truth of formula as part of a soundness proof. What my account adds to this is an emphasis on the role of sense, and on the way in which focusing on the role of sense-clarification in eliciting the self-evidence of basic laws manages to avoid the problems inherent in the semantic and elucidatory interpretations.

It is, of course, quite consistent for Burge to maintain that eliciting self-evidence is done by sense clarification. The bigger difference between our views is how we take this sense-clarification to be carried out. Burge argues for a connection between what he calls "pragmatic considerations" and the recognition of self-evidence (1998/2005a), and in an earlier article (1990/2005b), he argues for a similar connection between pragmatic considerations[6] and sense-clarification. His reasoning, in brief, is as follows – the self-evidence of a proposition is more objective than mere obviousness, and rather more akin to being independently recognizable as true by anyone who fully understands it. This full understanding may, however, require one to first appreciate certain inferential connections (Burge, 1998/2005a, p. 341). Logical theory-building helps us appreciate these inferential connections, and pragmatic considerations are part and parcel of such theory building (Burge, 1998/2005a, pp. 340–341).

The view I put forth here is committed to no such connection. I grant it's plausible that gaining the requisite level of understanding to recognize self-evidence requires appreciating some of the inferential connections in which that thought figures. It's also plausible that logical theory-building requires pragmatic considerations. But it's far less plausible that identifying inferential connections requires those elements of theory-building in which the pragmatic considerations come into play. In Frege's view, axioms must be objectively basic, and unprovable. But for identifying inferential connec-

---

[6] As he uses the term, "pragmatic considerations" encompasses general theoretical considerations of fruitfulness, independence, and simplicity of axioms – I shall follow this usage here

tions, it is of no consequence which laws we take to be objectively basic. If the role of pragmatic considerations is in identifying such basicness, then it has no role in helping us see inferential connections, and thus no role in eliciting the self-evidence of basic laws by clarifying their sense.

6 Conclusion

The semantic interpretation holds that Frege's philosophical commitments *necessitated* semantic theorizing to establish the correctness of his laws and inference rules, while the elucidatory interpretation holds that his philosophical commitments *precluded* the same. The sense-clarification interpretation denies both – Frege could have, but didn't (and needn't have) done semantics systematically, whereas he could and did make use of semantic arguments in certain parts of *Grundgesetze*, such as in §10 and §31. The sense-clarification interpretation makes good sense of the varied way in which Frege discusses his laws, while giving equal due to the present but limited role of semantic theorizing, and the elucidatory element in Frege's system.

References

Burge, T. (2005a). Frege on Knowing the Foundation. In *Truth, Thought, Reason: Essays on Frege*. Oxford: Oxford University Press. (Original work published 1998)

Burge, T. (2005b). Frege on Sense and Linguistic Meaning. In *Truth, Thought, Reason: Essays on Frege*. Oxford: Oxford University Press. (Original work published 1990)

Cook, R. (2013). Appendix: How to Read Grundgesetze. In P. Ebert & M. Rossberg (Trans.), *Basic Laws of Arithmetic*. Oxford: Oxford University Press.

Frege, G. (1960). *The Foundations of Arithmetic: A Logico-mathematical Inquiry into the Concept of Number* (J. L. Austin, Trans.). New York: Harper Brothers. (Original work published 1884)

Frege, G. (1977). Begriffschrift. In J. van Heijenoort (Ed.), *From Frege to Gödel: A Sourcebook in Mathematical Logic*. Cambridge: Harvard University Press. (Original work published 1879)

Frege, G. (1979a). Comments on Sense and Meaning. In H. Hermes, F. Kambartel, & F. Kaulbach (Eds.), *Posthumous Writings*. Oxford: Basil Blackwell. (Original work published 1892–1895)

Frege, G. (1979b). Logic in Mathematics. In H. Hermes, F. Kambartel, & F. Kaulbach (Eds.), *Posthumous Writings*. Oxford: Basil Blackwell. (Original work published 1914)

Frege, G. (1979c). On Euclidean Geometry. In H. Hermes, F. Kambartel, & F. Kaulbach (Eds.), *Posthumous Writings*. Oxford: Basil Blackwell. (Original work published 1899–1906)

Frege, G. (1980). Letter to Huntington (H. Kaal, Trans.). In G. Gabriel, H. Hermes, F. Kambartel, C. Thiel, & A. Veraart (Eds.), *Frege's Philosophical and Mathematical Correspondence*. Oxford: Basil Blackwell. (Original work published 1902)

Frege, G. (1984a). Function and Concept. In B. McGuinness (Ed.), *Collected Papers on Mathematics, Logic, and Philosophy*. Oxford: Basil Blackwell. (Original work published 1891)

Frege, G. (1984b). On the Foundations of Geometry: First Series. In B. McGuinness (Ed.), *Collected Papers on Mathematics, Logic, and Philosophy*. Oxford: Basil Blackwell. (Original work published 1903)

Frege, G. (2013a). *Basic Laws of Arithmetic* (P. Ebert & M. Rossberg, Trans.). Oxford: Oxford University Press. (Original work published 1893)

Frege, G. (2013b). *Basic Laws of Arithmetic* (P. Ebert & M. Rossberg, Trans.). Oxford: Oxford University Press. (Original work published 1903)

Heck, R. G., Jr. (2007). Frege and Semantics. *Grazer Philosophische Studien, 75*, 27–63.

Ricketts, T. (1986a). Frege, the Tractatus, and the Logocentric Predicament. *Noûs, 19*, 3–15.

Ricketts, T. (1986b). Objectivity and Objecthood: Frege's Metaphysics of Judgment. In J. Hintikka & L. Haaparanta (Eds.), *Frege Synthesized: Essays on the Philosophical and Foundational Work of Gottlob Frege*. Dordrecht: D. Reidel Publishing Company.

Ricketts, T. (2010). Concepts, Objects and the Context Principle. In M. Potter & T. Ricketts (Eds.), *The Cambridge Companion to Frege*. Cambridge: Cambridge University Press.

Weiner, J. (2005). Semantic Descent. *Mind, 454*, 321–354.

Clarifying Sense: Frege's Commentary on His Symbolism

Alexander Yates
St Andrews/Stirling
Scotland
E-mail: ary4@st-andrews.ac.uk

Logics of Generalized Classical Truth Values

DMITRY ZAITSEV

Abstract: This paper represents relatively fresh trend in the tideway of the generalized truth values project, termed "Logics of generalized classical truth values". A sufficiently detailed description of this project can be found in (Zaitsev & Grigoriev, 2011; Zaitsev & Shramko, 2013), that is why I just delineate it in Section 1 here. In particular, I will introduce three logics of generalized classical truth values, namely $\mathbf{FDE}_{ref}$, $\mathbf{FDE}_{inf}$, and $\mathbf{FDE}_{scl}$. Section 2 is devoted to axiomatization of $\mathbf{FDE}_{scl}$ and provides a completeness proof for it. Finally, in Section 3 some promising extensions are considered.

Keywords: Generalized truth values, semi-classical negation, logic of generalized classical values

1 Introduction: generalized classical truth values

The term and relevant conception of *generalized truth values* as subsets of a given basic set is due to Y. Shramko, J. M. Dunn, and T. Takenaka, and was introduced in (Shramko, Dunn, & Takenaka, 2001). In fact, it is in keeping with a more general logical framework of generalized valuational systems started in seminal works of J. M. Dunn and N. Belnap and onward and upward from there. The core idea of this framework, namely, the concept of generalization procedure, if depicted briefly, presupposes the power-setting of an initial set of truth-values. To get full-fledged generalized valuational system, one has to generalize valuational function as well to be the map from the set of formulas into generalized set of values. Due to the resent results of Y. Shramko and H. Wansing (2005; 2006; 2011; 2014), a noticeable progress in developing the generalization procedure as the method of philosophical investigations in logic was achieved.

New opportunities for further development of this project opened up by virtue of recent results obtained by D. Zaitsev, O. Grigoriev, and Y. Shramko who tended to regard classical truth values not as Fregean logical (atomic) obects but rather as comlex collages. In so doing, we started with the idea of taking classical truth values to be also a result of certain generalization procedure. Namely, we consider two different notions of truth and falsity. On the one hand, truth values as viewed from an epistemological perspective play important role in determining valid arguments. In this context, truth can be interpreted as a property preserved from premises to conclusion in correct reasoning. The truth values so interpreted are labeled as "inferential values". On the other hand, truth values can be considered ontologically in Fregean tradition to be abstract objects denoted by sentences. This approach allows understanding truth values as "referential values".

In (Zaitsev & Shramko, 2013) we endenizened the notion of *Cartesian classical truth values* denoting the set $\mathbf{4}^{t,1}$ being a direct product of two sets $\mathbf{2}^t$, $\mathbf{2}^1$, where $\mathbf{2}^t = \{T, F\}$ is a set of classical truth values interpreted referentially as ontological entities and $\mathbf{2}^1 = \{1, 0\}$ represents a set of classical inferential (epistemical)truth values. Thus, $\mathbf{4}^{t,1}$ is a four-element set of generalized classical truth values consisting of

T1 = $\langle T, 1 \rangle$: a sentence is both ontologically and epistemically true;
T0 = $\langle T, 0 \rangle$: a sentence is ontologically true and epistemically false;
F1 = $\langle F, 1 \rangle$: a sentence is ontologically false and epistemically true;
F0 = $\langle F, 0 \rangle$: a sentence is both ontologically and epistemically false.

If now we define an ordering relation on $\mathbf{4}^{t,1}$ via combination of two natural orderings on $\mathbf{2}^t$ and $\mathbf{2}^1$, we arrive at a distributive lattice $FOUR^{t,1}$:

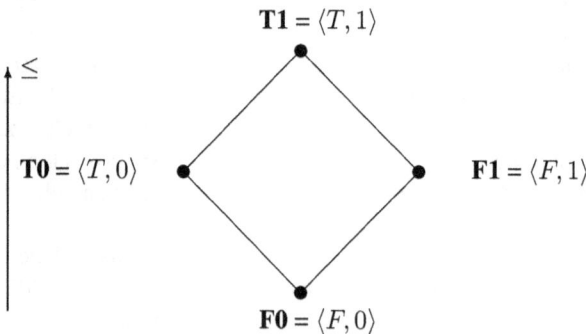

Figure 1: Lattice $FOUR^{t,1}$

To get the new valuational system, it is enough to define a valuation v^4 as a map from the set of propositional variables into $4^{t,1}$ and extend it to compound formulas. To proceed further, interpret conjunction and disjunction via meet and join on elements of v^4.

Definition 1
$A \models B \Leftrightarrow \forall v^4(v^4(A) \leq v^4(B))$.

Definition 2
(1) $v^4(A \wedge B) = v^4(A) \sqcap v^4(B)$;
(2) $v^4(A \vee B) = v^4(A) \sqcup v^4(B)$.

As for negation, it is natural to introduce it through an appropriate operation of involution. The minimum logical requirement for such an operation is to turn truth into falsity and *vice versa*. And there are two suitable candidates, each of which meets the above requirement at a minimum. The first one provides the change-over from one referential component to the other while the inferential component remains intact. The situation with the second is dual: it interchanges only between inferential truth and falsity completely ignoring the referential component. We label these operations by a'^t and a'^1 correspondingly (for an arbitrary $a \in FOUR^{t,1}$).

Interestingly, as was shown in (Zaitsev & Shramko, 2013), these operations turn to be semi-Boolean complementations. We call a complementation Boolean w.r.t. $4^{t,1}$, if it meets the following requirement:
(1) $a \sqcup \sim_b a = \mathbf{T1}$;
(2) $a \sqcap \sim_b a = \mathbf{F0}$.
For our operations a'^t and a'^1 it can be easily shown that

Proposition 1
(1) $a'^t{}'^t = a'^1{}'^1 = a$;
(2) $a'^t{}'^1 = a'^1{}'^t = \sim_b a$.

Thus, now we are in a position to introduce the notion of *semi-Boolean complementation* generically, and we have two examples of such an operation at hand.

Definition 3
*Let L be a lattice with $\top$ as the top and $\bot$ as the bottom, and let $'$ be an unary operation defined on L. Then this operation is called a **semi-Boolean complementation** iff for any $a \in L$ the following condition holds:*

$$a \sqcup a' = \top \Leftrightarrow a \sqcap a' \neq \bot.$$

The two semi-Boolean (and hence, semi-classical) complementations introduced above are presented in Table 1. It is straightforward that their superposition gives classical complementation.

a	a'^t	a'^1
T1	F1	T0
T0	F0	T1
F1	T1	F0
F0	T0	F1

Table 1: Semi-Boolean complementations in $FOUR^{t,1}$

Now we can extend Definition 2 on the negative clause:
(3) $v^4(\neg_t A) = v^4(A)'^t$;
(4) $v^4(\neg_1 A) = v^4(A)'^1$,
and name them as referential negation and inferential negation correspondingly.

Proceeding further, we considered three propositional languages and three corresponding consequence logics:

$\mathcal{L}_t \quad A ::= p \mid A \wedge A \mid A \vee A \mid \neg_t A \qquad \textbf{FDE}_{ref}$;
$\mathcal{L}_1 \quad A ::= p \mid A \wedge A \mid A \vee A \mid \neg_1 A \qquad \textbf{FDE}_{inf}$;
$\mathcal{L}_{t,1} \quad A ::= p \mid A \wedge A \mid A \vee A \mid \neg_t A \mid \neg_1 A \quad \textbf{FDE}_{scl}$.

Unfortunately, at the time of writing (Zaitsev & Shramko, 2013) we did not have an adequate deductive formalization of the logics $\textbf{FDE}_{ref}$, $\textbf{FDE}_{inf}$, and $\textbf{FDE}_{scl}$ and left this task for future work. In the next section I will provide such an axiomatization for the logic $\textbf{FDE}_{scl}$.

2 Completeness for FDE$_{scl}$

Definitely, it was not a soundness but rather a completeness that causes problems. Below I will zero in on the completeness proof but let me first reconsider the interpretation of generalized classical truth values in a more familiar way as a result of a power setting formation.

Namely, let the set $\textbf{II}$ include two basic values standing for truth from different spheres, i.e. $\textbf{II} = \{T, 1\}$, and $\mathcal{P}(\textbf{II}) = \{\{T, 1\}, \{T\}, \{1\}, \varnothing\}$. If we consider the standard set-inclusion on the elements from $\mathcal{P}(\textbf{II})$ as an ordering relation, it quite predictably gives rise to the same lattice $FOUR^{t,1}$.

Now it seems reasonable not only to define an entailment relation via set-inclusion but associate conjunction and disjunction with the standard set-theoretic operations, which are the operations of meet and join in $FOUR^{t,1}$ as well.

Summing up we may state the following Proposition.

Proposition 2
$1 \in v(\neg_1 B) \Leftrightarrow 1 \notin v(B); \quad t \in v(\neg_t B) \Leftrightarrow t \notin v(B);$
$t \in v(\neg_1 B) \Leftrightarrow t \in v(B); \quad 1 \in v(\neg_t B) \Leftrightarrow 1 \in v(B);$
$1 \in v(A \wedge B) \Leftrightarrow 1 \in v(A) \text{ and } 1 \in v(B);$
$t \in v(A \wedge B) \Leftrightarrow t \in v(A) \text{ and } t \in v(B);$
$1 \in v(A \vee B) \Leftrightarrow 1 \in v(A) \text{ or } 1 \in v(B);$
$t \in v(A \vee B) \Leftrightarrow t \in v(A) \text{ or } t \in v(B).$

A first degree consequence system **FDE**$_{scl}$ is determined by the following axiom schemas and rules of inference:

A1. $A \wedge B \vdash A$
A2. $A \wedge B \vdash B$
A3. $A \vdash A \vee B$
A4. $B \vdash A \vee B$
A5. $A \vee (B \wedge C) \vdash (A \vee B) \wedge C$
A6. $\neg_1 \neg_1 A \vdash A$
A7. $A \vdash \neg_1 \neg_1 A$
A8. $\neg_t \neg_t A \vdash A$
A9. $A \vdash \neg_t \neg_t A$
A10. $\neg_1 \neg_t A \vdash \neg_t \neg_1 A$
A11. $\neg_t \neg_1 A \vdash \neg_1 \neg_t A$
A12. $\neg_1 A \wedge \neg_t A \vdash B$

R1. $A \vdash B, A \vdash C / A \vdash B \wedge C$
R2. $A \vdash C, B \vdash C / A \vee B \vdash C$
R3. $A \vdash B, B \vdash C / A \vdash C$
R4. $A \vdash B / \neg_1 \neg_t B \vdash \neg_1 \neg_t A$

For a completeness proof we use a machinery of theories with some amplifications. Let theory be a set of formulas closed under $\vdash$ and adjunction as usual. A theory γ is *prime* iff the following holds: if $A \vee B \in \gamma$, then $A \in \gamma$ or $B \in \gamma$. A theory γ is $\neg$-*complete* (*complete* w.r.t. abstract negation) iff either $A \in \gamma$ or $\neg A \in \gamma$. A theory γ is $\neg$-*consistent* (*consistent* w.r.t. abstract negation) iff it is not the case that both $A \in \gamma$ and $\neg A \in \gamma$. A theory γ is

$\neg$-*normal* iff it is both $\neg$-complete and $\neg$-consistent, i.e. $A \in \gamma \Leftrightarrow \neg A \notin \gamma$. Correspondingly, a theory γ is $\neg$-*abnormal* if $A \in \gamma \Leftrightarrow \neg A \in \gamma$.

In what follows, we need two special kinds of theories: α-theories and β-theories.

A theory α is α-*theory* iff it is $\neg_t$-*normal* and $\neg_1$-*abnormal*, that is $A \in \alpha \Leftrightarrow \neg_t A \notin \alpha \Leftrightarrow \neg_1 A \in \alpha$.

A theory β is β-*theory* iff it is $\neg_1$-*normal* and $\neg_t$-*abnormal*, that is $A \in \beta \Leftrightarrow \neg_1 A \notin \beta \Leftrightarrow \neg_t A \in \beta$.

Define canonical valuation in terms of prime α- and β-theories.

Definition 4
$1 \in v^c(p) \Leftrightarrow p \in \beta, \quad t \in v^c(p) \Leftrightarrow p \in \alpha.$

Lemma 1 *For any arbitrary formula A,*
$1 \in v^c(A) \Leftrightarrow A \in \beta, \quad t \in v^c(A) \Leftrightarrow A \in \alpha.$

Proof. Routine simultaneous induction on the length of a formula. As an example, consider only clauses with negations.

Let A be $\neg_1 B$ and the lemma holds for B.

Ad 1: $1 \in v^c(\neg_1 B) \Leftrightarrow 1 \notin v^c(B)$ [Prop. 2]
$\Leftrightarrow B \notin \beta$ [Ind. hyp.]
$\Leftrightarrow \neg_1 B \in \beta$ [β-theory]

Ad t: $t \in v^c(\neg_1 B) \Leftrightarrow t \in v^c(B)$ [Prop. 2]
$\Leftrightarrow B \in \alpha$ [Ind. hyp.]
$\Leftrightarrow \neg_1 B \in \alpha$ [α-theory]

Let A be $\neg_t B$ and the lemma holds for B.

Ad t: $t \in v^c(\neg_t B) \Leftrightarrow t \notin v^c(B)$ [Prop. 2]
$\Leftrightarrow B \notin \alpha$ [Ind. hyp.]
$\Leftrightarrow \neg_t B \in \alpha$ [β-theory]

Ad 1: $1 \in v^c(\neg_t B) \Leftrightarrow 1 \in v^c(B)$ [Prop. 2]
$\Leftrightarrow B \in \beta$ [Ind. hyp.]
$\Leftrightarrow \neg_t B \in \beta$ [α-theory]

Logics of Generalized Classical Truth Values

$\square$

To proceed further, we need another lemma stating a useful property of **FDE**$_{scl}$ consequence relation.

Lemma 2 *For any $A, B \in \mathcal{L}_{t,1}$,
if $A \nvdash B \Rightarrow A \wedge \neg_t A \nvdash B$ or $A \wedge \neg_1 A \nvdash B$.*

Proof. The proof is straightforward. Assume $A \nvdash B$ and, on the contrary, $A \wedge \neg_t A \vdash B$ and $A \wedge \neg_1 A \vdash B$. Then, by R2. and A5., $A \wedge (\neg_t A \vee \neg_1 A) \vdash B$. As $A \vdash A \wedge (\neg_t A \vee \neg_1 A)$, applying R1. to last two consequnces we get $A \vdash B$ – a contradiction. $\square$

This lemma will be crucial for proving generalized version of Lindenbaum's Lemma, which is within the given context worthy of the name Double Lindenbaum's Lemma.

Lemma 3 *For any $A, B \in \mathcal{L}_{t,1}$,
if $A \nvdash B$, then there exists a prime α-theory χ with $A \in \chi$ and $B \notin \chi$ or there exists a prime β-theory χ with $A \in \chi$ and $B \notin \chi$.*

Proof. Let $A \nvdash B$ then, by Lemma 2, $B \notin \{C : A \wedge \neg_1 A \vdash C\}$ or $B \notin \{C : A \wedge \neg_t A \vdash C\}$.

If $B \notin \{C : A \wedge \neg_1 A \vdash C\}$ then $\chi_0 = \{A, \neg_1 A\}$, otherwise $\chi_0 = \{A, \neg_t A\}$.

We enumerate all the formulae of $\mathcal{L}_{t,1}$: $C_1, C_2, \ldots$ and construct a consequence of sets inductively starting with χ_0.

If $A \wedge \neg_1 A \in \chi_n$ then, if $B \notin \chi_n + \{C_n, \neg_1 C_n\}$, then $\chi_{n+1} = \chi_n + \{C_n, \neg_1 C_n\}$, otherwise $\chi_{n+1} = \chi_n$.

If $A \wedge \neg_t A \in \chi_n$ then, if $B \notin \chi_n + \{C_n, \neg_t C_n\}$, then $\chi_{n+1} = \chi_n + \{C_n, \neg_1 C_n\}$, otherwise $\chi_{n+1} = \chi_n$.

It is not a difficult task to show that χ defined as the union of all χ_n is *prime theory* either of α- or β-*type* (applying A6.–A11. and R3., R4.).

$\square$

Now we have what it takes to provide a completeness proof.

Theorem 1 *For any $A, B \in \mathcal{L}_{t,1}$,
if $A \models B$, then $A \vdash B$.*

Proof.
1. Assume $A \models B$, that is $\forall v(v(A) \subseteq v(B))$, and hence $\forall v \forall x(x \in v(A) \Rightarrow x \in v(B))$.
2. $\forall x(x \in v^c(A) \Rightarrow x \in v^c(B))$, i.e. $(t \in v^c(A) \Rightarrow t \in v^c(B))$ and $(1 \in v^c(A) \Rightarrow 1 \in v^c(B))$ – *from 1*.
3. Assume $A \nvdash B$.
4. There exists a prime α-theory χ with $A \in \chi$ and $B \notin \chi$ or there exists a prime β-theory χ with $A \in \chi$ and $B \notin \chi$ – from 3 by Double Lindenbaum's Lemma.
5. Now assume that there exists a prime α-theory χ with $A \in \chi$ and $B \notin \chi$.
6. $A \in \alpha$ and $B \notin \alpha$ – $\exists_{elim}$ *from 5*.
7. But immediately we get if $A \in \alpha$, then $B \in \alpha$ – *from 2*, contradicting 6.
8. It is not the case that there exists a prime α-theory χ with $A \in \chi$ and $B \notin \chi$, by contradiction, and hence, there is a prime β-theory χ with $A \in \chi$ and $B \notin \chi$ – from 4, by $\vee_{elim}$.
9. $A \in \beta$ and $B \notin \beta$ – $\exists_{elim}$ *from 8*.
10. And finally, if $A \in \beta$, then $B \in \beta$ – again *from 2*, leading to contradiction.
11. Hence, $A \vdash B$. □

3 Future lines of work: the FDE_{scl}'s connectives

Besides philosophical implications there are some extra useful by-products, namely semi-classical negations as such. In fact we received much more food for thought than we reckoned with. There are a number of unary connectives worthy of consideration and our semi-negations among them. To make this observation systematic consider two groups of conditions that arbirtary unary connective $\sharp$ may satisfy.

I.
($\underline{1}$) $1 \in v(A) \Leftrightarrow 1 \in v(\sharp A)$
($\overline{1}$) $1 \in v(A) \Leftrightarrow 1 \notin v(\sharp A)$
($\underline{2}$) $t \in v(A) \Leftrightarrow t \in v(\sharp A)$
($\overline{2}$) $t \in v(A) \Leftrightarrow t \notin v(\sharp A)$

II.
($\underline{3}$) $t \in v(A) \Leftrightarrow 1 \in v(\sharp A)$
($\overline{3}$) $t \in v(A) \Leftrightarrow 1 \notin v(\sharp A)$
($\underline{4}$) $1 \in v(A) \Leftrightarrow t \in v(\sharp A)$
($\overline{4}$) $1 \in v(A) \Leftrightarrow t \notin v(\sharp A)$

Logics of Generalized Classical Truth Values

It is obvious that both groups contain pairwise mutually exclusive conditions. For the group **I** these are (1) and $(\bar{1})$, (2) and $(\bar{2})$. For the group **II** – (3) and $(\bar{3})$, (4) and $(\bar{4})$. The remaining combinations of conditions within each group correspond to the following connectives:

$(\bar{1})$ and (2) – semi-negation$_1$ ($\neg_1$);
(1) and $(\bar{2})$ – semi-negation$_2$ ($\neg_2$);
$(\bar{1})$ and $(\bar{2})$ – Boolean negation ($\sim$);
(1) and (2) – identity transformation (i).

It is worth noticing that there is no relevant negation in this happy crew. It reveals itself among connectives of the second group:

$(\bar{3})$ and (4) – 'left turn' ($\twoheadleftarrow$);
(3) and $(\bar{4})$ – 'right turn' ($\twoheadrightarrow$);
$(\bar{3})$ and $(\bar{4})$ – deMorgan negation ($\neg_r$);
(3) and (4) – semantic shift ($\star$).

Strange labels 'left turn' and 'right turn' stand for functions corresponding to superposition of star and semi-negations. More specifically 'left turn'($\twoheadleftarrow$) may be expressed via $\neg_1 \star$ or $\star \neg_2$ and 'right turn' ($\twoheadrightarrow$) corresponds to $\neg_2 \star$ or $\star \neg_1$. Thus if one treats unary function as a map of P(**II**) into itself conditions from different groups can not be mixed. At the same time we may easily piece together corresponding connectives.

This whole manifold of functions forms two diamonds of bipartite unary connectives.

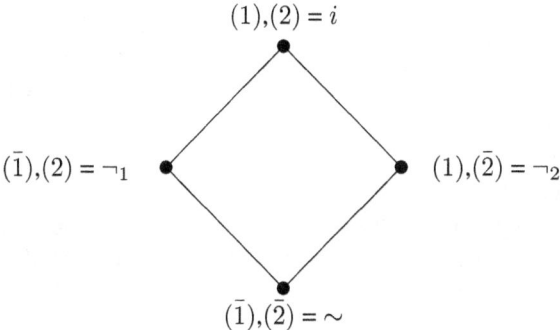

Figure 2: Diamond **I**

Dmitry Zaitsev

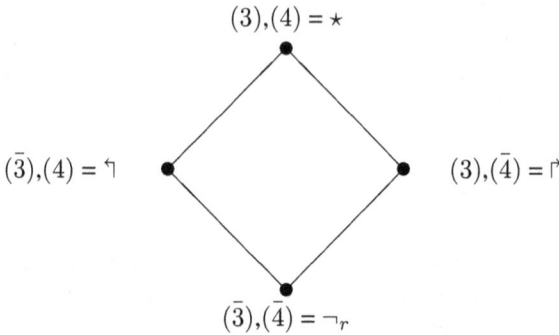

Figure 3: Diamond **II**

The various extensions of positive logic by these connectives are not all equally interesting for axiomatisation. One more direction for future work is to consider logic with two or three unary connectives. For star, relevant negation, and Boolean negation the answer is known, however there are other interesting and promising combinations.

Yet another interesting issue concerns the functional completeness of **FDE**$_{scl}$. In (Zaitsev & Shramko, 2013) we assumed that the set $\{\star, \neg_1, \neg_2\}$ is fuctionally complete w.r.t. the set of unary connectives. This guess was justified in (Wintein & Muskens, 2014). The authors claimed that the set consisting of De Morgan negation, semantic shift, conjunction and a new binary connective @ is functionally complete. As @ is concerned, it can be exhaustively characterized in the following way:

$1 \in v(A@B) \Leftrightarrow 1 \in v(B)$;

$t \in v(A@B) \Leftrightarrow t \in v(A)$.

Intuitively $A@B$ can be interpreted as "A is true, and next (and thus) B is acceptable", which opens up new vistas of future research.

References

Shramko, Y., Dunn, J. M., & Takenaka, T. (2001). The Trilattice of Constructive Truth Values. *Journal of Logic and Computation*, *11*(6), 761–788.

Shramko, Y., & Wansing, H. (2005). Some Useful 16-valued Logics: How a Computer Network Should Think. *Journal of Philosophical Logic*, *34*(2), 121–153.

Shramko, Y., & Wansing, H. (2006). Hyper-contradictions, Generalized Truth Values and Logics of Truth and Falsehood. *Journal of Logic, Language and Information*, *15*(4), 403–424.

Shramko, Y., & Wansing, H. (2011). *Truth and Falsehood. An Inquiry into Generalized Logical Values*. Berlin: Springer.

Shramko, Y., & Wansing, H. (2014). Truth Values. In E. N. Zalta (Ed.), *The Stanford Encyclopedia of Philosophy* (Summer 2014 ed.). http://plato.stanford.edu/archives/sum2014/entries/truth-values/.

Wintein, S., & Muskens, R. A. (2014). From Bi-facial Truth to Bi-facial Proofs. *Studia Logica*, 1–14.

Zaitsev, D., & Grigoriev, O. (2011). Two Truths – One Logic. *Logical Investigations*, *17*, 121–139.

Zaitsev, D., & Shramko, Y. (2013). Bi-facial Truth: A Case for Generalized Truth Values. *Studia Logica*, *101*(6), 1299–1318.

Dmitry Zaitsev
Lomonosov Moscow State University
Russia
E-mail: zaitsev@philos.msu.ru

www.ingramcontent.com/pod-product-compliance
Lightning Source LLC
Chambersburg PA
CBHW071328190426
43193CB00041B/956